Springer Complexity

Springer Complexity is an interdisciplinary program publishing the best research and academic-level teaching on both fundamental and applied aspects of complex systems – cutting across all traditional disciplines of the natural and life sciences, engineering, economics, medicine, neuroscience, social and computer science.

Complex Systems are systems that comprise many interacting parts with the ability to generate a new quality of macroscopic collective behavior the manifestations of which are the spontaneous formation of distinctive temporal, spatial or functional structures. Models of such systems can be successfully mapped onto quite diverse "real-life" situations like the climate, the coherent emission of light from lasers, chemical reaction-diffusion systems, biological cellular networks, the dynamics of stock markets and of the internet, earthquake statistics and prediction, freeway traffic, the human brain, or the formation of opinions in social systems, to name just some of the popular applications.

Although their scope and methodologies overlap somewhat, one can distinguish the following main concepts and tools: self-organization, nonlinear dynamics, synergetics, turbulence, dynamical systems, catastrophes, instabilities, stochastic processes, chaos, graphs and networks, cellular automata, adaptive systems, genetic algorithms and computational intelligence.

The two major book publication platforms of the Springer Complexity program are the monograph series "Understanding Complex Systems" focusing on the various applications of complexity, and the "Springer Series in Synergetics", which is devoted to the quantitative theoretical and methodological foundations. In addition to the books in these two core series, the program also incorporates individual titles ranging from textbooks to major reference works.

Understanding Complex Systems

Founding Editor: Scott Kelso

Future scientific and technological developments in many fields will necessarily depend upon coming to grips with complex systems. Such systems are complex in both their composition – typically many different kinds of components interacting simultaneously and nonlinearly with each other and their environments on multiple levels – and in the rich diversity of behavior of which they are capable.

The Springer Series in Understanding Complex Systems series (UCS) promotes new strategies and paradigms for understanding and realizing applications of complex systems research in a wide variety of fields and endeavors. UCS is explicitly transdisciplinary. It has three main goals: First, to elaborate the concepts, methods and tools of complex systems at all levels of description and in all scientific fields, especially newly emerging areas within the life, social, behavioral, economic, neuro- and cognitive sciences (and derivatives thereof); second, to encourage novel applications of these ideas in various fields of engineering and computation such as robotics, nano-technology and informatics; third, to provide a single forum within which commonalities and differences in the workings of complex systems may be discerned, hence leading to deeper insight and understanding.

UCS will publish monographs, lecture notes and selected edited contributions aimed at communicating new findings to a large multidisciplinary audience.

More information about this series at http://www.springer.com/series/5394

Amparo Alonso-Betanzos • Noelia Sánchez-Maroño
Oscar Fontenla-Romero • J. Gary Polhill
Tony Craig • Javier Bajo • Juan Manuel Corchado
Editors

Agent-Based Modeling of Sustainable Behaviors

Springer

Editors
Amparo Alonso-Betanzos
Facultad de Informática
Universidad de A Coruña
A Coruña, Spain

Noelia Sánchez-Maroño
Facultad de Informática
Universidad de A Coruña
A Coruña, Spain

Oscar Fontenla-Romero
Facultad de Informática
Universidad de A Coruña
A Coruña, Spain

J. Gary Polhill
Information and Computational Science
The James Hutton Institute
Aberdeen, Scotland, UK

Tony Craig
Socio-Economic and Geographical Sciences
The James Hutton Institute
Aberdeen, Scotland, UK

Javier Bajo
Departamento de Inteligencia Artificial
Universidad Politécnica de Madrid
Madrid, Spain

Juan Manuel Corchado
Departamento de Informática y Automática
Universidad de Salamanca
Salamanca, Spain

ISSN 1860-0832 ISSN 1860-0840 (electronic)
Understanding Complex Systems
ISBN 978-3-319-46330-8 ISBN 978-3-319-46331-5 (eBook)
DOI 10.1007/978-3-319-46331-5

Library of Congress Control Number: 2016959004

Printed on acid-free paper

This Springer imprint is published by Springer Nature
The registered company is Springer International Publishing AG
The registered company address is: Gewerbestrasse 11, 6330 Cham, Switzerland

Preface

Sustainability is a broad field, encompassing different aspects from technological, business, economic, environmental, and social sciences (among others), aiming at finding ways for humans to live more harmoniously with their environments, preferably without prejudicing the opportunities for humans in the future to do so. One of the most pressing issues at present is the potential of anthropogenic greenhouse gas emissions to increase global mean temperatures to a point where there are larger regions on Earth with unsuitable habitats for wildlife and/or poor-quality agricultural land. There are other issues around depletion of natural resources more generally (e.g., water, soil, biodiversity, phosphorus, and fuels), not to mention issues beyond purely environmental concerns, such as the provision of energy, jobs, growth, equality of opportunity, and ensuring human lifestyles themselves are psychologically enjoyable.

Sustainability is important because, by definition, our future depends on it. With a widespread perception that our lifestyles are currently unsustainable, finding ways to live more sustainably is critical to our future: we cannot maintain our quality of life as human beings, the diversity of life on Earth, or Earth's ecosystems unless we embrace it. It should be a priority in civil planning, environmental consultancy, agriculture, economics, corporate strategies, health assessment and planning, law, and politics. Understanding and striving for sustainability will pave the way for active and effective policies in environmental protection, while allowing for social and economical development.

Sustainability is complex because it can be perceived within different contexts, such as environmental, social, psychological, and economic [1]. These contexts could include the sustainability of economic sectors, ecosystems, countries, municipalities, neighborhoods, worker behavior patterns, transportation, private life patterns, and lifestyles. Fundamentally, sustainability is a problem involving humans making decisions. As such, studies of environmental sustainability issues can no longer ignore the human factor: there are now very few (if any) "pristine" ecosystems unaffected by human activity. Environmental sustainability is therefore a question bound by other issues influencing human behavior: such as prosperity, comfort, expectations, and governance.

The Industrial Revolution set a historical starting point for the connection between economic growth and environmental degradation, at least in terms of carbon emissions, which are observed to rise significantly from 1850 onward. Indeed, many of the global environmental problems we face today are associated with efforts to increase prosperity and improve the human condition (better healthcare, food security, shelter, and mobility), appearing with and to a large extent enabled by the Industrial Revolution. It was the beginning of major technological transitions from the use of hand tools to power tools and ultimately high technology enabling production on very large scales and robust economic development. However, we are now in the position that if we would like to maintain our lifestyles, we need to put technology, economy, and social sciences to work toward balancing human development and welfare and the needs of the environment, while also addressing the challenges posed to that balance of population growth and technological progress. If we are to develop a sustainable society, we must be willing to reexamine the conceptual foundation upon which our currently unsustainable society is built.

Traditional quantitative modeling of human behavior has, of necessity, relied on assumptions of human reasoning and access to information that those in the social sciences who study real humans have disputed. The supposed rift in the social sciences between qualitative and quantitative strands has led to skepticism about the ability of the social sciences to contribute meaningfully to studies in the natural sciences that are increasingly recognizing the need to include human behavior in the system. Once, calculus was all we had to model systems. However, the exponential growth of computational power over the past few decades has meant that it is now feasible to build simulations of social systems representing individual humans and their interactions with each other and their environment. These simulations need not rely on making assumptions for the sake of simplicity [2] or on studies of the conditions prevailing under equilibria that are never reached.

Interestingly, the simulation of human behaviors in computers has evolved separately from disillusioned economists experimenting with new ideas. Artificial intelligence (AI) has been preoccupied with this for several decades and, particularly in its early years, drew heavily on psychology for inspiration [3]. Many in the agent-based community now draw on, or are inspired by, AI architectures used to represent human decision-making and planning, not least of which include beliefs-desires-intentions [4–7], case-based reasoning [8–10], and other more general rule-based systems. That said, it is not uncommon to find agent-based modelers using simple heuristic rules or even utility maximization (though usually under conditions of limited information).

However human decision-making is simulated, developing the means by which we can explore scenarios under which people live more sustainably is arguably an urgent task, especially when more traditional approaches are simply not capable of addressing the social complexity of the transition. In particular, the concepts of "path dependence" in complex systems and "non-ergodicity" in dynamical systems mean that the doors to various visions of sustainable futures [11–14] could even now be being closed.

The purpose of this book is to gather together the latest work on simulating sustainable behaviors using agent-based modeling. Different applications of sustainability related with interesting areas as transportation, traffic management, and agricultural sustainability are also explored.

Programs that are effective in changing people's behaviors are to be promoted in order to make a graceful transition to a more sustainable future. Unfortunately, education and/or campaigns alone cannot be expected to change the individual behaviors, as various studies have already found [15–17]. Lack of knowledge and unsupportive attitudes are two possible barriers to adopting more sustainable behaviors and lifestyles, but there are also others to be accounted for, such as cultural practices, social interactions, and human feelings. In the chapter by Schaat et al., the authors describe two models for agent-based social simulations (ABSS), Consumat and SiMA-C, as a means of incorporating psychological and social factors in agent behavior. The authors examine work related to the use of ABSS in the study of sustainable behaviors. The chapter includes a review of Consumat as a tool for evaluating the influence of social policies on the adoption of sustainable behaviors. The chapter also includes a detailed explanation of the inner workings of SiMA-C and provides an example of its use in studying sustainable behaviors. Finally, it elaborates on the influence of psychological factors on decision-making in SiMA-C.

By contrast, the two chapters by Sánchez-Maroño et al. and Polhill et al. use decision trees as a supposedly transparent way to model the agents' decision-making in a project working with field researchers in environmental psychology. Their model, of everyday pro-environmental behavior in the workplace, is primarily aimed at demonstrating how data from appropriately designed questionnaires on sustainable behaviors can be used in a model to simulate the dynamics of norms. The first of the two chapters describes the processes used to derive the decision trees, comparing various methods and exploring the differences obtained. The work described in the second chapter not only suggests that different pro-environmental behaviors respond to different norms in different ways, it also argues that the topology of the social network has an effect on how norms work. The implications of this for traditional social science would be that reports of the effects of norms in a population under study should be accompanied with an account of the social network. For empirical agent-based modeling, it also means that social network topology is important data when constructing the model. The work demonstrates how agent-based modeling can collaborate with more traditional social sciences for the mutual benefit of both disciplines.

Household behavior is one of the most pertinent areas to study, as environmental pressure from households is projected to significantly increase by 2030 [18]. Better understanding of the relationships between policies implemented by governments and household decisions will improve guidance to policy makers on effective and efficient environmental policies, while addressing social concerns. One of the areas affected is waste management, as policies in this area have been successful in diverting increasing amounts of valuable materials from landfill, reducing associated environmental impacts. Despite this success, waste management is still anticipated to be a major challenge in the coming decade. Aside from various policy instruments, the literature on waste generation and recycling [19, 20] examines the role of sociodemographic, attitudinal, and contextual characteristics in households' decisions over waste management activities. Other issues still to be explored or expanded upon include whether there are interaction effects between policy variables and sociodemographic and attitudinal attributes; if there are such effects, it is important for policy makers to be aware of how and to what extent household and community characteristics can influence the success or failure of different policies. The chapter by Scalco et al. is devoted to the problem with waste. According to the authors, the increasing complexity of current non-recycled waste treatment makes waste prevention the most desirable outcome. In such a framework, recycling household waste becomes crucial, as it would both reduce waste and save resources. Household behavior is integral to the success of a recycling program. This chapter presents theoretical concepts related to recycling behaviors such as social norms and integrates them into a computational approach by formalizing, in this case, the theory of planned behavior. The resulting agent-based model is used to investigate the determinants of recycling behavior, focusing on the question of what is needed to encourage more of it.

The European Commission has launched a program for employment and social solidarity, aimed at contributing to the achievements of EU 2020 goals in employment, social affairs, and equal opportunities area [21]. Other countries, such as Japan or the USA, have similar programs to help employers find qualified applicants with disabilities, enforcing laws in the latter country by the Equal Employment Opportunity Commission [22]. There is an enormous employment gap between disabled population and nondisabled population that varies between 20 and 40 % in some countries, while is much larger, around as much as 80 %, in others. Thus, as disability benefit expenditures also tend to rise in the most developed economies, governments should converge toward activation policies that could ensure transitions to open labor markets that can promote people with disabilities. If disability policies are adequately designed, they can contribute to social inclusion and sustainable employment opportunities. The chapter presented by López Barriuso et al. introduces an innovative agent-based platform that uses 3D models of the environment to perform accurate simulations. This platform is specifically oriented toward facilitating the integration of people with disabilities in the workplace. The chapter is mainly focused on the description of the platform including specific sections about the locating infrastructure, the agent-based model and the environment. Two case studies are presented to demonstrate the technical and conceptual validity of the platform: one based on an environment of dependent people and the other dealing with an emergency situation.

The current transportation scenario leaves much room for improvement in several aspects regarding efficiency, safety, costs, and sustainability, as transportation accounts for approximately 25 % of total greenhouse gas emissions in the European Union. While emissions from other areas have been decreasing in general, those from transport continued to rise until 2008, when transport emissions started to decrease due to oil prices, increased efficiency of passenger cars, and slower growth in mobility. Despite this more recent decreasing, in 2012, transport emissions were still 20.5 % above 1990 levels and would need to fall by 67 % by 2050 in order to meet the 2011 Transport White Paper target reduction of 60 % compared to 1990. Among other measures, intelligent vehicles and traffic management is one development in personal transportation that is expected to make travel safer, more cost effective, and greener. Three chapters in this book describe research efforts in this area. The sustainability benefits of such systems could be significant, improving transportation safety, reducing traffic congestions, avoiding and reducing traffic accidents, increasing energy efficiency, and decreasing greenhouse gas emissions. The chapter by Jeffery Raphael et al. describes the application of ABM to optimizing traffic signal timings, by modeling critical elements such as vehicles and traffic control devices as autonomous agents. In this application, the MAS paradigm offers a flexible and inexpensive method for modeling the stochastic nature of the problem, allowing different models to be tested, which are then easier to maintain and scale in a real situation. Specifically, traffic control is modeled as a coordination problem, using actions to achieve coordination among traffic signal agents. One of the main characteristics of the proposed approach is that the authors propose an auction-based controller that does not need a vehicle agent. Experimental results under different traffic conditions demonstrate the interest of the proposal.

The chapter by Martin Schaefer et al. describes AgentDrive, a platform that supports development and testing of new coordination algorithms for intelligent vehicles in various levels of abstraction. As the platform is agent based, it allows the possibility of managing heterogeneous agents in any scenario. Besides a high-level description of the architecture of the platform, scalability properties are highlighted, as they are necessary for simulation of real traffic conditions. In addition, in this work, the platform is used for developing a lane-changing assistant technology, with experimental results enabling safer and swifter lane changing than the traditional non-coordinating approaches. The chapter by Francesco Barile et al. focuses on the problem of city parking and describes an automated system where software agents negotiate between the supply side and demand side of the parking allocation problem. A simulation of the automated negotiation system is presented, and notable aggregate social welfare benefits are found to be associated with its implementation. Understanding and managing the dynamics of parking is clearly an important issue for city managers in terms of allocating a scarce resource (parking places). Moreover, from a sustainability perspective, efficient allocation reduces emissions and air pollution, and also systems such as the one described here may have considerable utility in the design of future transportation systems which may incentivize, for example, car pools or shared ownership schemes.

In farming, new and more efficient farming methods will allow farms to consolidate, transforming the old models of ownership and exploitation to achieve gains in agricultural productivity and economic efficiency, with the help of government policies and technological tools. Although sustainable farming operations are site specific, individualistic, and dynamic, some general underlying characteristics of successful sustainable agricultural operations are beginning to emerge from diverse experiences. Using these characteristics, we can understand how to organize and manage sustainable farms. The chapter by Navarrete Gutiérrez et al. describes the application of a combined model, using agent-based model (ABM) and a life cycle assessment (LCA) to simulate the evolution of the agricultural system of the Grand Duchy of Luxembourg under different conditions given by policy-driven actions. On the one hand, the goal of the ABM is to represent the farmers via agents that take decisions about the crops that will be planted on their farms and the associated rotation schemes to be applied in order to maintain the health of the soil. The model provides as output the changes in land use arising from exogenous drivers, in terms of hectares of land planted with each crop. The ABM model is used to estimate the volume of agricultural commodities produced by the farmers at a given time under certain policy-driven scenarios. On the other hand, the LCA is employed to quantify the environmental impacts of these products across their whole life cycle. The LCA measures the environmental consequences, at a global scale, of different decisions made by farmers. This methodology can provide, as outputs, different levels of mid-point and end-point environmental assessments. These assessments could be a valuable tool to inform farmers on the potential impact of their activities.

The sector of energy production and management has become an important pillar for social and economic progress in modern societies. Recently, with the emergence and growth of smart grids, energy management systems have become more complex, but on the other hand, they provide a huge potential for the optimization of the energy consumed and produced. This allows the creation of new systems that boost sustainable energy management. The chapter by Lopes et al. presents the application of an agent-based software system to simulate the negotiating process of bilateral contracts between the main participants of the electricity markets (generating companies, retailers, etc.). Moreover, a risk-preference concession-making strategy was also included in the process. The agents, pursuing a predefined strategy, are able to prepare offers and counteroffers, according to different levels of risk attitude, in order to reach mutually beneficial agreements. Finally, the chapter presents a case study aiming to analyze the role of contracts for difference (CFD) as a financial tool to prevent price volatility.

On a similar topic, the chapter by Klaimi et al. presents an approach based on a multi-agent system and intelligent storage systems for energy management and control in smart grids, by balancing electric power supply and minimizing energy bill, while considering residential consumers' preferences and comfort levels. The aim of the system is to point to more responsible energy consumption while establishing lower contract prices. The proposal introduces in the smart grid four types of agents: the grid agent, the storage agent, the prosumer agent, and the consumption agent. They must control generation, load, and storage assets primarily from the outlook of power flows. Furthermore, the energy management system is split into two layers: the proactive layer and the reactive layer. The first one is responsible for the prediction of energy production and consumption. The second one is responsible for planning and negotiating consumption at shorter periods and helps to buy energy with a minimal cost. Finally, the approach has been evaluated in a simulated scenario using JADE.

As has been observed in other introductory materials to edited publications of applications of agent-based modeling to complex issues [23], the flexibility of the agent-based modeling approach in terms of the diversity of domains to which it can meaningfully be applied is ably demonstrated by the contributions to this book. One obvious observation, perhaps, is the lack of standardization in the approaches to developing the agent-based model. Though this, perhaps, is an inevitable consequence of applying agent-based models to such a diversity of fields, it is a matter that will need to be addressed if agent-based modeling is to become more established as a tool for analysis and development of robust policies aimed at transitioning societies to more sustainable futures. A key aspect of this is in building the means for the various stakeholders in a model to have confidence in the results it shows. Various chapters take a different approach to this, from the formalization of established social theory to the use of transparent, data-driven decision-making algorithms for the agents. However, in all agent-based models, a matter of notable significance is what might be termed their "ontological realism." The explicit representation of the important actors and processes in a system, without recourse to oversimplifying assumptions for the sake of analytical tractability, is an important strength of agent-based modeling. It is this that facilitates dialogue both with qualitative social scientists skeptical of traditional quantitative approaches and policy makers and stakeholders who are looking for hard facts and quantified uncertainty. In the field of sustainability research, such dialogue is essential in bringing about the sustainable future we all so desperately need.

A Coruña, Spain — Amparo Alonso-Betanzos
A Coruña, Spain — Noelia Sánchez-Maroño
A Coruña, Spain — Oscar Fontenla-Romero
Aberdeen, UK — J. Gary Polhill
Aberdeen, UK — Tony Craig
Madrid, Spain — Javier Bajo
Salamanca, Spain — Juan Manuel Corchado

References

1. Pol, E.: Environmental management: a perspective from environmental psychology. In: Bechtel, R., Churchman, A. (eds.) Handbook of Environmental Psychology, pp. 55–84. Wiley, New York (2002)
2. Moss, S.: Relevance, realism and rigour: a third way for social and economic research. CPM Report No. 99-56, 30 November 1999. http://cfpm.org/papers/Moss%20-%20R%20R%20and%20R.pdf (1999). Accessed 18 May 2016
3. Boden, M.A.: Computer Models of Mind: Computational Approaches in Theoretical Psychology. Cambridge University Press (1988)
4. Rao, A.S., Georgeff, M.P.: BDI agents: from theory to practice. In: Proceedings of the First International Conference on Multiagent Systems ICMAS-95. pp. 312–319 (1995)

5. Baptista, M., Martinho, C.R., Lima, F., Santos, P.A., Prendinger, H.: Improving learning in business simulations with an agent-based approach. J. Artif. Soc. Soc. Simul. **17**(3), 7. http://jasss.soc.surrey.ac.uk/17/3/7.html (2014). Accessed 18 May 2016
6. Nickles, M., Rovatsos, M., Schmitt, M., Brauer, W., Fischer, F., Malsch, T., Paetow, K., Weiss, G.: The empirical semantics approach to communication structure learning and usage: individualistic vs. systemic views. J. Artif. Soc. Soc. Simul. **10**(1), 5. http://jasss.soc.surrey.ac.uk/10/1/5.html (2007). Accessed 18 May 2016
7. Albino, V., Carbonara, N., Giannoccaro, I.: Coordination mechanisms based on cooperation and competition within industrial districts: an agent-based computational approach. J. Artif. Soc. Soc. Simul. **6**(4), 3. http://jasss.soc.surrey.ac.uk/6/4/3.html (2003). Accessed 18 May 2016
8. Aamodt, A., Plaza, E.: Case-based reasoning: foundational issues, methodological variations, and system approaches. AI Commun. **7**(1), 39–59 (1994)
9. Izquierdo, L.R., Gotts, N.M.: The implications of case-based reasoning in strategic contexts. Lect. Notes Econ. Math. Syst. **564**, 163–174 (2006)
10. Lee, K.C., Lee, N.: CARDS: Case-based reasoning decision support mechanism for multi-agent negotiation in mobile commerce. J. Artif. Soc. Soc. Simul. **10**(2), 4. http://jasss.soc.surrey.ac.uk/10/2/4.html (2007). Accessed 18 May 2016
11. Polhill, J.G., Gimona, A., Gotts, N.M.: Nonlinearities in biodiversity incentive schemes: a study using an integrated agent-based and metacommunity model. Environ. Model. Softw. **45**, 74–91 (2013)
12. Costanza, R.: Visions of alternative (unpredictable) futures and their use in policy analysis. Conserv. Ecol. **4**(1), 5. http://www.consecol.org/vol4/iss1/art5/ (2000). Accessed 18 May 2016
13. Quist, J., Vergragt, P.: Past and future of backcasting: the shift to stakeholder participation and a proposal for a methodological framework. Futures **38**(9), 1027–1045 (2006)
14. Rotmans, J., van Asselt, M., Anastasi, C., Greeuw, S., Meltors, J., Peters, S., Rothman, D., Rijkens, N.: Visions for a sustainable Europe. Futures **32**(9–10), 809–831 (2000)
15. Stern, P.: Towards a coherent theory of environmentally significant behaviour. J. Soc. Issues **56**(3), 407–424 (2000)
16. Geller, E.S.: The challenge of increasing proenvironment behaviour. In: Bechtel, R., Churchman, A. (eds.) Handbook of Environmental Psychology, pp. 525–540. Wiley, New York (2002)
17. Jackson, T.: Motivating Sustainable Consumption: A Review of Evidence on Consumer Behaviour and Behavioural Change : A Report to the Sustainable Development Research Network. Centre for Environmental Strategy, University of Surrey, Guildford (2005)
18. OECD: Environmental Outlook to 2030, OECD, Paris (2008)
19. Dijkgraafand, E., Gradus, R.H.J.M.: Cost-savings in unit-based pricing of household waste. The case of The Netherlands. Resour. Energy Econ. **26**, 353–371 (2004)
20. OECD: Impacts of Unit-Based Waste Collection Charges, OECD, Paris (2006)
21. PES approaches for sustainable activation of people with disabilities. Analytical Paper. The European Commission Mutual Learning Programme for Public Employment Services. DG Employment, Social Affairs and Inclusion (August, 2013)
22. United States Department of Labour: https://www.dol.gov/general/topic/disability/hiring
23. Polhill, J.G., Gimona, A., Aspinall, R.J.: Agent-based modelling of land use effects on ecosystem processes and services. J. Land Use Sci. **6**(2–3), 75–81 (2011)

Contents

Contributors

Hugo Algarvio Instituto Superior Técnico, Universidade de Lisboa, Lisboa, Portugal

Amparo Alonso-Betanzos Universidade da Coruña, A Coruña, Spain

Javier Bajo Department of Artificial Intelligent, Technical University of Madrid, Madrid, Spain

Francesco Barile Dipartimento di Matematica e Applicazioni, Università degli Studi di Napoli Federico II, Napoli, Italy

Enrico Benetto Luxembourg Institute of Science and Technology (LIST), Belvaux, Luxembourg

Marino Bonaiuto Centro Interuniversitario di Ricerca in Psicologia Ambientale (CIRPA), Sapienza Università di Roma, Rome, Italy

Mirilia Bonnes Centro Interuniversitario di Ricerca in Psicologia Ambientale (CIRPA), Sapienza Università di Roma, Rome, Italy

Luigi Bove Dipartimento di Ingegneria Elettrica e Tecnologie dell'Informazione, Università degli Studi di Napoli Federico II, Napoli, Italy

Michal Čáp Department of Computer Science, Artificial Intelligence Center, Czech Technical University in Prague, Prague, Czech Republic

Giuseppe Carrus CIRPA, Università Roma Tre, Rome, Italy

Andrea Ceschi Department of Human Science, University of Verona, Verona, Italy

Juan M. Corchado Department of Computer Science and Automation Control, University of Salamanca, Salamanca, Spain

Tony Craig The James Hutton Institute, Aberdeen, Scotland, UK

Stephan Dickert School of Business and Management, Queen Mary University of London, London, UK

Linköping University, Linköping, Sweden

Claudia Di Napoli Istituto di Calcolo e Reti ad Alte Prestazioni, C.N.R., Naples, Italy

Fernando De la Prieta Department of Computer Science and Automation Control, University of Salamanca, Salamanca, Spain

Adina Dumitru Universidade da Coruña, A Coruña, Spain

Oscar Fontenla-Romero Universidade da Coruña, A Coruña, Spain

Ferdinando Fornara CIRPA, Università di Cagliari, Rome, Italy

Jean-Marc Frayret Department of Mathematics and Industrial Engineering, École Polytechnique de Montréal, Montréal, QC, Canada

Ricardo García-Mira Universidade da Coruña, A Coruña, Spain

Corina Ilin Universitatea de Vest din Timisoara, Timisoara, Romania

Wander Jager Center for Social Complexity Studies, University of Groningen, Groningen, The Netherlands

Kees Keizer Rijksuniversiteit Groningen, Groningen, The Netherlands

Joelle Klaimi EDST, Centre-Azm, Laboratoire des Systèmes électroniques, Télécommunications et Réseaux (LaSTRe), Lebanese University, Beirut, Lebanon

ICD/ERA, Troyes University of Technology, Troyes, France

Fernando Lopes LNEG-National Research Institute, Lisboa, Portugal

Alberto López Barriuso Department of Computer Science and Automation Control, University of Salamanca, Salamanca, Spain

Fridanna Maricchiolo CIRPA, Università Roma Tre, Rome, Italy

Antonino Marvuglia Luxembourg Institute of Science and Technology (LIST), Belvaux, Luxembourg

Simon Maskell Department of Electrical Engineering & Electronics, University of Liverpool, Liverpool, UK

Leila Merghem-Boulahia ICD/ERA, Troyes University of Technology, Troyes, France

Tomás Navarrete Gutiérrez Luxembourg Institute of Science and Technology (LIST), Belvaux, Luxembourg

J. Gary Polhill Information and Computational Science, The James Hutton Institute, Aberdeen, Scotland, UK

Rana Rahim-Amoud EDST, Centre-Azm, Laboratoire des Systèmes électroniques, Télécommunications et Réseaux (LaSTRe), Lebanese University, Beirut, Lebanon

Faculty of Sciences, Lebanese University, Beirut, Lebanon

Jeffery Raphael Department of Computer Science, University of Liverpool, Liverpool, UK

Sameer Rege Luxembourg Institute of Science and Technology (LIST), Belvaux, Luxembourg

Sara Rodríguez-González Department of Computer Science and Automation Control, University of Salamanca, Salamanca, Spain

Silvia Rossi Dipartimento di Ingegneria Elettrica e Tecnologie dell'Informazione, Università degli Studi di Napoli Federico II, Napoli, Italy

Angela Ruepert Rijksuniversiteit Groningen, Groningen, The Netherlands

Noelia Sánchez-Maroño Universidade da Coruña, A Coruña, Spain

João Santana Instituto Superior Técnico, Universidade de Lisboa, Lisboa, Portugal

INESC-ID, Lisboa, Portugal

Riccardo Sartori Detpartment of Human Science, University of Verona, Verona, Italy

Andrea Scalco Department of Human Science, University of Verona, Verona, Italy

Samer Schaat Institute of Computer Technology, TU Wien, Vienna, Austria

Martin Schaefer Department of Computer Science, Artificial Intelligence Center, Czech Technical University in Prague, Prague, Czech Republic

Itad Shiboub Higher Institute of Comprehensive Professions, Nalut, Libya

Elizabeth I. Sklar Department of Informatics, King's College London, London, UK

Linda Steg Rijksuniversiteit Groningen, Groningen, The Netherlands

Jiří Vokřínek Department of Computer Science, Artificial Intelligence Center, Czech Technical University in Prague, Prague, Czech Republic

Psychologically Plausible Models in Agent-Based Simulations of Sustainable Behavior

Samer Schaat, Wander Jager, and Stephan Dickert

Abstract Agent-based modelling (ABM) proves successful as a methodology for the social sciences. To continue bridging the micro-macro link in social simulations and applying ABM in real-world conditions, conventional and often simplified models of decision-making have to be utilized and extended into psychologically plausible models. We demonstrate the contribution of such models to enhance validation and forecasts in social simulations with two examples concerned with sustainable behavior. We start with the Consumat framework to demonstrate the contribution of an established psychological plausible decision-making model in various scenarios of sustainable behavior. Then we use the SiMA-C model to explain how different psychological factors generate social behavior and show how a detailed model of decision-making supports realistic empirical validation and experimentation. A scenario of social media prompting of environmental-friendly behavior exemplifies the details of how individual decision-making is influenced by the social context. Both examples, Consumat and SiMA-C, emphasize the importance of psychological realism in modelling behavioral dynamics for simulations of sustainable behavior and provide explanations on the psychological level that enable the development of social policies on the individual level.

S. Schaat (✉)
Institute of Computer Technology, TU Wien, 1040 Vienna, Austria
e-mail: samer.schaat@tuwien.ac.at

W. Jager
Center for Social Complexity Studies, University of Groningen, Groningen, The Netherlands
e-mail: w.jager@rug.nl

S. Dickert
School of Business and Management, Queen Mary University of London, London, UK

Linköping University, Linköping, Sweden
e-mail: s.dickert@qmul.ac.uk

A. Alonso-Betanzos et al. (eds.), *Agent-Based Modeling of Sustainable Behaviors*, Understanding Complex Systems, DOI 10.1007/978-3-319-46331-5_1

1 Towards Psychologically Plausible Social Simulations

As is common in modelling approaches, ABSS (Agent-Based Social Simulations) strive to abstract and reduce information. The first step to tackle the micro-macro gap in ABSS is done abstractly (following a top-down approach in science) and has brought important methodological insights in showing how the paradigm of agents—as heterogeneous individuals with frequent interactions—is able to tackle emergent social phenomena. However, oversimplification in ABSS is increasingly criticized (e.g. [1, 2]) since their unrealistic assumptions often impede their evaluation and application in real-world conditions. Additionally, in reducing social simulations on the interactional aspect, conventional approaches often neglect how social interactions are generated. These weak points are recently tackled by different approaches, which consider psychological processes that generate social interactions. Such psychologically plausible agents in ABSS can overcome these limitations and bridge the gap between psychological and sociological perspectives by providing a platform for their integration.

The discussion between sociological, economical, and psychological approaches is reflected in the discourse between social simulations and cognitive simulations. Similarly to sociology, conventional ABSS approaches argue that macro-level phenomena are better explored using abstract micro-specifications. Typical arguments against sophisticated decision-making models include that sometimes an agent's properties at the psychological level are assumed to be constant and that often the processes that give rise to the agent's action (and their alternatives) are not relevant [3, p. 430]. That is, in such cases it does not matter why the agent decided the action, but only that the agent does it in the given context.

Recently different approaches in ABSS and cognitive science (e.g., [4, 5]) argue against the usage of simplified, psychologically implausible decision models in ABSS. Typical arguments are that many macro-level phenomena are caused by the dynamics between the micro and macro level, which is not captured by (too) simplified models [4]. Another key argument is that replicating macro-level phenomena in conventional ABSS does not imply that the relevant micro-specifications are considered in the model. In extreme cases, "right" phenomena are replicated due to "wrong" assumptions. Regarding validation, simplified abstract micro-specifications are difficult to compare against established psychological models or even against empirical observations [5, p. 17]. This limits the usage of conventional ABSS in real world applications (e.g., for policy testing). ABSS supports understanding the dynamics in social simulations and identifying emergent, i.e., unpredictable, phenomena. Hence ABSS is helpful as a starting point to identify turbulent states in complex systems but is limited in addressing them. For instance, to be able to compare the model's parameter to empirical data and in turn inform policies of how to tackle the phenomenon, psychologically plausible models are necessary. In the end, a tradeoff is always necessary, since on the one hand a model implies abstraction, but on the other hand we cannot know the causes of a phenomenon a priori, but we can tackle the structure that generate these causes, which is the human mind. This is particularly the case with counter-intuitive emergent phenomena. One possibility for a trade-off is to use basic building blocks of cognitive models as foundations and, dependent on the concrete research question at hand and insights about emergent behavior from prototypical simplified ABSS, to decide on their detail of specification and extension.

The integration of social and psychological aspects with psychologically plausible models in agent-based social simulations support deeper explanations and better forecasting. In particular, it provides deeper process explanations which contribute to a more adaptive management of less predictable (complex) developments. As general in science, top-down or bottom-up approaches in integrating social and psychological levels in simulations are possible. A top-down approach would start with a simplified social simulation that informs the necessity and detail of specifications in the psychological layer. However, often (wrong) explanations are possible on the social level. Additionally, often docking from the social level to the psychological level is difficult to do, in particular due to the counter-intuitive aspect of emergence. The bottom-up approach would start with representing a psychologically plausible representation of human decision-making, being the source of all behavioral phenomena, and identifies the (often unexpected) causes of the social phenomena. Here simplification would be done after exploring the model in simulations. A bottom-up approach—starting with assumed building blocks—would also resemble a generative approach, which is one of the core-theme in ABSS.

Overall, reaching the limits of social simulations, one conclusion is that deeper explanations and realistic forecasts require sophisticated models of human decision-making.

Next, we demonstrate the Consumat approach as an established psychologically plausibly model to explore sustainable behavior. After that we use the SiMA-C approach to demonstrate details of considering the social context in individual decision-making using an example of prompting to switch to environmental-friendly energy providers in social media.

2 Simulations of Sustainable Behavior with the Consumat Model

The Consumat approach has been developed as a generic conceptual framework to guide the development of social simulation models that include different human needs and decision strategies.

The basic drivers of behavior in the Consumat framework are *needs* and the fulfilment of needs resulting in *satisfaction*. Humans have multiple needs as indicated by Maslow [6] and Max-Neef [7]. In developing a simple social simulation model we often want to reduce the number of needs, hence we make a basic distinction between three main need forces: *existence*, *social* and *personality. Existence* relates to having means of existence, food, income, housing and the like. Agents act in order to avoid depletion of these resources over time. *Social* relates to having interactions with others, belonging to a group, and having a social status. *Personality* relates to satisfying one's personal taste with respect to overall life values and norms, such as environmental protection, altruism, or enjoyment of life. Depending on how important these needs are for an agent (personality) and how satisfied/depleted these needs are, the agents have a motivation to perform a particular behavior out of a range of possible behaviours (or behavioural options).

To perform a particular behavior, an agent possesses *abilities*, which relate to its capacity to actually use particular behavioral options. This relates to e.g. income, land possession, availability of tools, cognitive capacity and other possible abilities that are a prerequisite for performing a particular behavior.

Decision-making on what behavior to perform is a key element of the Consumat approach [8]. The uncertainty and satisfaction of the agent drive its type of decision-making. The lower the satisfaction of an agent, the higher the motivation to elaborate on alternative behaviors that improve satisfaction. And the more uncertain the agent is, e.g., due to the complexity of the decision and the many options to choose from, the more likely it will use other agent's behavior as a source of information. In the consumat approach this is conceptualized in four basic decisional strategies; repetition, imitation, deliberation and inquiring. In case of low uncertainty and high satisfaction, agents engage in *repetition*, which is the mechanism behind habitual behavior. A high uncertainty combined with high satisfaction results in *imitation*. When satisfaction is low, the agents are more motivated to invest effort in improving their situation. Hence when they are certain but dissatisfied they will engage in *deliberation*, which is a form of optimizing or homo-economicus kind of behavior. Dissatisfaction combined with uncertainty results in *inquiring*, where the behavior of comparable others is evaluated and copied when expected satisfaction increases. The interactions between the agents during social decision-making can focus on a more normative influence (imitation) or informative (inquiring), thus capturing the two main social interaction mechanisms as distinguished in the social sciences (e.g., [9]). Social decision-making is usually directed at similar others, where similarity is related to abilities. Agents have a memory for behavioral opportunities and other agents' behavior and abilities, which is only updated if cognitively demanding strategies are being used. Figure 1 gives an overview of the Consumat architecture.

The Consumat provides a generic framework that can be applied to different domains. Depending on the domain and the available data, the Consumat approach can guide the development of a specific social simulation model. In the following sections, we will briefly discuss a number of applications of the Consumat approach in different domains that are relevant in the context of sustainable development.

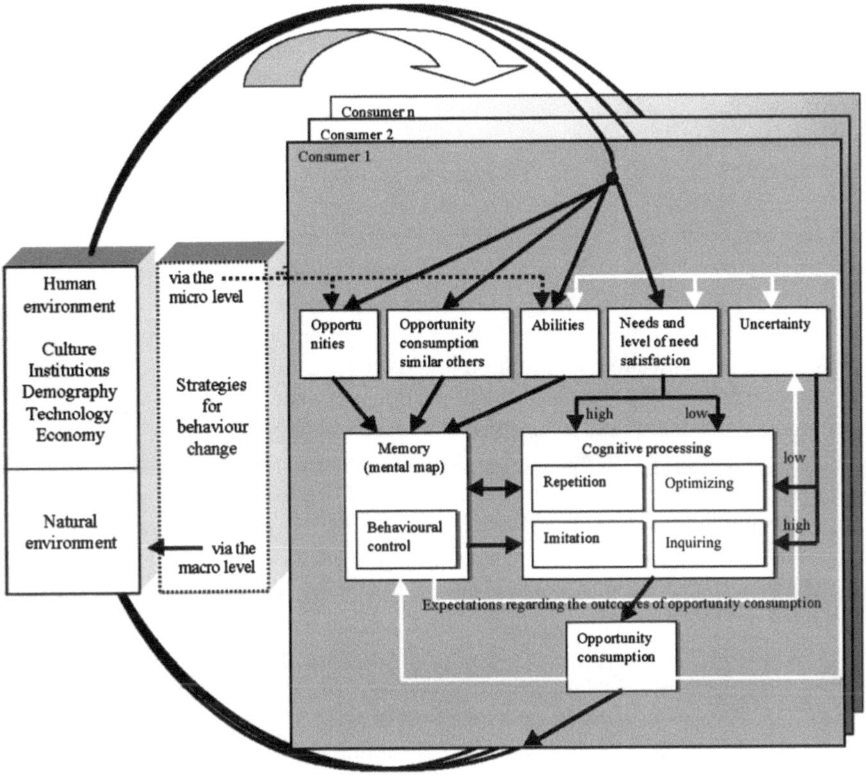

Fig. 1 Overview of the consumat framework

2.1 *Simulating Consumer Behavior*

The Consumat approach has been used to study consumer behavior in a variety of settings. Examples are the diffusion of "green" products [10], changes in sustainable life styles in Italy [11] and sustainable transport choices by US households [12]. In the following we will discuss very briefly four applications related to household lighting, the diffusion of electric cars, farmer crop choices and a transition in an artificial society. A relevant difference here is that household lighting is a domain where consumer choice is less important (financially, socially), whereas the diffusion of electric cars is a more complicated domain, including technology and charging network development.

2.1.1 Household Lighting: The Diffusion of LED's

In the last decade, governments and environmental groups have tried to promote energy-efficient lighting to consumers. With the rise and subsequent lowering of retail prices of LED-lamps, which are extremely durable and energy-efficient, the reasons for continuing use of the incandescent light bulbs are diminishing. The newer LED lamps are more economically attractive to the general consumer and can be more aesthetically pleasing compared to incandescent lamps because their colour spectrum can be configured. Despite these advantages and the pressure of governments and interest groups, consumer adoption of LED lights has proven to be low. Many governments around the world have even resorted to a (partial or complete) ban on incandescent lighting in an effort to reduce energy-use.

The behaviour of consumers in the lighting market is interesting, because it seems as if consumers need to be convinced to act in their own interest. In exploring possible scenarios to stimulate the diffusion of LED a Consumat application addressed consumer behaviour in this domain [13]. Based on empirical data a Consumat model was formalised. When a light broke down the agent replaced it by a LED, Halogen, CFL/ or incandescent light. The agent focused here on price, energy efficiency, colour discrepancy and ramp-up time. The importance of these aspects depended on the function of the light: functional as in a garage, versus spherical in the living room.

Depending on satisfaction and uncertainty, the agent would engage in (1) repetition: the broken lamp is replaced with exactly the same type, (2) imitation: the agent will select a close peer and pick a lamp of the inventory of this other agent (unless being anti-conformist), (3) optimization: look at all available lamps and select the best lamp considering all criteria, and (4) inquiring: select a close peer and select the best lamp from the other agent's inventory (unless being anti-conformist).

After a series of experiments replicating historic developments, a number of experiments addressed the stimulation of the adoption of LED's. This involved subsidising LED, giving 2 LED's for free, taxing incandescent lights and a yearly campaign (opinion boost). Considering this setting we experimented with a projected time horizon of 10 years. Figure 2 shows the results for this experiment.

The "no intervention" scenario shows that, while LED lamp prevalence does steadily climb, the penetration is slow. This behaviour shows that a new product appearing on the market alone does not strongly influence the consumer, basically because most consumers replace a broken lamp with an identical one (repetition). Whereas the financial and opinion boost strategies have some effect, the simulation shows that offering 2 LED lamps for free has the strongest impact. However, after the initial boost this provides, the same slow continuation is seen as with the "no intervention" progression. This is due to the updating of opinions of agents only when a lamp breaks. Combined with the longevity of LED lamps, this means the opinions of the agents are very slow to shift in favour of LED lamps.

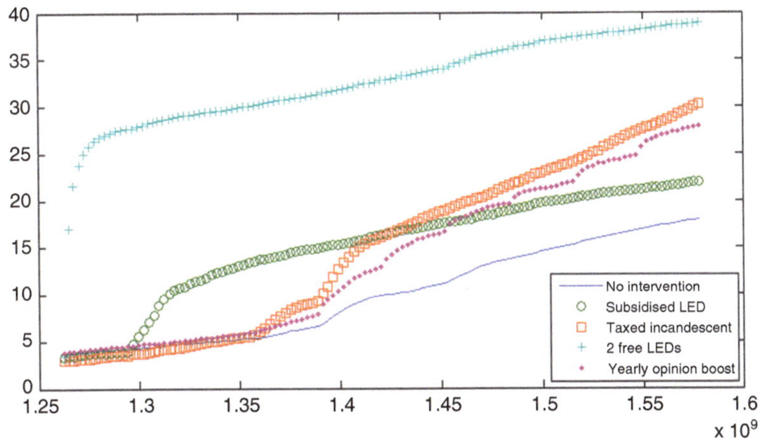

Fig. 2 Proportion of LED lamps in the house for different policies

This application of the Consumat approach demonstrated how it can be applied to explore the efficacy of different policies, acknowledging the possible impact of consumer decision-making strategies.

2.1.2 Diffusion of Electric Cars

A next application of the Consumat addresses the diffusion of electric cars, specifically plug-in hybrid vehicles (PHEVs) and full battery electric vehicles (BEVs). The transition from fuel cars to electric cars is a large-scale long-term process involving many consumers and other stakeholders. To explore how policies may stimulate the transition towards a more sustainable transport system a Consumat application, called STECCAR, has been developed [14, 15]. In STECCAR, detailed data of 2974 representative Dutch respondents were used to parameterise an agent architecture based on the Consumat architecture. The data includes individual characteristics of the respondents, their current vehicle, and driving behavior, as well as perceptions and evaluations of various attributes of full electric cars, attitudes towards full electric technology, the likelihood of adopting a full electric car, and the adopter type regarding innovative cars.

Experiments were conducted with different policies, such as taxing fuel cars, subsidising BEV and PHEV cars and expanding the network of fast chargers. The results demonstrated that single policies do not result in very large effects. Interestingly, combining different policies resulted in an overall effect that was larger than the sum of the individual effects. From an emissions perspective it makes a difference if PHEV's or BEV's dominate the market. Hence experiments were conducted supporting both types of cars versus BEV only. In Fig. 3 you see the diffusion of PHEV's and BEV's over a period of 20 years.

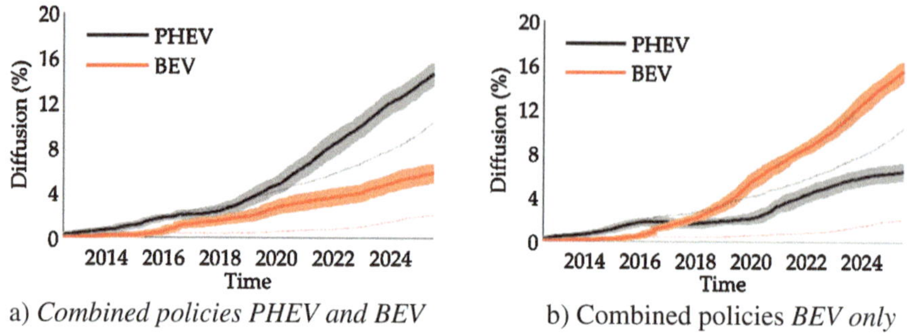

Fig. 3 Diffusion of PHEVs and BEV for stimulating both (**a**) or only BEV (**b**). The *solid line* represents the average diffusion over ten runs, the *lighter area* represents the standard deviation at each moment. The average diffusion process during the default scenario is added for reference (*dotted line*)

The results of this simulation exercise indicate that—given consumer decision-making processes—a diffusion process can be a relative slow process, and different policies could be implemented ensemble in order to create a stronger aggregated effect. Also it is clear that a good timing is important (sequencing of policies) and a clear long-term determination, because of the slow effects. Experiments indicate that effective policy requires a long-lasting implementation of a combination of monetary, structural and informational measures. The strongest effect on emission reduction requires an exclusive support for full battery electric cars, and no support for hybrid cars. As a general conclusion this study showed the possibility of studying technology development, economic policy and behavioral effects in combination using a social simulation model.

2.1.3 Simulating Farmer Behaviour and Land Use

The consumat approach is also been used beyond the domain of consumer behavior. In an early application the overharvesting of common-pool resources was explained from a behavioural dynamic perspective [16]. Responses to floods were also being explored using the Consumat by Brouwers and Verhagen [17]. In particular a number of applications have been developed targeting farmer behaviour. For example, Speelman [18] address the type of crop and individualistic versus cooperative behaviour of farmers in Mexico in relation to e.g. developments in the international market. Acosta-Michlik and Espaldon [19] and Mialhe et al. [20] studied the vulnerability of farmers in the Philippines for climate change.

Van Duinen et al. [21] studied how farmers in the south-west delta area of the Netherlands would adopt to an increasing risk of drought and salinization of the soil. Due to climate change droughts are expected to occur more frequently and to become more severe in the future, threatening crop production. Farmers need to adapt to these possible draughts in order to secure their income. However, among farmers there is a lot of uncertainty about the future frequency and severity of droughts. A survey based on a potential sample of 1474 members of a Dutch agricultural organization (the LTO) was conducted to elicit farmers' drought risk perceptions, adaptive behavior and socio-economic characteristics. These data were used to create an empirically based agent population based on the Consumat framework. The aim of the simulation study was investigating the effects of empirical social networks and different behavioral rules on farmers' choices under drought risk and its impacts on several macroeconomic indicators such as the rate of adaptation and income of the agricultural sector.

Farm agents operate in an environment consisting of the biophysical and socio-economic sphere. Their decisions depend on and affect both of these environments. In simulation experiments the degree of adopting irrigation technologies was being studied for a period of 30 years. This was being done for conditions of a stable climate and for a KNMI'06 climate change scenario developed by the Royal Dutch Meteorological Institute. Moreover, a comparison was made between the traditional economically optimizing agent versus agents that would employ all four decisional strategies as implemented in the Consumat. Figure 4 shows the results are presented for the resulting 4 conditions, each condition representing 30 model runs.

The *rate of adopters* gives insight in the dynamics of the adaptation process, see Fig. 4. The rate of adopters is measured as the share of farm agents who adopt irrigation on at least one of his fields.

The results indicate that when farm agents rely on heuristics and social networks the adoption process follows a different path than under optimization. The adoption process takes more time as farmers rely on information flows within their social network to inform their decisions. When agents do not observe a satisfactory number of adopters in their social network they remain uncertain and are reluctant to adopt. Once a critical threshold of number of farmers has adopted the technology, the

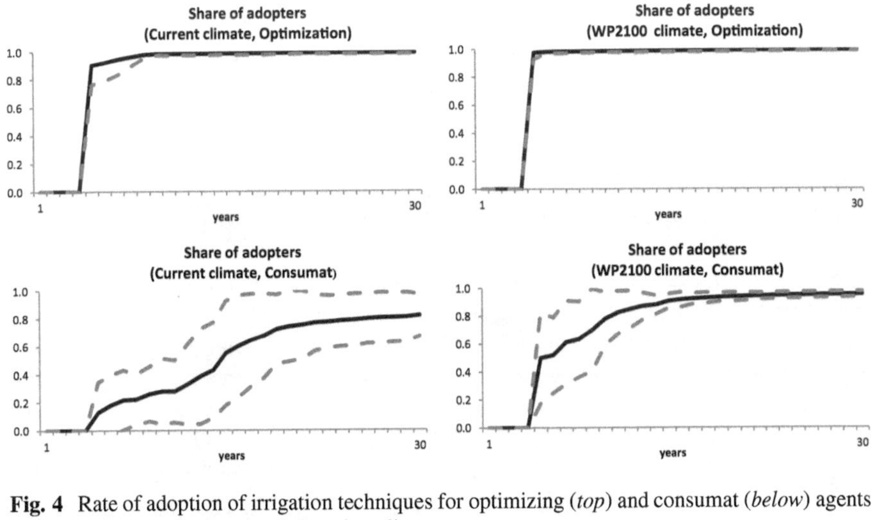

Fig. 4 Rate of adoption of irrigation techniques for optimizing (*top*) and consumat (*below*) agents under conditions of current and changing climate

diffusion of the technology takes off. Uncertainty about the effects of climate change slows down the adoption process. Hence, in case of climate change, more severe draught events reduce the uncertainty and stimulate a faster diffusion of irrigation technology.

This application of the Consumat approach demonstrates the relevance of including behavioural decision-making strategies in models, as they give a different, and possibly for policy making relevant perspective on how farmers respond to climate change.

2.1.4 Transitions in a Virtual Society

Formalising behavioural theory also offers a perspective on integrating social and environmental sciences. An early demonstration of this possibility is done by Jager et al. [8], who explored how psychologically more realistic agents would manage an "artificial world". The basic question was if such formalizations of behaviour would result in different human–environment interactions compared to standard economically optimizing agents. An artificial world called Lakeland was constructed based on a simple integrated model comprising two natural resources: a fish stock in a lake and a nearby gold mine. Mining would pollute the lake and have a negative effect on the fish stock. The model also included an economical submodel, allowing the agents to sell fish and gold, and to buy food and status-enhancing products. The model comprised only 16 agents that were equipped with four needs: (1) subsistence, to be satisfied with fish or gold, (2) identity, expressed as the relative amount of money an agent owns in comparison to a subset of agents having about similar abilities, (3) leisure, referring to the share of the time spent on leisure, and (4) freedom, associated with the total amount of money owned. Following the Consumat approach, the satisfaction and uncertainty experienced by these agents led them to employ different decisional strategies.

In an experiment we started with all agents being fisher, and opening the gold mine at $t = 1$. Because the agents differed in their abilities—their skills in both fishing and mining—the agents could decide on changing behaviour. We formalised both the homo economicus, that would exclusively engage in optimising, and the homo psychologicus, that would employ all four decisional strategies from the consumat approach.

For both the homo psychologicus and the homo economicus condition, 100 simulation runs were performed. In Fig. 5 the proportion of time spent fishing and mining for both the homo economicus and homo psychologicus are presented. The transition from a fishing to a mining society was more complete for the psychologically realistic agents. Due to processes of imitation and social comparison, many more agents started working in the mine than was optimal from an economical point of view. This instigated extra pollution of the lake, which led to a decreasing fish stock. As a consequence, the relative harvest from the mine was larger, thereby propagating the completion of the transition. These results confirmed the idea that macro level indicators of sustainability, such as pollution and fish harvest, are strongly and predictably affected by behavioural processes at the micro level.

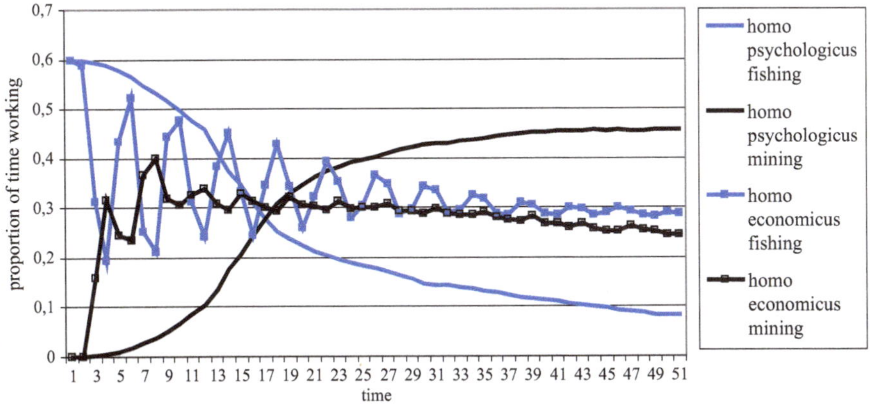

Fig. 5 Time spent fishing and mining for the homo economicus and psychologicus conditions

This Consumat application demonstrates that the incorporation of a microlevel perspective on human behavior within integrated models of the environment yields a better understanding of the processes involved in environmental degradation. In particular, the large-scale transition toward mining was explained by processes of imitation and social comparison.

2.2 *Future Challenges and Perspectives*

The Consumat has been developed as a conceptual tool to facilitate the modelling of human behaviour. The key contributions reside in the coupling of different behavioural principles in a single structure. This allows for modelling behavioural dynamics such as habitual behaviour. For example, if agents are satisfied, they can continue their behaviour habitually despite of the presence of better alternatives. When their satisfaction decreases due to worse outcomes they may reconsider their behaviour. Also when they become uncertain due to friends changing behaviour, they may engage in social processing. These are routes that may cause a behavioural change, and if the new behaviour is satisfactory a new habit may emerge. Aggregating these behavioural processes over a population of heterogeneous agents will result in outcomes that may deviate significantly from the rational actor approach. Especially when dealing with a transition towards a sustainable society, it is important for policy development to take into account such behavioural dynamics. A deeper understanding of these dynamics may contribute to a better identification of possible barriers for change, and may provide insights that contribute to more effective policies. In providing students and managers an environment that contributes to their understanding of these complex dynamics, the consumat has been implemented in a gaming context (see [22]). Here players have to manage a transition towards sustainability in the energy sector, playing the roles of companies and political parties that have to interact with a consumat based artificial population.

3 Examination of Environmental-Friendly Behavior with SiMA-C

When examining psychological reasons of sustainable behavior, the mental architecture SiMA (Simulation of the Mental Apparatus and Applications) [23] is a suitable means as a point of departure, since it provides a grounded explanation model of human behavior. Many commonalities with the Consumat approach can be identified (e.g. the motivational system and the consideration of different modes of decision-making), where the SiMA approach elaborates on some aspects, e.g. a process model of emotion and a detailed memory system.

3.1 *SiMA Approach*

The SiMA approach aims to develop a functional equivalent model of human information processing. Different than conventional cognitive architectures higher-level cognition and decision-making is based on low-level mechanisms, such as bodily needs and emotion, since it is assumed that such an embodied, holistic and functional model is better able to account for human decision-making.

The functional model (see Fig. 6) has four inputs, two for drives and two for perception, and an output for the decided actions. Drives represent the agent's desires stemming from bodily needs. Perception is distinguished in body-perception and environment perception. Drives and perception are compared and associated with the agent's experience to give a basis how the system may satisfy its desires in the perceived external world. In the defense track desires may lead to conflicts

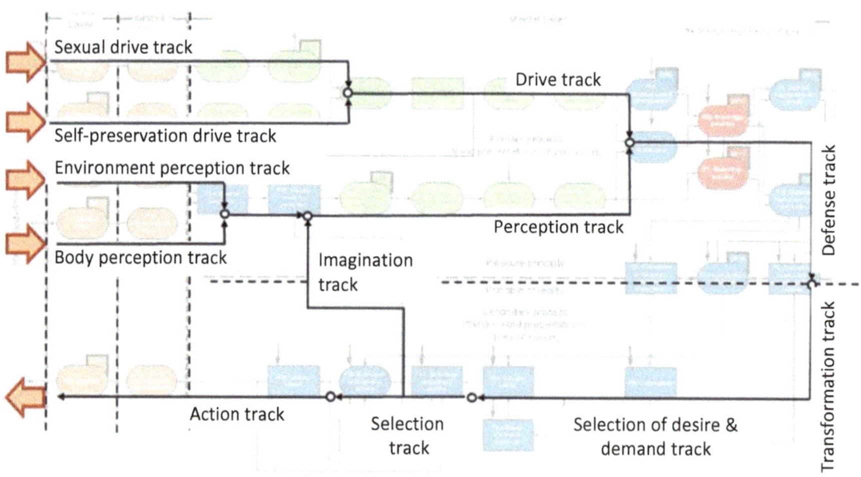

Fig. 6 SiMA model at the track level [24]

with internalized rules. The conflicts are handled by so called defense mechanisms, which may transform desires to social accepted ones. Next, goals are generated from desires and the affordances of the external environment. Finally, the best evaluated goal is chosen, a plan and action is selected that satisfies the different demands in the current situation best. In the action track this action is finally executed.

3.2 SiMA-C Model

The SiMA model provides a platform to transform concepts from psychoanalysis, psychology, and neuroscience into a computational model. However, for applications in social simulations a simplified version, SiMA-C (SiMA-Compressed), is developed [25, 26]. A basic psychological assumption is that decisions are fundamentally affected by emotion, which—in the context of decision-making—can be regarded as a valuation mechanism, e.g., a means of cost-benefit calculations (e.g., [27, 28]). Another psychological assumption considers the relevance of social norms in motivating behavior and their constitution by emotion. In this regard it is possible to distinguish norms that inform us about what is typically approved/disapproved (injunctive norms) and those that inform us about what is typically done (descriptive norms; [9]).

The SiMA-C model solves the problem of goal selection based on valuated memories that are activated by conflicting internal and external sources, representing demands and affordances, and possible reflections of this selection using emotion as a representation of an agent's current state of pleasure, displeasure, and conflict [26]. Model processing is triggered by a change in bodily needs or the external world, which causes data activation by two sources: Demands and affordances (see Fig. 7). Hence the activation process determines possibly relevant data for the current internal or external situation.

Two types of demand sources are distinguished: physical (bodily demands represented by drives and pain from body perception) and psychological (activated memorized norms). Both demand sources activate memorized goals and norms—directly by triggered activation or indirectly by spreading of activation—that are expected to bring pleasure by satisfying these demands. Goals may also be memorized with expected displeasure, if an (activated) goal would prevent the fulfillment of a demand or if a goal's object is expected to bring harm.

Beside demands, affordances are the second source of data activation. They activate memories that are similar to the current external world and—indirectly via spreading of activation—norms that are valid for the current external situation.

The relevance of activated goals for the current internal and external situation is determined by different valuation processes, triggered by the different activation sources, and considering the goal's memorized valuation for these sources. Hence a goal may have different valuations of expected pleasure and displeasure regarding the different activation sources. The current relevance of the single valuation is determined by the caused activation from that source. Besides memorized normative

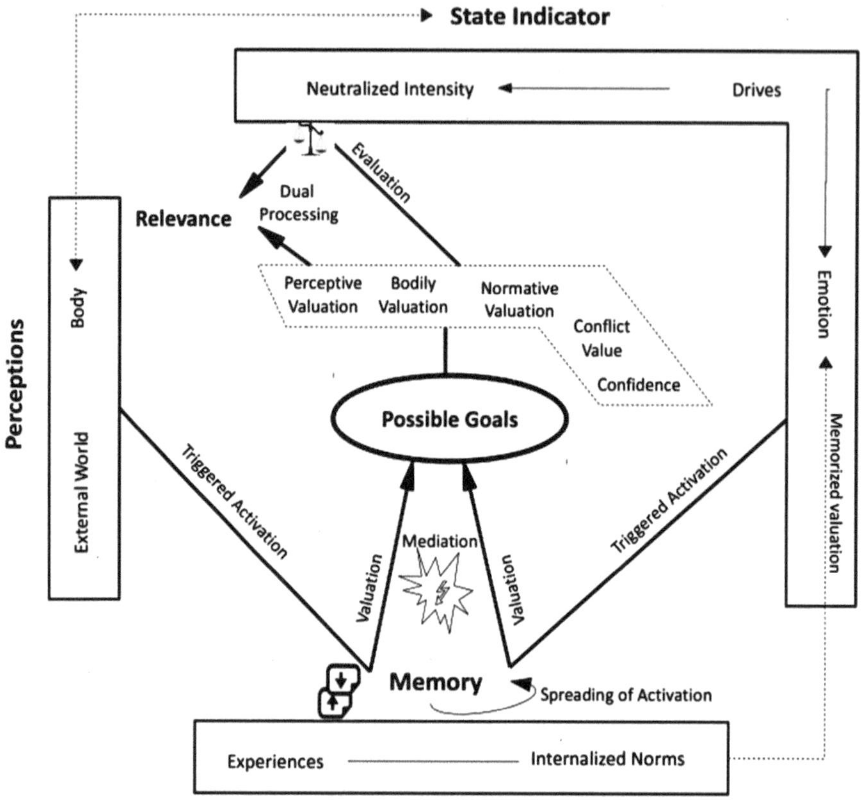

Fig. 7 SiMA-C model overview [26]

valuation, activated by normative sources, and bodily valuation, activated by bodily sources, perceptive valuation, activated by external perception are distinguished. The first two types of valuations correspond to an expected short-term fulfillment of bodily or normative demands, the latter uses a goal's memorized summary valuation, which provides integrative holistic information about all aspects of how a goal changed the agent's state, considering long-term expectations and context. The resulting data from the valuation process can be termed effective or affective, in an etymological sense, meaning that this data should have an impact (on the decision).

Contradicting valuations in a goal (e.g., expected pleasure for normative demands, but expected displeasure for bodily demands) cause conflicts. Additional conflicts are caused by contradictions between demands and affordances or between contradicting goals. For example, if an internalized norm is activated, goals unrelated to the norm would prevent its fulfillment. Therefore, they would be marked by a conflict. Two types of conflicts are distinguished: normative and reality conflicts, with ownership conflicts being a specialization of the latter. The different kind of conflicts are addressed in processes of mediation, which operate by changing the different valuations in a goal. The result can be called arranged data.

Parallel to these processes, the generated displeasure from the different demands and activated data, the currently experienced pleasure and conflict intensity together form the agent's state indicator, representing the agent's current emotional state. A share of the displeasure from drives is used as so-called neutralized intensity for reflective processes, regulating the grade of dual processing.

Based on the reactive processes described so far, evaluative reasoning processes, which relate the different valuations to the state indicator, are possible. The separation between valuation and evaluation corresponds to a dual processing model of the human mind (e.g., [29]). The degree of evaluation is dependent on its necessity in case of ambiguities between goals, and the agent's neutralized intensity. The overall process corresponds to weighing the different valuations in integrating them to a single relevance value. This evaluation corresponds to a multi-criteria aggregation aiming for displeasure minimization and pleasure maximization, while considering a goal's conflicts and the agent's confidence in the valuations. However, the guiding principle is that of a satisficing, not an optimizing agent (cf. [30]). Overall, the evaluation process results in determining the most relevant goal for the current internal and external conditions.

3.3 Applying the SiMA-C Model in Social Media Prompting of Environmental-Friendly Behavior

The applicability of using the SiMA-C model to examine the mechanisms and impact factors of environmental-friendly behavior is explored in the CogMAS (Cognitive Multi-Agent System Supporting Marketing Strategies of Environmental-Friendly Energy Products) project [25]. We chose to use a scenario of social media communication to explore the requirements and assumptions for the SiMA-C model in the CogMAS context. In a next step we structured and concretized this description into a data model for the social media scenario. The analyzed parameters are mapped into survey questions to gather empirical data to calibrate the model. Simulations show that the assumptions behind our model are indeed capable of generating and determining the targeted behavior (e.g., switching to green electricity in a social media scenario, see Sect. 3.3.2). By demonstrating that the behavior of the simulation model stands in line with the survey data, we show that the model is a valid explanation of the people's decision-making process. Going beyond the possibilities of a survey, the simulations is used to address additional questions (see the section "Experimentation").

3.3.1 Gathering Empirical Data for Agent Parametrization

The empirical data for the agent parametrization was based on our general assumptions that norms and emotions are key factors in social decision-making (i.e., decisions that are influential and influenced by the social context one operates in).

We reasoned that this should also be the case for environmental decision-making such as the choice between green and grey energy providers. The goal of collecting this empirical data was to be able to use concrete parameters which—in combination with the data they activate and valuate- influence the SiMA-C agent's decision to switch to green electricity or not.

For this, a survey on green electricity was designed and implemented to explore the reasons that encourage the decision for green energy consumption. The survey consisted of a questionnaire about the general attitude towards the environment, social norms and personality indicators of the participants themselves. Additionally, we included a brief experiment of the awareness of social media as a suitable communication channel about switching from grey to green energy.

To identify empirical evidence for decision factors of agents, a number of characteristics and attitudes were defined and tested in the survey. To give an overview, based on the required parameters for the SiMA-C model, the following seven categories of possible influencing factors were formed: (1) Personality factor (i.e., intuitive vs. reflexive decision-making), mapped to personality parameters in the SiMA-C agent; (2) initial internal state (e.g., happy, sad, hungry, . . .), mapped to the agents' state indicator; (3) previously activated data (i.e., daily plans and relevant activities), mapped to the agents' initial activation values of memories; (4) attitude, trust and prior knowledge about green energy, (5) emotions related to green energy (e.g., pride, envy, guilt), mapped to associated memories, confidence, and summary valuation of green energy; (6) social norms related to environmental behaviour (i.e., descriptive and injunctive norms), mapped to memorized norms and their strength and implemented by different ways in activating norms; and (7) self-efficacy in relation to environmental protection, mapped to the agents' personality parameters. In the next step, exploratory analyses were conducted to see whether any of these variables can predict participants' choice of green versus grey energy consumers.

Regression analyses revealed that the perceived ease of switching energy providers, personal involvement, and higher descriptive norms (i.e., how many people in one's social environment already use green energy) explain 43 % of the variance in participants' current energy choice (i.e., which energy source they currently use). These predictors correctly identify 75 % of green and 90 % of grey energy consumers.

Additionally, people's intentions to switch from grey to green energy can be predicted by participants' attitude towards green energy (i.e., more favorable attitudes are connected with stronger intentions to switch), personal involvement, and injunctive norms (i.e., whether one *should* switch). These three predictors can explain 28 % of the variance in intentions to switch from grey to green energy, and they correctly forecast 86.4 % of those people who are willing to switch and 57.3 % of those who are not willing.

Finally, these variables as well as emotion variables (such as pride, envy, and guilt) were used to create different clusters of energy consumers, which can later be used as a blueprint for specific agents. To do so, three analyses were done with different variable combinations. In the first analysis the variables ease of switching, involvement and descriptive norms were used to form the following four homogenous clusters on base of the current energy resource: (Cluster 1) Grey energy consumer with high involvement, (2) Grey energy consumer with low involvement, (3) Green energy consumer with low descriptive norms, and (4) Green energy consumer with high descriptive norms. In the second analysis, the variables attitude, involvement, injunctive norms, pride, envy and guilt were used to form four homogenous clusters on basis of the willingness to switch: (5) High willingness to switch with weak emotions, (6) High willingness to switch with strong emotions and high descriptive norms, (7) Mixed group with weak emotions, and (8) High willingness to switch with strong emotions and low descriptive norms. Finally, the third analysis used the variables attitude, involvement, injunctive norms, descriptive norms, ease of switching, pride, envy and guilt to form three homogenous clusters on base of the willingness to switch: (9) Mixed group with weak emotions, (10) High willingness to switch, and (11) Low willingness to switch with low involvement.

To sum up, the results show three main decision factors which will influence the agent´s acting in the simulation. These include the attitude towards green energy, personal involvement, and injunctive norms. An agent who is willing to switch therefore should have a positive attitude towards energy and prior knowledge. The agent should also be interested in this topic and willing to collect a variety of information about different energy providers. Furthermore, agents ought to place importance on behavior which is typically approved or disapproved in the society.

3.3.2 Simulation

To replicate the findings of the survey, agents were parametrized according to the clusters from the survey. We normalized the Likert scale (1–5) used in the survey to a range of 0–1 and parametrized the agents' variables according to the mapping described above. Using a social media scenario in simulations (where an agent gets a message from a friend about switching to green electricity, details see below), the goal's relevance to get green electricity is compared to the participants' tendency to get green electricity, which reflects a question in the survey (see Fig. 8).

We achieved a good replication of the empirical data without any model calibration: The tendency to get (or have) green electricity can be predicted by the SiMA-C model. However, analyzing the relevant variables in the clusters that are not replicated well (e.g., cluster 2 and 11), we assume that the interplay of valuation and evaluation (see Sect. 3.2) does not sufficiently cover the interplay of intuitive and reflective decision. This seems especially the case for considering variables of involvement.

Experimentation

The survey's experimental part showed some indication that manipulating the visual cues given by the message's background picture influenced participants' expressed

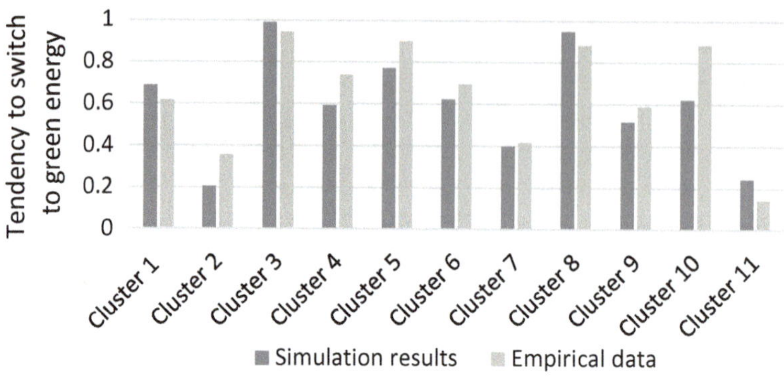

Fig. 8 Comparison of simulation and survey results in the CogMAS project

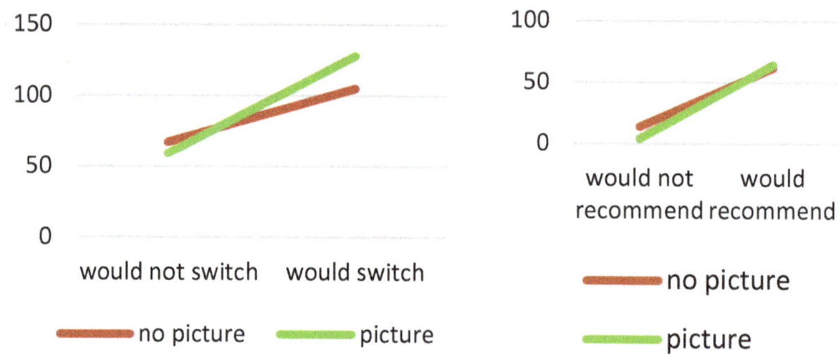

Fig. 9 Visual cues influence willingness to switch to and recommend green energy (Y-axes denote absolute frequencies)

willingness to switch to green electricity (see Fig. 9). Apparently, people are more willing to switch to green energy when they are prompted with a background picture that displays nature. However, displaying the picture has no significant impact on the willingness to recommend green energy to friends.

We further tested the impact of a background picture of nature in simulations and observed if the SiMA-C model is able to propose reasons for its impact on the decision. We extended the model to consider the impact of state-perception on decision-making. This functionality is triggered by perceiving an object (nature) or agent's (message-sender) state. For instance, another agent may indicate pleasure by corresponding bodily expressions. Based on the memorized relation and valuation of that object or the other agent (which can be parametrized according to the survey), the agent's pleasure and displeasure state (represented by the state indicator, see Sect. 3.2) is altered, which influence evaluation of goals. This process can be described as affective empathy.

Similar to the survey's findings about the presence of visual cues about nature, simulations show an impact on the decision when perceiving a message with a background picture of nature (see Fig. 10). Additionally, the state of nature in the

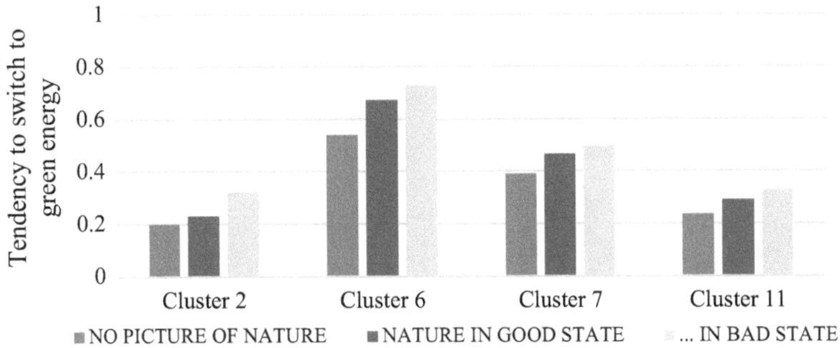

Fig. 10 Impact of perceiving different states of nature on the tendency to switch to green energy

picture only slightly changes the tendency to switch to green energy. Hence although it may seem counterintuitive, the state of nature in a background picture of a social media message does not seem to have an impact on the tendency to switch to green electricity (see Fig. 10, where an agent associates different states of displeasure and pleasure for the different pictures).

An analysis using our model provides candidate explanations: The manipulation of the agent's state (e.g., increase of displeasure due to the bad state of nature in a background picture) does not consider the concrete reasons of the current state, but only the abstract sources (normative, perceptive, bodily). In other terms: Unconscious affective empathy alone is not sufficient, the agent also has to reflect on the concrete reasons of its current state to decide which goal will have the best effect on its state. That is, the message has to emphasize the relation of the agent's state and the environments state and nudge the agent to reflect about that.

Model Walk-Through Using a Social Media Scenario

Next we will use different visualizations (generated in simulations) for a walk-through in the simulations to demonstrate the causal chain in decision-making and how the various impact factors of environmental friendly behavior are covered by the generic SiMA-C model [26]. For demonstration purposes, we use a concrete social media scenario that considers all mentioned impact factors [31]. Victoria is currently working and gets a social media message from her friend Caroline about switching to green electricity and how good this felt. The configuration of simulations includes creating memories. Victoria memorizes several internalized norms, e.g., to appreciate friendship, improve the world, and protect the environment. Part of the memories are also the knowledge about how she satisfies her needs, pleasureful thoughts and her relation to nature and to other people, e.g., to Caroline. Victoria's initial internal state is configured as hungry and a bit exhausted. But nevertheless Victoria is in an overall joyful state. Her conflict state is quite high (which does not exclude the joyfulness) since she has not confirmed to her own

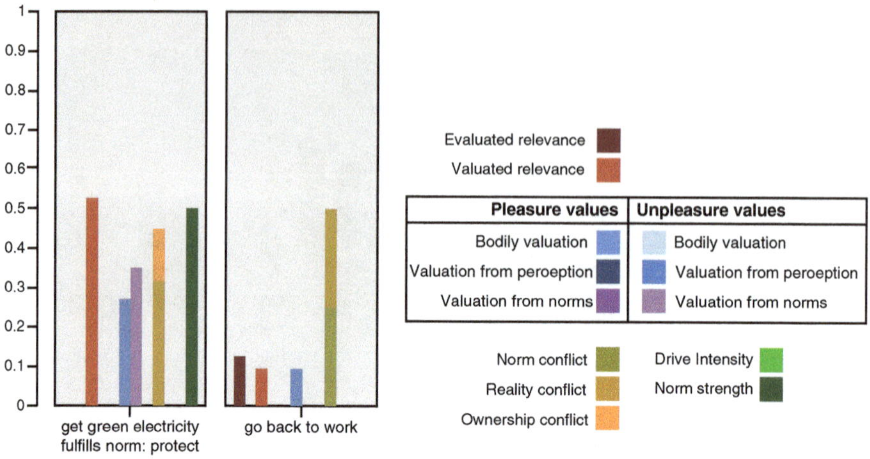

Fig. 11 Overview of goal selection in a social media scenario

norms recently. The personality factors of Victoria informs about her general norm conformity and how intuitive she is in general. Due to the simulation scenario, the goal to work is configured as pre-activated.

Figure 11 gives an overview of why the goal to get green electricity is highly ranked, given the agent's current internal and external situation.

The high relevance value in red is mainly due to valuation (light red), evaluation does not affect the overall goal relevance. Norms that are associated with green electricity are highly activated and impact the decision (green bar). Due to the effort to get green electricity mainly reality conflicts exist. Unsurprisingly the primary source of valuation are norms (in magenta) and the secondary source are direct activations from perception (blue). Figure 12 shows concretely how the goal to get green electricity is activated after perceiving the social media message.

Perceiving the prompt from Caroline to switch to green electricity directly activates the corresponding goal. Caroline is memorized as a close friend (i.e., high association strength and positive summary valuation), which is why her normative prompt is activated highly. An additional activation comes from the perceived high number of "Likes". These two aspects can be described as following a descriptive norm. Due to direct and indirect activation of the norm to protect environment (which is activated directly to the message and indirectly due its associations with the perceived message's background picture of nature and the prompt to get green electricity) the goal gets most of its activations from normative sources. This corresponds to following an injunctive norm. Hence these norm types are distinguished by the type of activation.

Due to the high activation by perception and memorized valuations perceptive and normative valuation has the highest impact. This reflects the relevance of an agent's attitude and norms (indicated by the survey), which is represented in the model by summary valuation and a configured pre-activation value.

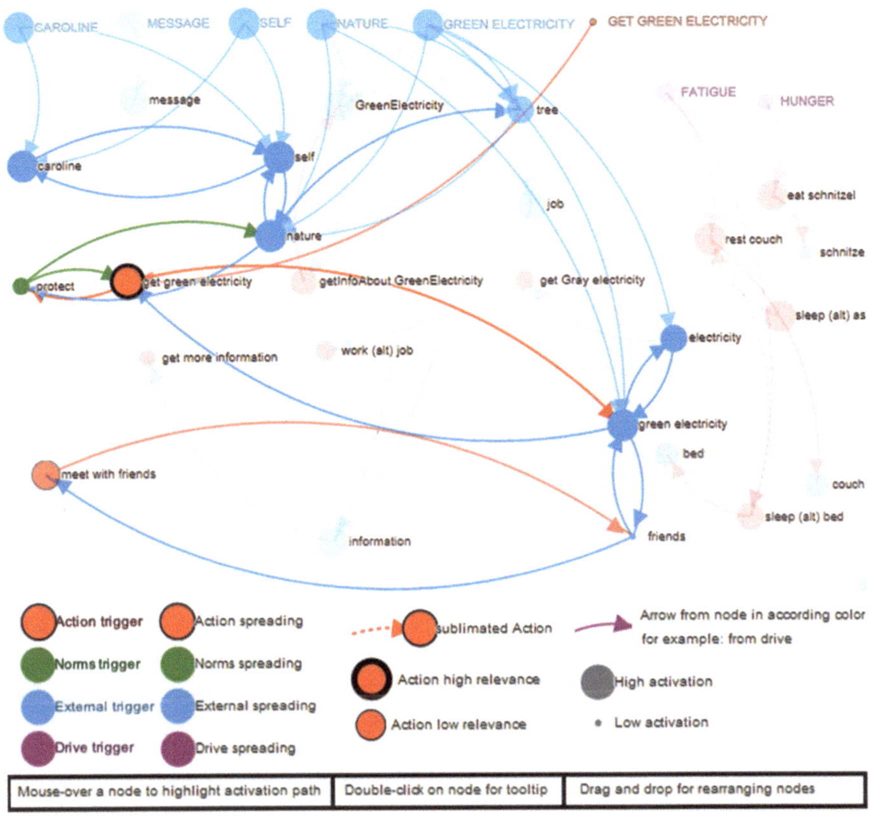

Fig. 12 Activation pattern in associative memory for the social media scenario (screenshot)

Due to the set personality parameters of Victoria, she is more of an intuitive type, thus valuation is the main source for the goal's relevance. Additionally the low uncertainty in valuations of different goals does not require an intensive reflection on their selection. However due to her hunger she has a high degree of neutralized intensity (which regulates dual processing, see chapter *SiMA-C model*) and reflects on her decision, i.e., conducts evaluation. As described, amongst others, evaluation also considers the conflict intensity of goals and consider the expected change of an agent's state. Since Victoria is initially in a conflicting state, which is increased by current conflicts due to the social media message, her normative valuations additionally impact the goal's relevance in evaluation. The current state, which is used in evaluation, has changed due to the current demands, activated data, and mediated conflicts. For instance, the high valuations of green electricity leads to an ownership conflict, since Victoria recognizes that Caroline, but not herself, has green electricity. Due to memorized association strength between Victoria and Caroline and positive summary valuations of Caroline, Victoria is able to identify with Caroline and imitate her behavior in getting green electricity. The resulting state indicator corresponds to the descriptive emotions of envy and guilt. Additionally the evaluation process is able to consider variables of involvement (e.g. confidence). However, as indicated in Victoria's case, the decision to get green electricity is mainly a reactive one due to activation and valuation of corresponding memories.

4 Discussion and Conclusion

Agent-based modeling (ABM) is able to bridge the gap between psychological and sociological perspectives by providing a platform for their integration. To model the dynamics of human behavior in such integrative simulations, psychologically plausible agents are required. This is especially the case for sustainable behavior, which can only be explained realistically on the psychological level under consideration of the social context.

We presented two examples of psychologically plausible agent models, Consumat and SiMA-C, and their application to examine sustainable behavior. By showing how social dynamics are generated with dynamics on the psychological level and how the social context influences individual decision-making, we integrate psychological and sociological perspectives. Both present examples of how psychologically plausible agent models provide explanations on the psychological level that enable the development of social policies on the individual level. In case of the Consumat approach we saw how the consideration of different decision strategies is able to explain dynamics in sustainable behavior. Using the SiMA-C model we showed how a detailed model of decision-making supports realistic empirical validation. In simulation experiments we show how the translation of emotion into a model of valuation and psychological dual processing is able to explore psychological causes of (often unexpected) observed behavior and inform policies for environmental-friendly behavior. Hence both models provide deeper explanations than conventional ABM.

The communalities between Consumat and SiMA-C show the importance of considering a generative approach to motivations and different modes of decision-making. This is especially important when the motives, and the external (environment) and internal (personality, agent's state) conditions of sustainable behavior are at stake. Hence, in explaining sustainable behavior, both, psychological aspects (such as different demands from needs and norms) and social aspects (such as social pressure) and their integration have to be considered. The Consumat and SiMA-C approach are both able to provide a framework to integrate psychological and sociological perspectives and enable the explanation of behavioral dynamics by the interplay of multiple model variables.

Consumat and SiMA-C are similar in basing behavior on motivations and in considering heuristics. The Consumat integrates these aspects more explicitly. For instance, a core process addressed in the Consumat framework—the selection of a suitable decision strategy given the situation—is not explicitly considered in SiMA-C. However, as general in the generative-functional SiMA-C model, the resulting interaction between its components is able to generate similar patterns of decision making as in Consumat. SiMA-C focusses on how these different strategies are generated by underlying mechanisms, most importantly emotion, a driver not explicitly addressed in the Consumat. Also within SiMA-C a more explicit modeling of the memory is formalized, allowing for the parameterizing of storage and retrieval of concrete memories. The higher degree of parametrization and memory-processing (e.g. spreading of activation) enables more detailed model exploration and experimentation, while it makes simulations (often unnecessarily) more complicated. Therefore, we expect that the Consumat model is more efficient in modeling the basic behavioral dynamics of a population that is interacting within an environmental relevant setting (e.g. resource use, consumption), whereas SiMA-C offers a possibility to experiment with strategies addressing the emotions and experiences of this population.

The compliance between the Consumat and the SiMA-C approach, and the partial difference in focus and level of details, provide possibilities to join the two approaches for deeper explorations of social phenomena. As Consumat provides a lightweight model, practical value in social simulations, and considers key factors of a psychological model, it can be used as a starting point to identify cases in simulations that require further exploration (e.g., because it identifies a wide variety of possible outcomes, or unexpected social barriers for change).

In such cases SiMA-C may be able to provide more insights of agents' decision-making processes in different personalities and concretize Consumat's variables where (assumedly) necessary. For instance, using a model of emotion in SiMA-C (and the dynamics of memory activation and valuation) detailed reasons for why agents choose goals can be provided. This also enables fine-grained parametrization with empirical data. Overall, Consumat enables a platform for broad policy testing, and may direct further explorations with SiMA-C, when it is assumed that the focus of SiMA-C provides detailed variables (for policies) to dock on. This would also support individually tailored policies and simulation experiments that address particular key-agents (e.g. opinion leaders) in a more precise way. Overall, as future work we aim to join the contributions of the Consumat and SiMA-C approaches to enhance exploration of social behavior in environmental contexts and harness the flexibility of using two compliant models with partial different focus and different levels of details.

References

1. Castelfranchi, C.: Cognitive architecture and contents for social structures and interactions. In: Cognition and Multi-Agent Interaction. Cambridge University Press, New York (2006)
2. Viale, R.: Methodological Cognitivism: Vol. 1: Mind, Rationality, and Society. Springer, Berlin (2012)
3. Gilbert, N.: When does social simulation need cognitive models? In: Sun, R. (ed.) Cognition and Multi-Agent Interaction. Cambridge University Press, New York (2006)

4. Jager, W., Janssen, M.: The need for and development of behaviourally realistic agents. In: Multi-Agent Simulations, I.I. (ed.) Lecture Notes in Computer Science, vol. 2581, pp. 36–49. Springer, Heidelberg (2003)
5. Sun, R.: Cognition and Multi-Agent Interaction. Cambridge University Press, New York (2006)
6. Maslow, A.H.: Motivation and Personality. Harper and Row, New York (1954)
7. Max-Neef, M.: Development and human needs. In: Ekins, P., Max-Neef, M. (eds.) Real-life Economics: Understanding Wealth Creation. Routledge, London/New York (1992)
8. Jager, W., Janssen, M.A., De Vries, H.J.M., De Greef, J., Vlek, C.A.J.: Behaviour in commons dilemmas: Homo economicus and Homo psychologicus in an ecological-economic model. Ecol. Econ. **35**, 357–379 (2000)
9. Cialdini, R.B., Goldstein, N.J.: Social influence: compliance and conformity. Anu. Rev. Psychol. **55**, 591–621 (2004)
10. Janssen, M.A., Jager, W.: Stimulating diffusion of green products—co-evolution between firms and consumers. J. Evol. Econ. **12**(3), 283–306 (2002)
11. Bravo, G., Vallino, E., Cerutti, A.K., Pairotti, M.B.: Alternative scenarios of green consumption in Italy: an empirically grounded model. Environ. Model. Software **47**, 225–234 (2013)
12. Natalini, D., Bravo, G.: Encouraging sustainable transport choices in american households: results from an empirically grounded agent-based model. Sustainability **6**(1), 50–69 (2014)
13. Schoenmacker, G.H.: Agent-based consumer modeling. Master thesis, Department of Artificial Intelligence, Faculty of Mathematics and Natural Sciences, University of Groningen (2013). http://scripties.fwn.eldoc.ub.rug.nl/scripties/Kunstmatigeintellige/Master/2014/Schoenmacker.G./
14. Kangur, A.: STECCAR: simulating the transition to electric cars using the consumat agent rationale. Master's thesis, Institute of Artificial Intelligence, University of Groningen (2014) http://scripties.fwn.eldoc.ub.rug.nl/scripties/Kunstmatigeintellige/Master/2014/Kangur.A.M.A./
15. Kangur, A., Jager, W., Verbrugge, R., Bockarjova, M.: Hybrid or full electric, that's the question: an agent-based model exploring the diffusion of electric vehicles. Journal of Environmental Psychology. In: Jager, W., Ernst, A. (eds.) Special issue "Social simulation in environmental psychology" (accepted)
16. Jager, W., Janssen, M.A., Vlek, C.A.J.: How uncertainty stimulates over-harvesting in a resource dilemma: three process explanations. J. Environ. Psychol. **22**(3), 247–263 (2002)
17. Brouwers, L., Verhagen, H.: Applying the CONSUMAT model to flood management policies. In: Müller, J., Seidel, M. (eds.) 4th Workshop on Agent-Based Simulation, pp. 29–34. SCS, Montpellier (2003)
18. Speelman, E.N.: Gaming and simulation to explore resilience in contested agricultural landscapes. PhD thesis, Wageningen University, Wageningen, the Netherlands, p. 221. ISBN: 978-90-6173-904-9 (2014). See http://edepot.wur.nl/301294
19. Acosta-Michlik, L., Espaldon, V.: Assessing vulnerability of selected farming communities in the Philippines based on a behavioural model of agent's adaptation to global environmental change. Glob. Environ. Chang. **18**, 554–563 (2008)
20. Mialhe, F., Becuc, N., Gunnelld, Y.: An agent-based model for analyzing land use dynamics in response to farmer behaviour and environmental change in the Pampanga delta (Philippines). Agric. Ecosyst. Environ. **161**, 55–69 (2012)
21. Van Duinen, R., Filatova, T., Jager, W., Van der Veen, A.: Going beyond perfect rationality: drought risk, economic choices and the influence of social networks. Ann. Reg. Sci. pp 1–35 (2015)
22. Verhagen, H., Johansson, M., Jager, W.: Games and online research methods. Fielding, N.G., Lee, R.M., Blank, G. (eds) The SAGE Handbook of Online Research Methods, 2nd edn. Sage Publications Inc. (2016)
23. Dietrich, D., Fodor, G., Zucker, G., Bruckner, D.: Simulating the Mind—A Technical Neuropsychoanalytic Approach. Springer, Vienna (2009)

24. Schaat, S., et al.: ARS: An AGI Architecture. Proceedings of the 7th conference on Artificial General Intelligence, LNAIC, pp. 155–164. Springer, Switzerland (2014)
25. Schaat, S., et al.: Socio-cognitive agents for the simulation of environmental-friendly consumer decisions. In: International Conference on Affective Computing and Intelligent Interaction, pp. 851–856, Xian (2015)
26. Schaat, S.: SiMA-C: a foundational mental architecture. In: Proceedings of the Conference on Brain Inspired Cognitive Architectures, July 2016
27. Damasio, A.R.: Descartes' error and the future of human life. Sci. Am. **271**(4), 15–20 (1994)
28. Slovic, P., Finucane, M., Peters, E., MacGregor, D.G.: The affect heuristic. In: Gilovich, T., Griffin, D., Kahneman, D. (eds.) Heuristics and Biases: The Psychology of Intuitive Judgment, pp. 397–420. Cambridge University Press, New York (2002)
29. Kahneman, D.: Thinking, Fast and Slow. Penguin, London (2011)
30. Simon, H.A.: Rational choice and the structure of the environment. Psychol. Rev. **63**(2), 129–138 (1956)
31. Schaat, S., et al.: Emotion in consumer simulations for the development and testing of recommendations for marketing strategies. In: Proceedings of the 3rd Workshop on Emotions and Personality in Personalized Systems, pp. 25–32, Vienna (2015)

Interactions Matter: Modelling Everyday Pro-environmental Norm Transmission and Diffusion in Workplace Networks

J. Gary Polhill, Tony Craig, Amparo Alonso-Betanzos, Noelia Sánchez-Maroño, Oscar Fontenla-Romero, Adina Dumitru, Ricardo García-Mira, Mirilia Bonnes, Marino Bonaiuto, Giuseppe Carrus, Fridanna Maricchiolo, Ferdinando Fornara, Corina Ilin, Linda Steg, Angela Ruepert, and Kees Keizer

Abstract This chapter demonstrates an approach to the agent-based modelling of norm transmission using decision trees learned from questionnaire data. We explore the implications of adding norm dynamics implied in static questionnaire data and the influence social network topology has on the outcome. We find that parameters determining network topology influence the outcome in both hierarchical and co-worker networks in a simulated workplace. As an exercise in empirical agent-based modelling, this work highlights the importance of gathering data on interactions in field studies.

J.G. Polhill (✉)
Information and Computational Science, The James Hutton Institute, Aberdeen, Scotland, UK
e-mail: gary.polhill@hutton.ac.uk

T. Craig
The James Hutton Institute, Aberdeen, Scotland, UK
e-mail: tony.craig@hutton.ac.uk

A. Alonso-Betanzos • N. Sánchez-Maroño • O. Fontenla-Romero • A. Dumitru • R. García-Mira
Universidade da Coruña, A Coruña, Spain

M. Bonnes • M. Bonaiuto
Centro Interuniversitario di Ricerca in Psicologia Ambientale (CIRPA), Sapienza Università di Roma, Rome, Italy

G. Carrus • F. Maricchiolo
CIRPA, Università Roma Tre, Rome, Italy

F. Fornara
CIRPA, Università di Cagliari, Rome, Italy

C. Ilin
Universitatea de Vest din Timisoara, Timisoara, Romania

L. Steg • A. Ruepert • K. Keizer
Rijksuniversiteit Groningen, Groningen, The Netherlands

A. Alonso-Betanzos et al. (eds.), *Agent-Based Modeling of Sustainable Behaviors*, Understanding Complex Systems, DOI 10.1007/978-3-319-46331-5_2

1 Introduction

The LOCAW (Low Carbon at Work: Modelling Agents and Organisations to achieve Transition to a Low Carbon Europe)[1] project concerned everyday pro-environmental behaviours in the workplace and their relationship with such activities at home. It focused on three areas of pro-environmental behaviour: waste management, energy consumption and transport. There were six case studies in locations throughout Europe: two in heavy industry, two service companies, a university and a municipality. Agent-based modelling was used in the four case studies not associated with heavy industry to simulate scenarios from back-casting workshops.

Everyday behaviour has been a focus of social science research since the early 1920s (e.g. [1]), and interest in it has been growing rapidly since the 1980s following the work of researchers such as Lefebvre [2] and de Certeau [3]. However, social and (physical) environmental drivers of behaviour are also important in facilitating the insertion of environmental awareness into the everyday—a domain more often associated with habit. In seeking to break the environmentally destructive aspects of habituated lifestyles without draconian regulation, there is an increasing focus on working with communities and engaging social norms. Socially cohesive communities have been found to be more supportive of pro-environmental behaviour [4, 5], and norms have been shown to have a role in predicting some pro-environmental behaviours, such as recycling [6] and household energy use [7].

Agent-based modelling of norms in the context of social control are reviewed by Hollander and Wu [8], who cite Verhagen [9] as couching social control in terms of achieving goals at the social level whilst allowing individual freedom, and Therborn [10] as arguing that social norms are a key tool in addressing the challenge of the social control problem in multi-agent systems. Though they state that normative architectures in agent-based systems have traditionally used Belief-Desire-Intention (BDI [11]) models of decision-making, there has also been a great deal of work on norms using more stylised models, often strongly influenced by game theory. The work of Axelrod [12] has been particularly influential in this area. Elsenbroich and Gilbert [13] provide a comprehensive review of approaches to the simulation of social norms in agent-based models, observing five categories thereof (p. 187), each with potentially different senses of what a norm is:

1. Environmental models, in which an environment contains resources and norms are represented in terms of property rights; the benefits of social norms are assessed by comparing outcomes in models using different decision-making algorithms (p. 90). The environment may also be used as a medium of agent interaction (stigmergy), rather than having the agents interact with each other directly (p. 92).

[1] http://www.locaw-fp7.com/.

2. Game theory models include the work of Axelrod [12], where norms are used as part of the decision-making model of the agents to encourage co-operation in social dilemmas, particularly the prisoner's dilemma.
3. Diffusion models. These are simple imitation models, where norms diffuse through a social network (or space) by agents copying what other agents are doing. The implementation of the imitation typically involves a relatively trivial equation weighting the individual agent's preferences against the behaviours of the other agents it can observe.
4. Social influence and learning models build on diffusion models by adding more deliberation on the part of the agent about whether or not a behaviour is adopted. Elsenbroich and Gilbert [13, pp. 116–120] give opinion dynamics and Schelling's [14] (and Sakoda's [15]) models of segregation as examples of this class, both of which could arguably be in the class of diffusion models, though they point out [13, p. 130] that there is a blurred boundary between diffusion models and social influence models.
5. Cognitive models of norms. These use richer cognitive architectures (such as BDI) from the (distributed) artificial intelligence and multi-agent systems literatures to represent reasoning about norms.

Although Hollander and Wu's [8] norm life cycle model is based on empirical results and theory, there has, interestingly, been relatively little work on the empirical agent-based modelling of norms, as the lack of a relevant category in Elsenbroich and Gilbert [13] suggests. The work presented here attempts to contribute to this area using agents with decision algorithms that include norm transmission (an area Hollander and Wu [8, para. 3.9] note is lacking in agent-based modelling literature), use local injunctive and descriptive norms, and have been configured using questionnaire data.

A growing interest in empirical agent-based modelling [16, 17] means there is a demand for methods to calibrate algorithms in such models from field data. For now, researchers are using standard tools in social science for this purpose: Smajgl et al. [17, p. 838] mention participant observation, surveys, interviews, census data, field experiments and role-playing games among others. Role-playing games are central to the maturing Companion Modelling approach [18] of the so-called 'French School' of agent-based modelling [19]. However, since empirical agent-based modelling typically involves interdisciplinary collaboration of social and computer scientists, we argue that there is a need to build on and develop other tools in social science, and in particular, those that focus on gathering evidence about interactions.

This chapter is concerned with the empirical agent-based modelling of norm transmission as a driver of pro-environmental behavioural change. We fit decision trees to questionnaire data that predict respondents' reported behaviour based on their answers to other questions on individual and normative drivers of pro-environmental behaviour. The questionnaire included questions on norm transmission (how likely the respondent is to tell their colleagues to behave more pro-environmentally), which, together with the questions on normative drivers of behaviour, creates an implied dynamic. The agent-based model simulates that dynamic, and we are interested in the effect the workplace social network has on the effect of the dynamic on the predicted behaviours of the agents. The workplace social network is interesting because interactions are mediated through formal hierarchical relationships and less formal co-worker relationships. Our hypothesis is that parameters affecting the topologies of these interactions affect the predicted behaviours of the agents, where these are affected by norms.

In what follows, we describe the questionnaire, how the responses to it were interpreted into an agent-based model, and experiments with the model exploring network topology parameters. We find that these significantly affect the results of the model, and conclude that empirical agent-based modelling studies (and indeed other empirical work in the social sciences) would benefit from a greater focus on development and adoption of methods for gathering social network data.

2 Method

2.1 Questionnaire Survey

The questionnaire survey included a series of questions about respondents' demographic and psychological characteristics, four questions on norm transmission, followed by several questions on everyday pro-environmental behaviour. The demographic questions included sex, age, education and level in the organisation (top manager, management, supervisory or organisational role). The psychological questions covered various areas, including values [20]. Data on biospheric, egoistic, altruistic and hedonic values were collected using a scale in which -1 represents 'opposed to my values', 0 stands for 'irrelevant' and 1–7 signifies an increasing degree of importance. Other psychological questions included efficacy, worldviews, norms and identity [21, 22]. The questions on norms were most relevant for the purposes of this article.

The norms questions were divided along two dimensions: descriptive versus injunctive norms, and local versus general norms. Local norms are concerned with norms in the workplace for the purposes of LOCAW, whilst general norms cover the respondents' neighbours, city and nation. Local norms are of primary interest for the research in this chapter, and reference to norms henceforth will be to the questions on local rather than general norms.

Descriptive norms focus on people's perceptions of what others are doing, with questions taking the form, "Most of my *colleagues* act pro-environmentally at work," with *colleagues* being replaced by four workplace relationships: subordinates, co-workers, supervisors, and members of the management team (These last two were grouped together, as there was little difference in their interpretation by respondents). Descriptive norms are what would be represented in the diffusion models in Elsenbroich and Gilbert's [13] classification above.

By contrast, injunctive norms pertain to the respondent's perception of what others think they should do, with questions taking the form, "Most of my *colleagues* think I should act pro-environmentally at work." Responses to both descriptive and injunctive norms questions were recorded using Likert scales from 1 (totally disagree) to 7 (totally agree). Hollander and Wu [8] refer to descriptive and injunctive norms as "passive" and "active" transmission respectively (para. 38), but the terms "descriptive" and "injunctive" are applied by Cialdini et al. [23] to contrast norms of what "is" and what "ought" to be.

Knowing what your colleagues think you should do requires them to communicate it to you. Norm transmission questions were added to the questionnaire at the request of the modelling team. Norm transmission in this context may be seen itself as a 'meta' pro-environmental behaviour insofar as it is effective in encouraging colleagues to behave more pro-environmentally. Agent-based modelling is generally held to be suited to contexts in which heterogeneity of and interactions among agents are of importance in determining macro-level outcomes [24]. Although some data on interactions were covered by the section on norms, these questions treat the respondent as a passive observer of their (social) environment; without data on norm transmission, there would have been no empirical basis on which to model agents taking actions to encourage others to behave more pro-environmentally.

Four questions on norm transmission were asked, each taking the form: "How often do you encourage [your *colleagues*] to act pro-environmentally at work?" These and other behaviours questions typically pertain to the respondent's perceived frequency with which they carry out certain everyday environmentally relevant behaviours. Questions not pertaining to norm transmission covered three areas (transport, energy use and waste) in two domains (workplace and at home), but we focus on the workplace in this chapter. The questions used with the model described here were answered using Likert scales from 1 (never) to 7 (always).

2.2 *Decision Tree Learning*

The demographic and psychological characteristics questions were used as the basis for explanatory variables for decision-tree learning; the everyday behaviour questions (including norm transmission) were treated as response variables. The process by which the decision trees were constructed is described in an accompanying chapter to this book [25], which compares various approaches to pre-processing the data before building the trees. Of the methods described therein, we have adopted here the CFS-MDL-DT approach without values clustering. CFS is Correlation-based Feature Selection [26], which is used to determine which explanatory variables should be sent to the decision-tree learning algorithm; MDL is Multi-interval Discretisation [27]; DT is the C4.5 decision tree algorithm [28]. Although Sánchez-Maroño et al. [25] point out that using values clusters creates different types of agent who behave in different ways, the main aim of this chapter is to see how social network characteristics affect the behaviour of the model. Heterogeneity of agents in this model is determined by their attributes and how these affect the route taken through the decision trees.

The model features trees for four everyday pro-environmental behaviours, and three norm transmissions (to subordinates (NT.1), co-workers (NT.2) and supervisors (NT.3)). The four behaviours are:

- BW.6: When you commute or drive for work purposes, how often do you drive in an energy efficient way (looking ahead and anticipating on traffic and brake and accelerate quietly and change to a higher gear as soon as possible)?
- BW.17: During the year when you are at work, how often do you turn on the heating at your workspace?
- BW.19: During the year when you are at work, how often do you turn on the air-conditioning at your workspace?
- BW.22: How often do you separate your plastic from the regular garbage at work?

These behaviours cover the three areas of behaviour with which LOCAW was concerned (transport, energy and waste), but were primarily chosen for the variety of injunctive and descriptive norms forming the conditional tests at branches in their trees. BW.6 is unaffected by descriptive or injunctive norms; BW.17 is affected by descriptive norms from all kinds of workplace relationship; BW.19 only by descriptive norms from supervisors; and BW.22 injunctive norms from co-workers and descriptive norms from subordinates.

The three norm transmission behaviours are also affected by injunctive and descriptive norms according to the decision trees. Norm transmission to subordinates is affected by descriptive norms from subordinates; norm transmission to co-workers by injunctive norms from subordinates; and norm transmission to supervisors by injunctive and descriptive norms from supervisors.

Figure 1 shows the decision tree for BW.6 and Fig. 2 the corresponding implementation as NetLogo code. The `get-value-of` procedure is described in the submodels section of the ODD in Sect. 2.3, and the source code is in the Appendix to this chapter.

The data used apply to the 'all-country' case study in [25], which amalgamates results from questionnaires sent to the Italian, Romanian and Spanish case studies.

2.3 *Model Design*

Decision trees for selected pro-environmental behaviours (including all norm transmission behaviours) in the questionnaire were implemented in a NetLogo model WERC-M Q (Worker-Environment Reinforcement Choice Model: Questionnaire). The process of adding dynamics to a static questionnaire involved (a) adding a social network to the agents; (b) interpreting the questions on local norms and norm transmission included in the questionnaire, where these appeared in the explanatory variables of the decision tree:

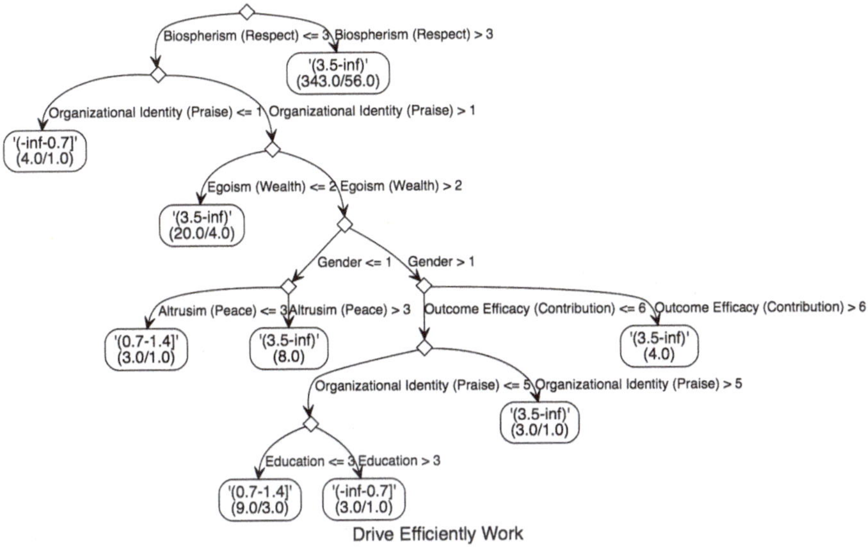

Fig. 1 An example decision tree for driving efficiently when commuting and/or driving on business (BW.6). Decision nodes are represented as *diamonds*, with the branches labelled according to the condition for following them. Leaf nodes show the range of responses from discretisation in quotes

```
to-report dt-drive-efficiently
report ifelse-value (get-value-of "VB.1" "BW.6" <= 3) [
  ifelse-value (get-value-of "IWO.3" "BW.6" <= 1) [0 + random-float 0.7] [
    ifelse-value (get-value-of "VE.2" "BW.6" <= 2) [3.5 + random-float 3.5] [
      ifelse-value (get-value-of "Q1" "BW.6" <= 1) [
        ifelse-value (get-value-of "VA.2" "BW.6" <= 3) [0.7 + random-float 0.7] [3.5 + random-float 3.5]
      ]
      [
        ifelse-value (get-value-of "OE.1" "BW.6" <= 6) [
          ifelse-value (get-value-of "IWO.3" "BW.6" <= 5) [
            ifelse-value (get-value-of "Q3" "BW.6" <= 3) [0.7 + random-float 0.7] [0 + random-float 0.7]
          ]
          [3.5 + random-float 3.5]
        ]
        [3.5 + random-float 3.5]
      ]
    ]
  ]
]
[3.5 + random-float 3.5]
end
```

Fig. 2 Netlogo code corresponding to the implementation of the decision tree in Fig. 1

- For descriptive norms, the response on the questionnaire (to "most of my *colleagues* act pro-environmentally at work") was replaced with an observation based on the agent's immediate neighbours in their workplace social network according to the appropriate workplace relationship. Specifically, the agent looked at the mean response of their subordinates/co-workers/supervisor to the question on the behaviour.

- For injunctive norms, the response on the questionnaire ("most of my *colleagues* think I should act pro-environmentally at work") was replaced with the mean norm transmission of the appropriate workplace relationships.

WERC-M Q is described using Grimm et al.'s [29, 30] ODD protocol below.

2.3.1 Purpose

The purpose of the model is to predict agents' responses to questions about everyday pro-environmental behaviour and norm transmission in the LOCAW questionnaire when the implied dynamics of norm transmission are taken into consideration.

2.3.2 Entities, State Variables and Scales

The entities in the model are the worker agents, each of which has state variables corresponding to the explanatory variables as constant determinants of behaviour. Agents also have dynamic state variables corresponding to each of the norm transmission and selected everyday pro-environmental behaviours. For each of these dynamic state variables, there is a state variable containing the initial value recorded. These variables are all stored in a hash table from variable name to value under the attribute named `data`. Agents also have a `level` attribute, which is their level in the artificial institution the model creates (1 being the most senior level).

Agents are connected in a social network comprising two parts: a hierarchical organisation network, which is a directed graph with each link going from supervisor to subordinate (referred to in the model as a `manager` link), and an informal co-worker network, which is an undirected graph, each link of which represents a mutual co-worker relationship, and is referred to in the model as a `co-worker` link.

2.3.3 Process Overview and Scheduling

Two processes operate in the model in a repeated time step, nominally representing a single working day:

1. Determine behaviours
2. Transmit norms

2.3.4 Design Concepts

Basic Principles

The model is designed to transparently reflect decision trees learned from questionnaire data, and hence the decision trees are hard-coded. It is intended to reflect the questionnaire as strictly as possible, and so only predicts the responses to the questions, not whether the corresponding pro-environmental behaviour is performed. It deviates from the questionnaire only in assigning a network structure to the norm interactions among agents, and interpreting normative influences on behaviour on the basis of model state rather than respondents' answers.

Emergence

What emerges from the model is the change in responses to behavioural questions associated with the dynamics imposed by the network structure and interpretation of norm influences.

Sensing

Agents sense the pro-environmental behaviours of their neighbours in the social network (descriptive norm).

Interaction

Agents may transmit and receive injunctive norms to/from each other.

Stochasticity

Used in building the social network during initialisation, and each time step in determining whether or not different kinds of norm transmission occur. Randomness is also used to determine the actual response from the discretised ranges at the leaf nodes (see Fig. 2), and in determining the order in which agents are activated each time step.

Observation

Data were collected on the time series of the distribution of predicted response for each behavioural variable, and the mean and variance of the difference between agents' predicted response at the beginning and end of the run.

2.3.5 Initialisation

The input file to the model is a CSV file with one row for each respondent to the questionnaire in the case study (in this case, an aggregation of data from three case studies), and one column for each question containing that respondent's answer or 'NA' if no answer was given. WERC-M Q uses this file (the location of which is specified using the questionnaire-data-file parameter) to initialise the agents' data attribute as a hash table of variable name to value.

The agents are then connected in a workplace social network implementing three relationship types: subordinate, co-worker and supervisor. Supervisor and subordinate relationships are implemented through the directed `manager` link network, in which *A* is a subordinate of *B* iff *B* is a supervisor of *A*. The workplace management hierarchy is determined by randomly selecting a single agent to be the chief executive of the artificial institution, assigning that agent a `level` of 1, setting a global variable `max-level` to 1, and then constructing the organogram as follows:

```
 1. While any agents have an unassigned level:
 2.    Let M = 0
 3.    For each agent A with level = max-level:
 4.       Let N = sample from a Poisson distribution with mean fan-out-mean
                                                              (a parameter)
 5.       If N > 0:
 6.          If the number of agents with unassigned level > N:
 7.             Let N = the number of agents with unassigned level
 8.          End If
 9.          For each of the N nearest agents B with unassigned level:
10.             Set level (of B) = max-level + 1
11.             Set M = M + 1
12.             Make a manager link from A to B
13.          End For
14.       End If
15.    End For
16.    If M > 0:
17.       Set max-level = max-level + 1
18.    End If
19. End While
```

Note that the code on lines 16–18 check that some agents have been assigned a `level` before increasing `max-level`. This effectively causes resampling of the Poisson distribution until agents have been assigned to the next level, and hence for low `fan-out-mean`, the distribution of subordinates of each supervisor may not quite be Poisson distributed.

The co-worker network is constructed using an adaptation of Hamill and Gilbert's [31] social circles model. The latter is argued by its authors to generate social networks that are more realistic than Watts and Strogatz's [32] small-world algorithm, or Barabási and Albert's [33] scale-free algorithm. Agents are randomly positioned in a 2D space, and then connected together if the distance between them is less than a parameter (`reach`). The adaptation of the algorithm used here imposes the further constraint on the prospective connection that the absolute difference in `level` between the pair of agents must be less than or equal to 1. The selection of the nearest agents in line 9 of the manager network construction algorithm is intended to ensure that this constraint during the construction of the co-worker network is affected as little as possible by the construction of the manager network.

2.3.6 Input Data

There are no time series input data for this model.

2.3.7 Submodels

Determine Behaviours

For each of the four behaviours (BW.6, BW.17, BW.19, BW.22), compute the result of the corresponding decision tree for each agent, with the following modifications:

1. Where the decision tree uses a descriptive norm in a branch conditional, use the mean of appropriately connected agents for that behaviour.
2. Where the decision tree uses an injunctive norm in a branch conditional, use the mean of the corresponding norm transmission decision for appropriately connected agents.
3. Where the decision tree uses the level in the organisation in a branch conditional, return 4 (operational) if `level` = `max-level`; 3 (supervisor) if `level` = `max-level`−1 and `max-level` > 3; 1 (top manager) if `level` = 1 or (`level` = 2 and `max-level` > 3); and 2 (manager) otherwise.
4. Where an agent has NA as the value of the variable in the branch conditional, use 0 (as was the case when the decision trees were constructed).
5. Where the decision tree used one of the hedonic, egoistic, altruistic or biospheric values questions in the branch conditional, use −10 as the value if the response was −1 (as was done when the decision trees were constructed).

Transmit Norms

These are assumed to occur less frequently than the everyday behaviours, each of which (driving efficiently when using a car on business or commuting, turning on heating or air conditioning, recycling plastic) could be thought of as occurring roughly once per day. Further, the frequency may depend on the relationship. The most frequent was assumed to be co-worker norm transmission, with probability 0.1 (corresponding roughly to once every two working weeks if the behaviours in

Table 1 Parameter settings varied in the exploration of model dynamics

Parameter	Values
`fan-out-mean`	1.5, 2, 5, 10, 20
`reach`	2, 4, 6, 8, 10

step one are seen as occurring daily), and the least frequent norm transmission to supervisors, with probability 0.05. Norm transmission to subordinates was given a probability of 0.075. Once a norm transmission had been determined to occur, the corresponding decision tree was used to determine the response, in exactly the same way as for the other behaviours in 'Determine behaviours' above.

2.4 *Experiments with the Model*

Experiments were conducted to investigate the effect of the network parameters `fan-out-mean` and `reach` on the predicted responses of the agents after 200 timesteps. The values for each are shown in Table 1, and certain combinations of them are visualised in Fig. 3. Each parameter setting was repeated 40 times. There were therefore 1000 runs of the model in total.

For each run, the model records in the last time step the mean and variance of the distribution of difference between each non-cloned agent's final and initial response to each behaviour. Specifically, for each behaviour, the NetLogo expressions for the means and variances are computed as shown in (1) and (2),

$$\texttt{mean}\ [\textit{behaviour} - \textit{initial-behaviour}]\ \texttt{of workers} \tag{1}$$

$$\texttt{variance}\ [\textit{behaviour} - \textit{initial-behaviour}]\ \texttt{of workers} \tag{2}$$

where *`behaviour`* and *`initial-behaviour`* are substituted for a lookup in the agent's `data` table for the corresponding behaviour variables at the end and beginning of the run respectively; and `workers` is the set of interacting agents.

In these experiments, we are primarily interested in knowing which of the variables in Table 1 had the greatest effect on the dynamics in the model. Our results are assessed using a Kruskal-Wallis test with the null hypothesis that there is no functional relationship between network parameters and the mean and variance of the difference between the initial and final predicted behaviour of each behaviour for each agent.

3 Results

The results are reported in Table 2 (mean difference in behaviour) and Table 3 (variance). For BW.6 (drive efficiently), which is effectively operating as a control,

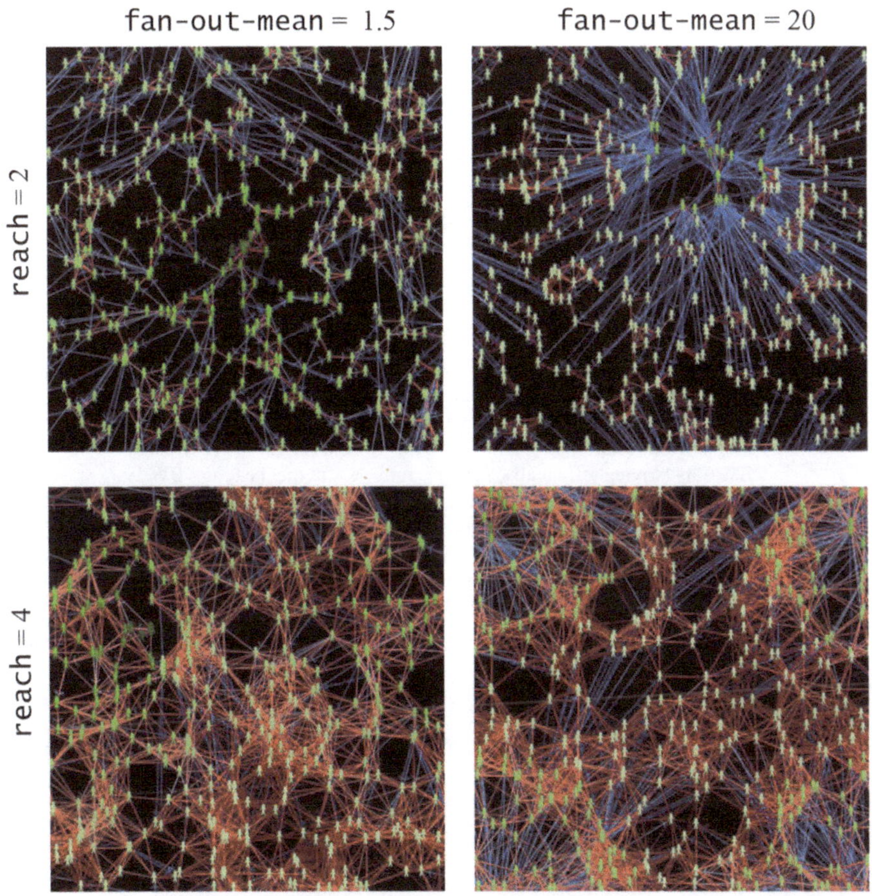

Fig. 3 Various visualisations of social networks derived from different parameters for the manager and co-worker networks. Agents are shown using *green 'person' icons*, with shade corresponding to the level in the organisation (darker is higher up, i.e. the agent's `level` attribute is closer to 1). *Blue links* show the manager network; *red links* show the co-worker network. Both are given a brighter shade of their respective colour if they have a higher mean norm transmission value along the link

it is apparent that neither network parameter has a significant effect on the difference between the initial and final predicted values from the BW.6 decision tree (Fig. 1). For behaviours BW.17 (turn on heating) and BW.22 (recycle plastic), which use the co-worker network, the `reach` parameter does have a significant effect on the mean difference, but not on the variance in the case of BW.22. As might be expected, where behaviours do not use co-worker norms, the `reach` parameter has no significant effect.

More interesting is the relationship between the use of the directed manager network in a behaviour, and the `fan-out-mean`. Norms from supervisors to

Table 2 Results for the Kruskal-Wallis tests of the effects of the network parameters `reach` and `fan-out-mean` on the mean difference between the initial and final predicted behaviour (rows) of each agent

Behaviour	Co-worker?	Manager?	reach	fan-out-mean
BW.6	–	–	–	–
BW.17	D	D±	***	–
BW.19	–	D–	–	*
BW.22	I	D+	**	**
NT.1	–	D+	–	***
NT.2	–	I+	–	***
NT.3	–	D – I–	–	***

The Co-worker? and Manager? columns show whether any of the explanatory variables in each behaviour's decision tree use descriptive (D) or injunctive (I) norms. In the directed Manager network, a + means norms from subordinates, and – (after D or I) norms from supervisors. BW.6 is driving efficiently, BW.17 turning on the heating, BW.19 turning on the air conditioning, and BW.22 recycling plastic. NT.1 is norm transmission to subordinates, NT.2 to co-workers, and NT.3 to supervisors
*$P<0.05$; **$P<0.01$; ***$P<0.001$

Table 3 As per Table 2, but for the variance of the distribution of differences of the initial and final predicted behaviours of each agent

Behaviour	Co-worker?	Manager?	reach	fan-out-mean
BW.6	–	–	–	–
BW.17	D	D±	***	**
BW.19	–	D–	–	*
BW.22	I	D+	–	***
NT.1	–	D+	–	***
NT.2	–	I+	–	***
NT.3	–	D – I–	–	***

subordinates (indicated by a D– or I– in the Manager? column of Tables 2 and 3) affect larger numbers of individuals than vice versa (D+ or I+), as in the model, individuals only ever have one supervisor. A larger `fan-out-mean` will increase the effect a single supervisor can have, whilst a smaller `fan-out-mean` will increase the effect a single subordinate can have.

In the case of BW.19 (turn on air conditioning), which only uses descriptive norms from supervisors to subordinates, the effect of `fan-out-mean` is weakly significant on both mean and variance (and indeed, given the number of significance tests done, a possible type I error). By contrast, NT.3 (norm transmission to supervisors), which uses descriptive and injunctive norms from supervisors to subordinates, has a strongly significant effect of `fan-out-mean`. BW.22, NT.1 (norm transmission to subordinates) and NT.2 (norm transmission to co-workers) all use only norms (descriptive in the cases of BW.22 and NT.1, and injunctive in the case of NT.2) from subordinates to supervisors, and yet still feature significant effects of `fan-out-mean` on both mean and variance. BW.17 is interesting in that

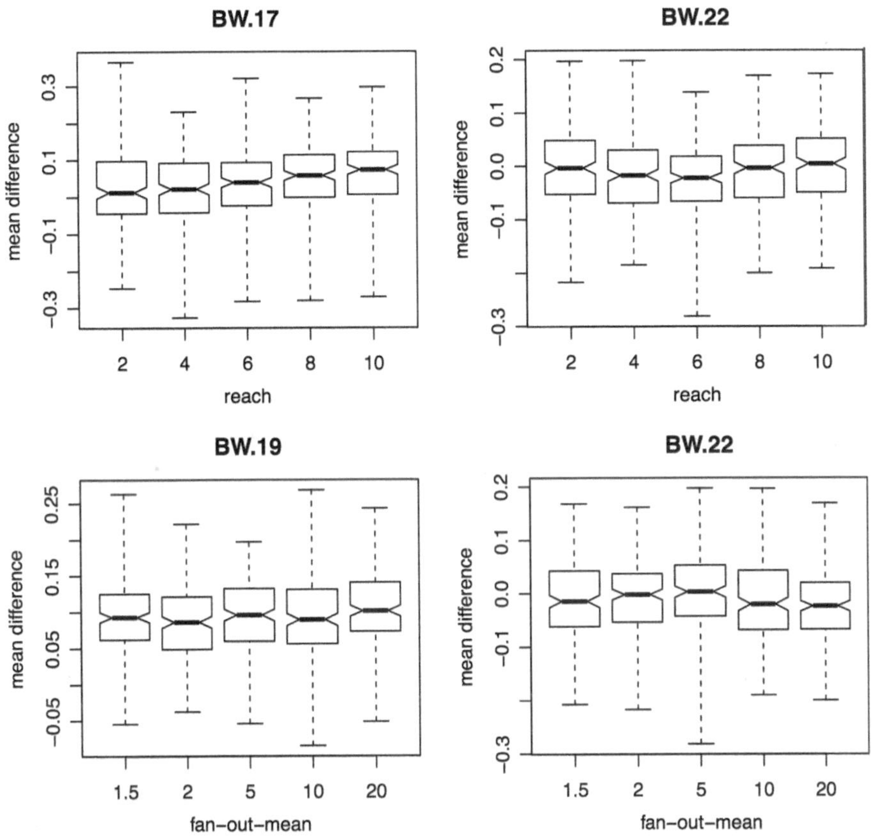

Fig. 4 Boxplots of the distributions of mean difference between initial and final predicted behaviour for those parameter and behaviour combinations for which a significant effect is shown in Table 2. The *thick lines* show the medians, whilst the *bottom* and *top of the boxes* are drawn at the first and third quartiles. The whiskers (*dashed lines*) extend to the minimum and maximum. The notches in the size of the box extend to ± 1.58 times the interquartile range divided by the square root of the sample size, as described in the help file for `boxplot.stats` in R, the statistical software used to draw the boxplots. As this text asserts, these notches show approximately a 95 % confidence interval for the purposes of comparing pairs of medians—if notches do not overlap, this is potential evidence of a significant difference in median

it features descriptive norms in both directions of the manager network, but no effect of `fan-out-mean` on the mean difference in predicted behaviour at the start and end of the run, though there is a significant difference in the variance.

There is therefore no clear pattern in whether norms from subordinates or supervisors are more or less effective in changing behaviour (for better or for worse) as the corporate hierarchy is adjusted. Neither is there any discernible pattern in whether injunctive or descriptive norms have a stronger effect, which is interesting since injunctive norms require norm transmission. Though this may be in part due to insufficient behaviours being studied to enable any such patterns to be observed, there are also questions of how norms are used by the decision trees themselves, and how sensitive they are to the attributes of agents being put in particular positions in the network. However, it should be no surprise that different behaviours will be sensitive in different ways to the topologies of norm interaction.

The Kruskal-Wallis test merely shows there is a functional relationship, but says nothing about the magnitude or direction of the effect. In Fig. 4, boxplots are used to show the distribution of the mean difference in behaviour for BW.17 and BW.22 in the case of `reach`, and for BW.19 and BW.22 in the case of `fan-out-mean` (these being the significant results in the first four rows of Table 2). Bearing in mind that responses vary from 1 to 7, the scaling on the y-axes of these graphs show that the magnitude of effect is small. Further, there is considerable overlap, especially when the extremes (whiskers) are included. However, Kaiser [34] proposes a nonlinear (logistic) formalisation of Campbell's [35] paradigm for assessing the attitude-behaviour gap. If this can be used to determine the probability of performing a behaviour from its reported frequency on a Likert scale, then, depending on the mean response level, small differences such as those reported here could have a larger effect on the performance of the actual behaviour.

BW.17 shows an increasing trend with respect to `reach`, as does BW.19 (though less clearly, which may explain the weak significance of this result) with respect to `fan-out-mean`. The wording of both these behaviours is such that an increase in response means less pro-environmental behaviour, since, respectively, they refer to turning the heating and air-conditioning on. The trends in the case of BW.22 (in which a positive difference does mean more pro-environmental behaviour—recycling plastic) are for a U-shaped curve in the case of the `reach` parameter, and an inverted-U for `fan-out-mean`. These shapes are also seen in the distributions of `fan-out-mean` against mean difference for each of the norm transmission behaviours (Fig. 5). Since BW.22 uses injunctive norms in the co-worker network, this may explain the U-shaped relationship between `reach` and mean difference observed in the top right boxplot in Fig. 4. The presence of U and inverted-U curves in the relationship between `fan-out-mean` and mean difference may reflect the observations above about the balance between `fan-out-mean` and the relative effectiveness of norms from subordinates and supervisors.

In general, the observations in Figs. 4 and 5 lend weight to the conclusions from the study of the results of the significance tests: the topologies of norm interactions affect different behaviours in different ways.

4 Discussion

Our hypothesis has been that, where behaviours are affected by various workplace norms, the topology of social interactions is influential. The results have indeed shown that (with the exception of BW.6 (drive efficiently), which was not found to be affected by norms using CFS-MDL-DT) the co-worker and manager network topology both have a small but significant effect on the dynamics implied by norm transmission and injunctive and descriptive norms in a static questionnaire pertaining to everyday pro-environmental behaviours in the workplace. The effect is context-sensitive: it changes according to the behaviour in question. Nevertheless, the implication for social and environmental psychology is that empirical findings should include a discussion of the structure of the social network, and for empirical agent-based modelling, that data are needed on the topology of social interactions as well as on the behaviours themselves.

Choi et al.'s [36] exploration of innovation diffusion (which is similar to norm diffusion insofar as it is imitative) found that random networks are a more difficult medium in which to transmit information from a low level of initial adoption, because the lack of cliques means there is not enough accumulation of adopters among the connections of later adopters for them to adopt. However, once the number of initial adopters is high enough, random networks diffuse much more rapidly than cliquey ones (ibid.). We have not studied the effect of changing the initial response—this could be done in future work—however, we did measure the global clustering co-efficient from the model (the proportion of triplets that are closed). This isn't quite the same as cliques (which are subgraphs with fully connected nodes), but will be related to it as more and larger cliques will mean more closed triplets. As Fig. 6 shows, there is no clear trend in the relationship between clustering coefficient and mean difference for BW.22 (one of the two behaviours for which `reach` had a significant effect), though the range of coefficients resulting from the parameters used in this study does not cover the full span of values from [0, 1].

Empirical findings [37] suggest that descriptive norms may be affected by relationship type. Specifically, co-workers performing pro-environmental behaviours have been found to make it more likely that individuals will also do so, whilst managers not performing pro-environmental behaviours have been found to make it more likely that individuals will also not do so. The data reported from our model

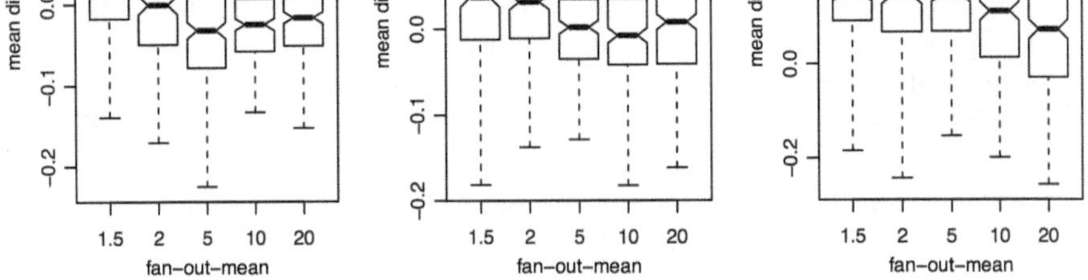

Fig. 5 As per Fig. 4, but for the bottom three rows of Table 2

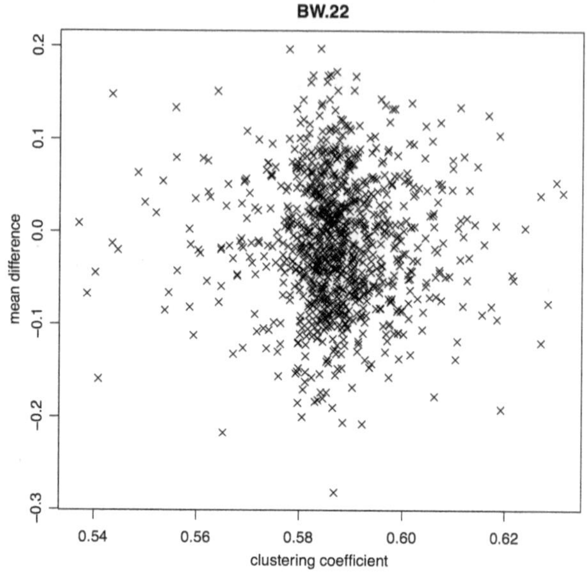

Fig. 6 Scatter plot of clustering coefficient against mean difference for BW.22 (recycling plastic)

do not allow us to determine whether this finding is upheld, but we do see in our results different dynamics in the effect of the `fan-out-mean` parameter of the mean difference and that of the `reach` parameter (where it uses only descriptive norms), the former having a monotonic effect on the mean difference, the latter a U or inverted-U shape to the relationship. The effects observed by Keiser et al. [37] should emerge from the model, rather than being imposed. Confirming this, and the differing dynamics in the hierarchical and co-worker networks, would be the possible subjects of future work.

As stated above, it would be better in empirical modelling work to gather data on the actual topology of interactions rather than generating the networks artificially. Though established methods, such as social network analysis [38] and theory, such as actor network theory [39], are available to explore interactions and networks in the social sciences, these tend to be applied by rather specialised communities. A search in the Thomson Reuters database of manuscripts published in 2012 returned less than 250 results for "snowballing" (a method used in the social network community), but over 130,000 results for the more standard tools of the social sciences: interviews and questionnaire surveys. Ours is far from being the first article in the modelling community to highlight the significance of interactions in a set of results. With a greater interest in empirical agent-based modelling [16], developing cost-effective approaches to gathering evidence to support modelling of interactions is a priority. We anticipate that such approaches, if widely used, would provide an interesting context in which to evaluate findings from studies in the social sciences, even when not involving modellers.

Gilbert [40] is critical of the use of survey data in agent-based models, pointing out that they fail to adequately address interactions with others and are a sample from a snapshot in time. Outright rejection of sampling approaches constrains the scope for collaborations needed to build empirical agent-based models, as individual researchers have their preferred tools. We have shown here that, with a few extra questions on norm transmission, it was possible to create an empirical agent-based model with dynamic behaviour that is based on questionnaire data. Although the results suggest that further data are needed on social network structure, in general we may reasonably expect some adaptation of traditional social science tools to be needed when they are to be applied (at least in part) to the development of agent-based models.

The model has used decision trees learned from questionnaire data as a means for representing agents' decisions. Although the structure of these trees is fixed, some dynamism in individual decision-making processes is enabled through the modification of explanatory variables. Here, these were purely those pertaining to descriptive and injunctive norms, but demographic characteristics (e.g. age), training, career progression and workplace re-organisation could all affect the route taken by an individual agent through the decision tree. There is no theoretical reason why decision trees could not have nodes depending on memory, which would provide a basis for implementing learning. However, gathering enough empirical data to build such a tree could prove challenging. The main advantage of decision trees in terms of their relationship with the empirical data is their transparency, and it is unsurprising that they have been used in other ABM work, empirical and otherwise (e.g. [41, 42]), with Verhagen [43] applying them to the learning of norms. An important issue with decision trees, where derived automatically from empirical data, is that their structure can be sensitive to the algorithm used to build it. Here, we have used an approach to decision tree construction found by Sánchez-Maroño et al. [25] mostly to minimise classification or validation error, though since we are primarily interested in demonstrating the principle of using decision trees to study norms in agent-based models, such errors are of less concern. Other researchers have found it is sometimes appropriate to refine the model qualitatively (e.g. [44, 45]), and this would be appropriate here, though if modelling all the behaviours in the questionnaire, modifying all the trees in this way would place a considerable burden on the research team. Again, our primary concern with demonstrating the principle has meant this step was not undertaken.

5 Conclusion

Agent-based modelling is predicated on the importance of individual heterogeneity and interactions in determining society-level outcomes. This work has shown that adding the norm dynamics implied in static questionnaire data affects the results thereof in ways that are sensitive to network topology parameters for both hierarchical and co-worker networks. Although this knowledge could be used to calibrate network construction parameters (particularly when combined with two-shot questionnaires), gathering data on networks and relationships is clearly an important part of an empirical agent-based modelling project. Indeed, the influence network properties have on the dynamics suggests that social network information should be relevant to all empirical work in the social sciences.

Acknowledgments This work was funded by the European Commission through Framework Programme 7 grant agreement number 265155 (Low Carbon at Work: Modelling Agents and Organisations to Achieve Transition to a Low-Carbon Europe), and by the Scottish Government Rural Affairs and the Environment Portfolio Strategic Research 2011–2016 Theme 4 (Economic Adaptation).

A.1 Appendix: Netlogo Code

The `get-value-of` procedure, called from the code shown in Fig. 2, implements the rules described in the submodels in the ODD in Sect. 2.3 as shown below. References to injunctive and descriptive norms are highlighted in `inverse video`. One detail not discussed in the text is the possibility that there are no co-workers, supervisors (for the top manager) or subordinates (for those at the bottom of the hierarchy). In this case, the treatment is the same as for 'NA': 0 is returned as the value of the variable. Another is that when assessing the descriptive norm, two of the behaviours (turning on the heating (BW.17) and turning on the air conditioning (BW.19)) are worded such that higher responses mean less pro-environmental behaviour. When reporting a descriptive norm (most of my colleagues behave pro-environmentally at work), the responses of colleagues for these behaviours need to be inverted so that they correspond to the sense used in the descriptive norm question.

```
to-report get-value-of [ var behav ]
 if (var="Q4") [
   ; Use level in virtual organisation
   report ifelse-value (level=max-level) [4] [
     ifelse-value (level=max-level - 1 and max-level>3) [3] [
       ifelse-value (level=1
                       or (level=2 and max-level>3)) [1] [2]
     ]
   ]
 ]
 if (length var>4 and substring var 0 4="NIL.") [
   ; Handle injunctive norms
   if var="NIL.1" [
     if not any? out-manager-neighbors
       with [is-number? table:get data "NT.3"] [
       report 0
     ]
     ; Most of my subordinates think I should behave
     ; pro-environmentally...
     report mean [table:get data "NT.3"] of out-manager-neighbors
       with [is-number? table:get data "NT.3"]
     ; ... so use norm transmission to supervisors of those I
     ; manage
   ]
```

```
      if var="NIL.2" [
        if not any? co-worker-neighbors
          with [is-number? table:get data "NT.2"] [
          report 0
        ]
        ; Return transmission from co-workers
        report mean [table:get data "NT.2"] of co-worker-neighbors
          with [is-number? table:get data "NT.2"]
      ]
      if var="NIL.3" or var="NIL.4" [
        if not any? in-manager-neighbors
          with [is-number? table:get data "NT.1"] [
          report 0
        ]
        ; Return transmission from supervisors
        report mean [table:get data "NT.1"] of in-manager-neighbors
          with [is-number? table:get data "NT.1"]
      ]
    ]
    if (length var>4 and substring var 0 4="NDL.") [
      ; Handle descriptive norms
      let desc-norm 0
      if var="NDL.1" [
        ; Return typical behaviour of subordinates (use behav)
        if not any? out-manager-neighbors
          with [is-number? table:get data behav] [
          report 0
        ]
        set desc-norm mean [table:get data behav] of
          out-manager-neighbors with [is-number? table:get data behav]
      ]
      if var="NDL.2" [
        if not any? co-worker-neighbors
          with [is-number? table:get data behav] [
          report 0
        ]
        ; Return typical behaviour of co-workers (use behav)
        set desc-norm mean [table:get data behav] of
          co-worker-neighbors with [is-number? table:get data behav]
      ]
      if var="NDL.3" or var="NDL.4" [
        if not any? in-manager-neighbors
          with [is-number? table:get data behav] [
          report 0
        ]
        ; Return typical behaviour of supervisors (use behav)
        set desc-norm mean [table:get data behav]
          of in-manager-neighbors
          with [is-number? table:get data behav]
      ]
      ifelse (behav="BW.17" or behav="BW.19") [
        ; Here 'behaving pro-environmentally' is behav=1
        report 8 - desc-norm
      ]
```

```
    [
      ; Here 'behaving pro-environmentally' is behav=7
      report desc-norm
    ]
  ]
  if table:get data var="NA" [
    report 0
  ]
  if substring var 0 1="V" and table:get data var=-1 [
    report -10
  ]
  report table:get data var
end
```

The code in Fig. 2 implements a behaviour (BW.6—drive efficiently) that does not use injunctive or descriptive norms. The procedure below implements the behaviour for recycling plastic (BW.22), which uses both injunctive and descriptive norms (highlighted in `inverse video`) in various branches:

```
to-report dt-recycle-plastic
report ifelse-value (get-value-of "Country" "BW.22"<= 2) [
 ifelse-value (get-value-of "IES.2" "BW.22"<= 3) [
   0.7+random-float 0.7
 ] [
   ifelse-value (get-value-of "WV.5" "BW.22"<= 4) [
     ifelse-value (get-value-of "IES.3" "BW.22"<= 6) [
       ifelse-value (get-value-of "IES.2" "BW.22"<= 5) [
         ifelse-value (get-value-of "IES.2" "BW.22"<= 4) [
           ifelse-value (get-value-of "VE.1" "BW.22"<= 0) [
             0.7+random-float 0.7
           ] [3.5+random-float 3.5]
         ]
         [2.8+random-float 0.7]
       ]
       [0.7+random-float 0.7]
     ]
     [3.5+random-float 3.5]
   ]
   [
     ifelse-value (get-value-of "VE.1" "BW.22"<= 2) [
       ifelse-value (get-value-of "IES.3" "BW.22"<= 5) [
         ifelse-value (get-value-of "IES.2" "BW.22"<= 6) [
           ifelse-value (get-value-of "WV.5" "BW.22"<= 6) [
             ifelse-value (get-value-of "NDL.1" "BW.22"<= 1) [
               ifelse-value (get-value-of "IES.2" "BW.22"<= 5) [
                 ifelse-value (get-value-of "VE.1" "BW.22"<= 1) [
                   ifelse-value (
                     get-value-of "IES.3" "BW.22"<= 4) [
                     ifelse-value (
                       get-value-of "NIL.2" "BW.22"<= 1) [
                       0.7+random-float 0.7
                     ] [1.4+random-float 0.7]
                   ]
                   [
```

```
                ifelse-value (
                  get-value-of "Q5" "BW.22" <= 1) [
                  1.4 + random-float 0.7
                ] [0.7 + random-float 0.7]
              ]
            ]
            [
              ifelse-value (
                get-value-of "IES.2" "BW.22" <= 4) [
                3.5 + random-float 3.5
              ] [0.7 + random-float 0.7]
            ]
          ]
          [0.7 + random-float 0.7]
        ]
        [3.5 + random-float 3.5]
      ]
      [
        ifelse-value (get-value-of "NDL.1" "BW.22" <= 1) [
          ifelse-value (get-value-of "SE.2" "BW.22" <= 3) [
            ifelse-value (get-value-of "VE.1" "BW.22" <= 0) [
              1.4 + random-float 0.7
            ] [0.7 + random-float 0.7]
          ]
          [3.5 + random-float 3.5]
        ]
        [0.7 + random-float 0.7]
      ]
    ]
    [0.7 + random-float 0.7]
  ]
  [
    ifelse-value (get-value-of "SE.2" "BW.22" <= 6) [
      ifelse-value (get-value-of "IES.3" "BW.22" <= 6) [
        ifelse-value (get-value-of "Q3" "BW.22" <= 4) [
          ifelse-value (get-value-of "VE.1" "BW.22" <= -10) [
            ifelse-value (get-value-of "NIL.2" "BW.22" <= 4) [
              ifelse-value (
                get-value-of "WV.5" "BW.22" <= 6) [
                3.5 + random-float 3.5
              ] [0.7 + random-float 0.7]
            ]
            [0.7 + random-float 0.7]
          ]
          [
            ifelse-value (get-value-of "VE.1" "BW.22" <= 1) [
              3.5 + random-float 3.5
            ] [2.8 + random-float 0.7]
          ]
        ]
        [
          ifelse-value (get-value-of "SE.2" "BW.22" <= 3) [
            0.7 + random-float 0.7
          ] [3.5 + random-float 3.5]
```

```
              ]
            ]
            [
              ifelse-value (get-value-of "NIL.2" "BW.22" <= 5) [
                ifelse-value (get-value-of "NIL.2" "BW.22" < = 3) [
                  0.7 + random-float 0.7
                ] [3.5 + random-float 3.5]
              ]
              [0.7 + random-float 0.7]
            ]
          ]
          [3.5 + random-float 3.5]
        ]
      ]
      [3.5 + random-float 3.5]
    ]
  ]
]
[3.5 + random-float 3.5]
end
```

References

1. Lukàcs, G.: History and Class Consciousness. Merlin Press, London (First pub. 1923) (1977)
2. Lefebvre, H.: Everyday Life in the Modern World. Penguin, London (1971)
3. de Certeau, M.: The Practice of Everyday Life. University of California Press, Berkeley (1984)
4. Bonaiuto, M., Bilotta, E., Bonnes, M., Ceccareli, M., Martorella, H., Carrus, G.: Local identity moderating the role of individual differences in natural resource use: the case of water consumption. J. Appl. Soc. Psychol. **38**(4), 947–967 (2008)
5. Uzzell, D., Pol, E., Badenas, D.: Place identification, social cohesion, and environmental sustainability. Environ. Behav. **34**(1), 26–53 (2002)
6. Nigbur, D., Lyons, E., Uzzell, D.: Attitudes, norms, identity and environmental behaviour: using an expanded theory of planned behaviour to predict participation in a kerbside recycling programme. Br. J. Soc. Psychol. **49**(2), 259–284 (2010)
7. Nolan, J.M., Schultz, P.W., Cialdini, R.B., Goldstein, N.J., Griskevicius, V.: Normative social influence is underdetected. J. Pers. Soc. Psychol. **34**, 913–923 (2008)
8. Hollander, C.D., Wu, A.S.: The current state of normative agent-based systems. J. Artif. Soc. Soc. Simul. **14**(2), 6 (2011). http://jasss.soc.surrey.ac.uk/14/2/6.html
9. Verhagen, H.J.E.: Norm autonomous agents. Ph. D. thesis, Department of Computer and Systems Sciences, The Royal Institute of Technology and Stockholm University, Sweden (2000)
10. Therborn, G.: Back to norms! On the scope and dynamics of norms and normative action. Curr. Sociol. **50**, 863–880 (2002)
11. Rao, A.S., Georgeff, M.P.: BDI agents: from theory to practice. In: Lesser, V. (ed.) ICMAS-95: Proceedings, first international conference on multiagent systems, 12–14 June 1995, San Francisco, California, pp. 312–319. AAAI Press, Menio Park, CA (1995)
12. Axelrod, R.: An evolutionary approach to norms. Am. Polit. Sci. Rev. **80**(4), 1095–1111 (1986)
13. Elsenbroich, C., Gilbert, N.: Modelling Norms. Springer, Dordrecht (2014)
14. Schelling, T.M.: Dynamic models of segregation. J. Math. Sociol. **1**, 143–186 (1971)
15. Sakoda, J.M.: The checkerboard model of social interaction. J. Math. Sociol. **1**, 119–132 (1971)

16. Janssen, M.A., Ostrom, E.: Empirically based, agent-based models. Ecol. Soc. **11**(2), 37 (2007). http://www.ecologyandsociety.org/vol11/iss2/art37/
17. Smajgl, A., Brown, D.G., Valbuena, D., Huigen, M.G.A.: Empirical characterisation of agent behaviours in socio-ecological systems. Environ. Model. Software **26**, 837–844 (2011)
18. Barreteau, O., et al.: Our companion modelling approach. J. Artif. Soc. Soc. Simul. **6**(2), 1. http://jasss.soc.surrey.ac.uk/6/2/1.html (2003)
19. Moss, S.: Alternative approaches to the empirical validation of agent-based models. J. Artif. Soc. Soc. Simul. **11**(1), 5 (2008). http://jasss.soc.surrey.ac.uk/11/1/5.html
20. Schwartz, S.H.: Universals in the content and structure of values: theoretical advances and empirical tests in 20 countries. In: Zanna, M.P. (ed.) Advances in Experimental Social Psychology Volume 25, pp. 1–65. Academic, San Diego (1992)
21. Steg, L., Perlaviciute, G., van der Werff, E., Lurvink, J.: The significance of hedonic values for environmentally relevant attitudes, preferences, and actions. Environ. Behav. **46**, 163–192 (2014)
22. van der Werff, E., Steg, L., Keizer, K.: The value of environmental self-identity: the relationship between biospheric values, environmental self-identity and environmental preferences, intentions and behaviour. J. Environ. Psychol. **34**, 55–63 (2013)
23. Cialdini, R.B., Reno, R.R., Kallgren, C.A.: A focus theory of normative conduct: recycling the concept of norms to reduce littering in public places. J. Pers. Soc. Psychol. **58**(6), 1015–1026 (1990)
24. Polhill, J.G.: Agent-based modeling of socio-economic processes related to the environment: example of land use change. In: Baveye, P., Mysiak, J., Laba, M. (eds.) Uncertainties in Environmental Modeling and Consequences for Policy Making, pp. 61–76. Springer, Dordrecht (2009)
25. Sánchez-Maroño, N., Alonso-Betanzos, A., Fontenla-Romero, O., Polhill, J. G., Craig, T.: Empirically-derived behavioral rules in agent-based models using decision trees learned from questionnaire data. In: Agent-Based Modeling of Sustainable Behaviors. Springer (2016)
26. Hall, M.A.: Correlation-based feature selection for machine learning. Ph.D. thesis, University of Waikato, New Zealand (1999)
27. Fayyad, U.M., Irani, K.B.: Multi-interval discretization of continuous-valued attributes for classification learning. In Bajcsy, R. (ed.) Proceedings of the 13th International Joint Conference on Artificial Intelligence. Chambéry, France, August 28 September 3, vol. 2., pp. 1022–1027. Morgan-Kaufmann (1993)
28. Quinlan, R.: C4.5: Programs for Machine Learning. Morgan Kaufman, San Francisco, CA (1993)
29. Grimm, V., Berger, U., Bastiansen, F., Eliassen, S., Ginot, V., Giske, J., Goss-Custard, J., Grand, T., Heinz, S.K., Huse, G., Huth, A., Jepsen, J.U., Jørgensen, C., Mooij, W.M., Müller, B., Pe'er, G., Piou, C., Railsback, S.F., Robbins, A.M., Robbins, M.M., Rossmanith, E., Rüger, N., Strand, E., Souissi, S., Stillman, R.A., Vabø, R., Visser, U., DeAngelis, D.L.: A standard protocol for describing individual-based and agent-based models. Ecol. Model. **198**, 115–126 (2006)
30. Grimm, V., Berger, U., DeAngelis, D.L., Polhill, J.G., Giske, J., Railsback, S.F.: The ODD protocol: a review and first update. Ecol. Model. **221**, 2760–2768 (2010)
31. Hamill, L., Gilbert, N.: Social circles: a simple structure for agent-based social network models. J. Artif. Soc. Soc. Simul. **12**(2), 3 (2009). http://jasss.soc.surrey.ac.uk/12/2/3.html
32. Watts, D.J., Strogatz, S.H.: Collective dynamics of 'small-world' networks. Nature **393**(6684), 409–410 (1998)
33. Barabási, A.-L., Albert, R.: Emergence of scaling in random networks. Science **286**(5439), 509–512 (1999)
34. Kaiser, F.G., Byrka, K., Hartig, T.: Reviving Campbell's paradigm for attitude research. Pers. Soc. Psychol. Rev. **14**, 351–367 (2010)
35. Campbell, D.T.: Social attitudes and other acquired behavioral dispositions. In: Koch, S. (ed.) Psychology: A Study of a Science Volume 6, pp. 94–172. McGraw-Hill, New York (1963)

36. Choi, H., Kim, S.-H., Lee, J.: Role of network structure and network effects in diffusion of innovations. Ind. Mark. Manag. **39**, 170–177 (2010)
37. Keizer, K., Lindenberg, S., Steg, L.: The spreading of disorder. Science **322**, 1681–1685 (2008)
38. Wasserman, S., Faust, K.: Social Network Analysis: Methods and Applications. Cambridge University Press, Cambridge (1994)
39. Latour, B.: Reassembling the Social: An Introduction to Actor-Network Theory. Oxford University Press, Oxford (2005)
40. Gilbert, N.: Agent-based social simulation: dealing with complexity. ESRC Research Methods Festival, St. Catherine's College, Oxford, 17–20 July 2006. http://www.ccsr.ac.uk/methods/festival/programme/witha/gilbert.pdf (2006)
41. Ekasingh, B., Ngamsomsuke, K., Letcher, R.A., Spate, J.: A data mining approach to simulating farmers' crop choices for integrated water resources management. J. Environ. Manage. **77**, 315–325 (2005)
42. Schreinemachers, P., Berger, T.: Land-use decisions in developing countries and their representation in multi-agent systems. J. Land Use Sci. **1**, 29–44 (2006)
43. Verhagen, H.: Simulation of the learning of norms. Soc. Sci. Comput. Rev. **19**(3), 296–306 (2001)
44. Oxley, T., Jeffrey, P., Lemon, M.: Policy relevant modelling: relationships between water, land use, and farmer decision processes. Integr Assess. **3**(1), 30–49 (2002)
45. Smajgl, A., Bohensky, E.: Behaviour and space in agent-based modelling: poverty patterns in East Kalimantan, Indonesia. Environ. Model. Software **45**, 8–14 (2013)

Empirically-Derived Behavioral Rules in Agent-Based Models Using Decision Trees Learned from Questionnaire Data

N. Sánchez-Maroño, A. Alonso-Betanzos, O. Fontenla-Romero, J. Gary Polhill, and T. Craig

Abstract With the increasing trend in exploring the use of agent-based models in empirical contexts, this paper reflects on the use of decision trees learned from questionnaire data as behavioral models for the agents. Decision trees are machine learning algorithms most commonly used in the data mining literature, especially for smaller datasets where other techniques such as Bayesian Networks cannot be applied. In agent-based modelling contexts, decision trees have the advantage over some other machine learning techniques in that the results are more transparent, and can be critiqued by domain experts without a background in computing or artificial intelligence. However, decision trees are sensitive to the way in which they are constructed, particularly with respect to preprocessing. We describe the processes by which the decision trees were derived in the context of a model of everyday pro-environmental behavior at work, comparing various preprocessing methods and exploring their differences.

1 Introduction

For a number of years there has been a growing interest in the empirical use of agent-based models (ABM). Janssen and Ostrom [30] outlined the challenges in the field introducing a special issue of Ecology and Society on the subject, noting that issues with data collection in the social sciences mean that good statistical fits to data are an insufficient benchmark for model acceptance in comparison with 'softer' criteria such as theoretical plausibility, stakeholder and expert opinion,

N. Sánchez-Maroño (✉) • A. Alonso-Betanzos • O. Fontenla-Romero
University of A Coruña, A Coruña 15071, Spain
e-mail: noelia.sanchez@udc.es; amparo.alonso.betanzos@udc.es; oscar.fontenla@udc.es

J.G. Polhill
Information and Computational Science, The James Hutton Institute, Aberdeen, Scotland, UK
e-mail: gary.polhill@hutton.ac.uk

T. Craig
The James Hutton Institute, Craigiebuckler, Aberdeen AB15 8QH, UK
e-mail: tony.craig@hutton.ac.uk

A. Alonso-Betanzos et al. (eds.), *Agent-Based Modeling of Sustainable Behaviors*, Understanding Complex Systems, DOI 10.1007/978-3-319-46331-5_3

and the generation of new knowledge or understanding. More recently, Smajgl and Bohensky [47] outlined a framework for the use of various sources of evidence to construct empirical agent-based models.

In the LOCAW (Low Carbon at Work: Modelling Agents and Organisations to Achieve Transition to a Low-Carbon Europe) Framework Programme 7 project, an agent-based modelling exercise was included to simulate formalisations of backcasting scenarios aimed at increasing the frequency with which everyday pro-environmental behaviors (such as switching lights off, using public transport, or recycling waste) are performed at work and in the home in four case study organisations: the University of A Coruña (research and administrative staff) in northwest Spain, Groningen Municipality in The Netherlands, ENEL Green Power (a privatised electricity supplier based in Rome), and Aquatim (the water company for Timişoara, Romania). The main quantitative data source for this was a questionnaire survey that included questions on standard psychological constructs theorised to drive pro-environmental behavior, and the frequency with which various such behaviors are performed. Sample sizes in each case study numbered from 100 to 200. We used decision trees for representing the behavior of the agents because they could be learned from these questionnaire results, providing an empirical foundation for the behavior of the agents; and because the trees themselves have a transparent structure that can be interpreted and understood by the psychologists in the project.

At the simplest level, decision trees are a data structure (formally, a binary tree) encoding series of nested if-then-else statements that are used to predict a value for a response variable. The boolean expressions in the 'if' clauses (which are the 'nodes' in the tree) are each tests of a single variable using comparison operators (<, =, >=, one-of, etc.). The contents of each then/else branch are either a further if-then-else statement or a value for the response variable. The variables tested in the boolean expressions may be referred to as explanatory variables for the response variable.

The advantages of such a data structure, particularly if it is able to provide a (statistically) good fit from the explanatory variables to the response variable, include the ease with which it is possible to interpret and understand how predictions are made. Moreover, there are several techniques for building these data structures automatically from datasets, the most well-known being C4.5 algorithm [38] and Classification and Regression Trees (CARTs) [5]. Algorithms for learning decision trees differ in such things as (a) deciding whether or not to 'split' a node (i.e. to use another if-then-else statement or report a predicted value for the response variable; (b) which variable to select for the boolean expression if a decision has been made to split; (c) how to choose the value at which to make the split; (d) how to treat the response variable (e.g. as a cardinal, ordinal or nominal number); (e) how to create a prediction (e.g. whether to fit a model in the 'leaf' nodes or to report a single predicted value) [53].

If the explanatory variables each form an axis in a multidimensional space, decision trees effectively divide that space with each test using hyperplanes orthogonal to the axis corresponding to the tested variable. For tests other than that in the root node of the tree, the hyperplanes will be bounded by those corresponding to tests in parent nodes. The space is thereby separated into a set of open or closed regions, each corresponding to a predicted value for the response variable. The orthogonality of the hyperplanes is a potential weakness of decision trees: the data may not be elegantly separable in this way. For this reason, data may need preprocessing prior to applying the decision tree learning algorithm. If transparency is an important reason for applying decision trees, then preprocessing techniques [18] need to be selected carefully to avoid detracting from this advantage. Decision trees also have the advantage that they can operate well on relatively small datasets in comparison with other machine learning algorithms [36].

In previous work, we have studied in depth the decision trees generated for the University of A Coruña (UDC) [42, 43] for two reasons: (a) it had the greatest number of respondents to the questionnaire of the four case studies; and (b) the co-location of the modellers and the psychologists in the organisation itself created the greatest opportunity for collaboration and meant more information was available about its everyday operation than was the case for the other case studies. To a lesser extent, two other case-study organizations in LOCAW have also been the subjects of earlier work: Aquatim [41] and ENEL Green Power [37]. In this chapter, with an interest in comparing different approaches to constructing decision trees and preprocessing the data, we want to use as much data as possible to enhance the quality of the resulting decision trees. We therefore decided to amalgamate data from the LOCAW case studies in Spain, Italy and Romania. As there may be differences in the behavior of the workers in these three countries and/or organizations, they were also analysed separately. We therefore effectively have four case studies: the three individual countries, and their amalgamation.

We did not include Groningen Municipality because it had a relatively low number of usable responses for decision tree analysis, and there are important cultural differences with respect to norms in the Netherlands than in the other more southern European cases studies in LOCAW. Data from the Municipality were therefore not included in the amalgamation. Further, the psychologists in each case study had the option of including questions that were only asked in that organisation. Our earlier work in UDC in particular made use of questions that were only asked in Spain. For the purposes of the work reported here, we could only use those parts of the questionnaire that were common to the three organizations.

In what remains of this chapter we will describe the method by which the trees were constructed and the results obtained applying different preprocessing techniques. The chapter is organized as follows: Sect. 2 enlightens relations to previous works. Section 3 describes the questionnaire used to obtain real data. Section 4 explains the methodology used, being Sects. 5 and 6 devoted to present two of the techniques used: feature selection and clustering, respectively. Section 7 shows the experimental results achieved. Finally, Sect. 8 concludes the chapter.

2 Background

One of the most important aspects of agent-based models is determining the algorithms to use for agent decision-making (the other issue being algorithms for interaction) [1]. There have been various empirical approaches to this in the literature, most notably role-playing games [9, 14], declarative modelling [19], laboratory experiments [15], besides other approaches such as recourse to formalisations of theory [28], use of algorithms from Artificial Intelligence [29] and, perhaps surprisingly given criticisms by some in the field of the representation of decision-making in mathematical analysis of social systems (e.g. [35]) recourse to utility maximisation [6]. As explained in the previous section, we have opted by using decision trees.

Gostoli [22] argues for the psychological foundations of decision trees for representing agent decision-making with reference to Gigerenzer and Goldstein's 'Take The Best' algorithm [20] for choosing cues for object recognition based on their Probabilistic Mental Models theory of decision-making from memory ([21] cited in Gigerenzer and Goldstein [20, p. 654]). Gigerenzer and Goldstein [20] showed by computer simulation that such 'fast and frugal' (in terms of computational speed and use of information) algorithms matched or outperformed rational approaches, demonstrating that rationality is not a necessary condition for reasoning in the real world. Indeed, Gigerenzer and Goldstein [20, p. 666] speculate with reference to [5] that their algorithm, which is designed to choose between just two alternatives, could be generalisable to other inference tasks.

Decision trees have been used extensively in agent-based models. Schreinemachers and Berger [44], for example, compare an agent-based model using decision-trees with another using an optimising approach based on mathematical programming. Their conclusion, that the differences between optimising and heuristic approaches are not particularly pronounced[1] is not so different from that of Gigerenzer and Goldstein [20], though interestingly they use this as a basis for preferring optimisation approaches, arguing that, by focusing on structural rather than individual cognitive factors as the source of societal inefficiency, optimisation is more policy-relevant [44]. However, for everyday behavior, which is much more likely to feature habituated routines than a sequence of carefully optimised decisions, such an argument is much less persuasive, even if structural factors are influential.

Other work using decision trees in agent-based models includes that of Smajgl and Bohensky [47], who use decision trees to develop a typology of households for use in their agent-based model of the impact of fuel price changes on vulnerable households in Indonesia.

[1]Empirical evidence in an agricultural context (as is that of [44]) that there is a difference has been found by Evans et al. [15], who show using laboratory experiments with human subjects compared with a utility-maximising agent-based model that humans produced more patchy landscapes. Since higher landscape fragmentation affects species distributions, this will be of interest to policymakers concerned with meeting the United Nations Convention on Biological Diversity's targets for halting biodiversity loss.

3 Materials: Questionnaire Data

Various quantitative and qualitative tools were used in LOCAW project to analyse the different organizations, including focus groups, interviews, and questionnaire surveys. We had the aim of making the ABM evidence-based, drawing on data supplied by field researchers and other supplementary information. Among the quantitative tools for data acquisition, the purpose of the questionnaire was to get data from personnel in each case study organization on the relationships between the individual demographic and psychological factors theorised to influence pro-environmental behavior and such behavior at work. The psychological factors in the questionnaire are based on the values-beliefs-norms (VBN) theory of environmental concern [50].

The questionnaire comprises five blocks of questions covering: (a) socio-demographic data; (b) hedonic, egoistic, altruistic and biospheric value profiles; (c) beliefs, including worldviews, efficacy and pro-environmental identity; (d) injuctive and descriptive local and general norms; (e) behaviors. Values can be seen as abstract concepts or beliefs concerning a person's goals and serve as guiding standards in his or her life. Schwartz identifies ten human value profiles [46], though some authors (see Chap. 1 in [23]) argue that just three may affect beliefs and behavior related to the environment: egoistic, altruistic and biospheric. For example, a person may reduce car use because the costs are too high (egoistic), because it endangers the health of fellow citizens (altruistic), or because it harms plants and animals (biospheric). More recent studies [49] have shown that hedonic value profiles also play an important role in the environmental domain. Following with the same example, a hedonic worker may commute by car because they find driving more pleasurable than (cheaper, healthier, or less polluting) alternatives.

Worldviews are measured using the New Human Interdependence Paradigm scale [8, 11], which assesses the degree to which respondents believe in the reciprocity and interdependence of human societies and the ecosystems they inhabit, and has been found to correlate with pro-environmental attitudes and behaviors. Efficacy pertains to beliefs about whether behaving pro-environmentally is possible and makes a difference. Specifically, self-efficacy concerns individuals' perceived personal control over their ability to undertake pro-environmental behavior [2], whilst outcome efficacy measures the extent to which people believe that their own pro-environmental behavior contributes to the solution of environmental problems [45]. The latter is particularly important in the context of the environment, as addressing the problems we face requires concerted global action. Environmental self-identity measures how people see themselves as being the kind of person who acts pro-environmentally [51].

Descriptive norms concern what the individual observes as being typically done by others, whilst injunctive norms are what the individual has been told to do by others [10]. Local norms have been distinguished from general norms, the former pertaining to normative social interactions that are associated with particular settings [17] (such as the workplace in this project), as opposed to more generally. More detail on the questionnaire, and the psychological theories behind it, is available in Deliverable 4.3 of the LOCAW project, which includes in an appendix a full listing of the questionnaire itself [40].

The behaviors questions covered a number of everyday practices at work and in the home, in three main categories:

- organization-related mobility,
- energy consumption,
- management and generation of waste.

A fourth 'metabehavior'—norm transmission (telling your colleagues to behave more pro-environmentally, which doesn't itself save any carbon emissions, but is intended to encourage others to)—was added at the request of the modellers in the project to enable the endogenous simulation of perceived injunctive norms using social networks in the agent-based model. As described in other work [41], the simulated social networks include vertical (hierarchical: between supervisors and subordinates) and horizontal (informal: among coworkers) components.

Questions of blocks (a–d) are considered as inputs for the decision trees, whereas questions in block (e) are the desired outputs. The questionnaire was made available online so that employees can easily access and complete it anonymously. Response rates were fairly typical for questionnaire surveys: for example, 300 responses were obtained for UDC, which has nearly 2000 employees to whom a link to the questionnaire was sent (including both administrative and teaching personnel). As is common with surveys, many respondents had only completed a small fraction of the questions, and their results had to be discarded as they are useless for the purposes of contributing to the construction of a decision tree. The final number of responses used for building the aggregate case study decision trees described herein is 397, of which 185 are from the UDC and the remainder equally distributed between Aquatim and ENEL (106 each). Results from some questions (in both the input and output sections of the questionnaire) were also discarded because most of the participants did not answer them. In the end, 66 questions were used as inputs, of which 7, 16, 22 and 21 related to blocks (a), (b), (c) and (d), respectively.

Tables 1 and 2 summarize the questions included in the questionnaire, based on studies of one of the research groups participating in the project [48]. Table 2 provides the values block (b) questions in detail, as these were treated somewhat differently than the others. Most questions in the questionnaire use seven-point Likert scales [31], from 'totally disagree' to 'totally agree' for various propositions in blocks (c) and (d); and 'never' through to 'always' for frequency of behaviors—though some of the behaviors were Boolean (e.g. 'Do you have personal control over the thermostat at your workspace?') or asked respondents for a number (e.g. 'How many kilometers per week do you on average commute by car?'). Questions related to value profiles (block b) have a slightly different scale, because in addition to a 1–7 scale ('important' 1–5, 'very important' 6, 'of supreme importance' 7) representing the importance of each value, options are included for 'Opposed to my values' (represented by -1) and 'Not important' (0).

Table 1 Questions that, together with the values questions in Table 2, form the inputs to the decision-making procedure

Socio-demographical (1–7)	Norms (36–57)	Beliefs (24–35, 58–66)
Country (1)	Descriptive-general (36–40)	Self-efficacy (24–26)
Gender (2)	Injuctive general (41–45)	Outcome-efficacy (27–29)
Age (3)	Descriptive local (46–49)	Worldviews (30–35)
Level of studies (4)	Injuctive local (50–53)	Environmental self-identity (58–60)
Organization level (5)	Personal (54–57)	Environmental organisational identity (61–63)
Exemplary role (6)		Organisational identification (64–66)
Job position (7)		

Parentheses are used to show the question number(s) of each input

Table 2 Questions related to values associated with their profile (column headings)

Altruist	Biospheric	Egoistic	Hedonic
Equality (8)	Respect earth (9)	Social power (10)	Pleasure (11)
World at peace (13)	Unity with nature (12)	Wealth (14)	Enjoying life (17)
Social justice (16)	Protecting environment (18)	Authority (15)	Self indulgent (22)
Helpful (20)	Preventing pollution (21)	Influential (19)	
		Ambitious (23)	

Brackets indicate the question number

The number of decision trees to learn are: 4 norm transmissions, 27 behaviors at work and 27 behaviors at home. However, the LOCAW project analysed (un)sustainable behavior and practices in the workplace and focused on the factors determining these actions in the organizations under investigation. The aim of this analysis was to explore how to promote more sustainable behavior at work; the home was of interest more in the degree to which there was 'spillover' between similar behaviors in these two main spheres of everyday life. Since the main focus of the project was the workplace, in this chapter we will deal with the first two desired outputs, that is, norm transmissions and behavior at work, leading to a total of 31 desired outputs.

The concern here is less with the trees themselves, but more with the methods by which they are produced, and the degree to which the results are (a) a good fit to the empirical data; (b) comprehensible to psychologists; (c) suitable for use in an agent-based model as the decision-making algorithm for the agents. Item (c) follows largely from (a) and (b), with an additional consideration largely pertaining to the justification for using ABM in the first place: to get interaction among the agents (without which an ABM is not really necessary), it is important that injunctive or descriptive local norms feature among the explanatory variables used in at least some of the decision trees.

4 Methodology

As a first attempt, we try to learn all the decision trees from the raw data, i.e. without preprocessing, and apply a stratified 10-fold cross-validation as a validation technique. Table 3 shows information about sixteen of the 31 decision trees derived (one from each of the four main output categories for each of the four case studies), including details on the training and estimated prediction accuracies. The second and third columns provide information on the size of the tree, in particular, the number of nodes and the number of attributes/features that have been used to form it. It is important to note that none of the decision trees derived uses all the 66 inputs available. Although larger trees using more features are generated when there are more samples (rows 'All'), but even in this case the decision tree with the most attributes used 53 (4. Waste—How often do you use recycled paper?). The fourth and fifth columns in Table 3 show accuracy performance results, the former for the training set and the latter for the test set. These two columns demonstrate that the decision trees are clearly overfitted: they perform adequately in the training

Table 3 Accuracy results for different behaviors using decision trees without any preprocessing

Country	*Size*	NumFeatures	AccTrain (%)	AccTest (%)
1. *Norm transmission (to subordinates)*				
All	151	43	90.7	42.1
Spain	87	35	81.6	21.1
Italy	55	24	87.7	31.1
Romania	15	7	95.3	82.1
2. *Mobility—how often do you commute by car*				
All	157	45	87.4	35.3
Spain	81	27	91.4	40.5
Italy	25	12	81.1	55.7
Romania	59	23	84.9	25.5
3. *Energy—how often do you have lights on at work when no-one is there*				
All	199	51	85.1	33.3
Spain	81	32	88.1	37.3
Italy	55	19	83.0	23.6
Romania	45	16	88.7	35.8
4. *Waste—how often do you use recycled paper*				
All	225	53	82.9	18.4
Spain	101	32	86.5	20.5
Italy	51	19	84	22.6
Romania	53	24	78.3	16.0

Size indicates the number of nodes and leaves of the tree, *NumFeatures* the number of features used to build the tree and *AccTrain* and *AccTest*, the accuracy for the train and test sets, respectively set, but they exhibit a very poor performance in the test set. The excessive size of the decision trees also corroborates this observation; the larger number of branches suggesting fitting to specific cases rather than general rules. An added problem is that such trees are very difficult for the psychologists to visualize and interpret. Moreover, our experimental results in a previous UDC study [43] evidenced that there is always a combination of discretization and feature selection methods that significantly improved the accuracy compared to not using these preprocessing methods. Therefore, we have applied these preprocessing techniques with a dual purpose:

- to obtain decision trees with better generalization capabilities—that is, they must not only adjust to the training set, but also perform properly on the test set;
- to derive smaller and thus easier to interpret decision trees. In this manner, decision trees can be analysed by experts to verify their theoretical consistency. Moreover, it is simpler to include them in the agent-based model and to analyse the possible consequences of a change in any of the variables involved.

Since the application of preprocessing techniques—especially where they are unfamiliar to the psychologists—is a potential source of reduction in transparency, the purpose of the study is to compare various combinations of the following preprocessing techniques:

- Feature selection (FS). Although the decision tree algorithms themselves select which variables are most relevant when deciding to split a node, selecting a subset of explanatory variables to use prior to applying the decision tree algorithm can improve its performance (measured by the fit of the resulting tree to data). Notice that the complete set of available samples was used, although this is not the usual procedure and it would not be fair for a comparative study of feature selection methods, but this is not the focus of this chapter. Given the low ratio samples/features, to obtain the most representative feature, it was necessary to use all samples available. Feature selection is defined as the process of detecting relevant features and discarding irrelevant and redundant features with the goal of obtaining a subset of features that accurately describe a given problem with a minimum degradation of performance [25]. Section 5 is dedicated to explain the existing methods of this preprocessing technique and it also explains the reasons that have led us to select the two methods that will be used in this work: Correlation-based Feature Selection [26] (CFS) and Support Vector Machine—Recursive Feature Elimination (SVM-RFE) [24].
- Discretization. Discretizing the response variable simplifies the prediction task, and can be used to alleviate some of the problems related with the use of the Likert scale employed by the questionnaire, such as differences among respondents situated in the extremes of the scale. In previous studies [43], simple binning techniques such as equal-width (EWD) and equal-frequency (EFD) [32] were adopted. These techniques require to be provided with the desired number of intervals (bins) by the user, and in an attempt to make this latter decision automatic, in this chapter we have opted for using Fayyad and Irani's MDL [16], determining the number of intervals by leave-one-out validation.

- Clustering of the values related questions. Clustering is an unsupervised machine learning technique that attempts to find distinct groups in the data. Using clustering approaches can create a single (nominal) variable from a number of cardinal variables, thereby simplifying considerably the input to a decision tree. However, the objective of this work is not to reduce the number of inputs; rather to group existing samples based on the values profiles, i.e., associate clusters to each profile: altruistic, biospheric, hedonic and egoistic. Clustering algorithms are as many and varied as those for constructing decision trees; however not many of them provide results that help to associate the cluster to a profile (further details in Sect. 6). Taking this into account, we have chosen k-means clustering [33], where the centroids generated can help us to see if there is a match with the profile, and we have had to rule out other methods, such as hierarchical clustering [34], in which this information is not available.

Figure 1 illustrates the decision making procedure of an agent. It shows the different techniques that are going to be tested and the different inputs that each one is using. Note that not all the steps are required, so if, for instance, the discretization step is discarded, the original output will be given to the decision tree for it to predict. Similarly, if feature selection is removed, the complete set of input variables is provided to the decision tree algorithm. All possible combinations (with/without feature selection, with/without discretization, with/without clustering) have been analysed in this work in order to achieve the objectives of obtaining better generalized decision trees, smaller and simpler to interpret by the experts. Due to the high number of combinations and behaviors, we have focused on the four behaviors

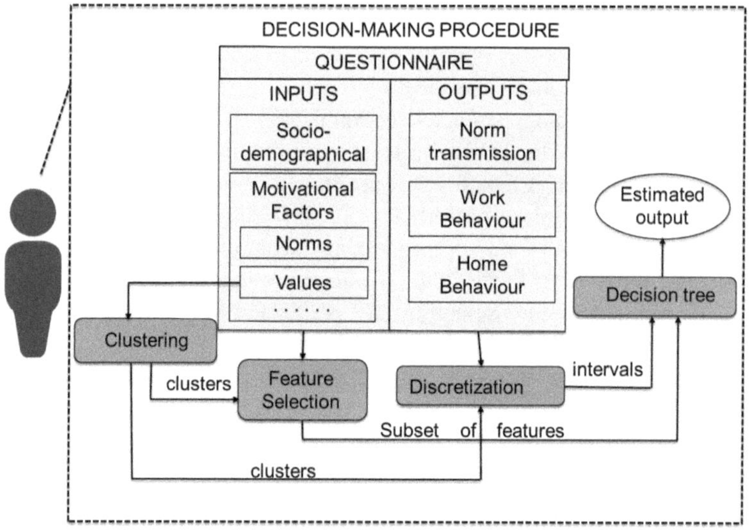

Fig. 1 Decision making procedure

shown in Table 3. Feature selection, discretization and decision trees (specifically, the J48 implementation for C4.5) algorithms were applied using Weka tool [27], whereas clustering was carried out in Matlab.

5 Feature Selection

Our goal is to obtain decision trees to be included as part of the decision-making process of the agents in our model. To ensure the theoretical-consistency of these decision trees they should be analysed by experts. Thus, it is important not only to derive decision trees that exhibit a good performance and have good generalization capabilities, but also that are as simple as possible to facilitate their analysis. Therefore, although the C4.5 algorithm may discard features (i.e., it may not use all the attributes available to construct the decision tree as can be seen in Table 3), a preprocessing step of feature selection can be included to improve the simplicity of these decision trees, whilst aiming also at obtaining better generalized trees.

Feature selection (FS) is the process of detecting the relevant features and discarding the irrelevant ones for a given data set. A correct selection of the features can lead to an improvement of the inductive learner, either in terms of learning speed, generalization capacity or simplicity of the induced model [3]. With regard to the relationship between a feature selection algorithm and the inductive learning method, three major approaches can be distinguished [3]:

- filters, which rely on the general characteristics of training data and carry out the feature; selection process as a preprocessing step with independence of the induction algorithm;
- wrappers, which involve optimizing a predictor as a part of the selection process;
- embedded methods, which perform feature selection in the process of training and are usually specific to given learning machines.

Feature selection methods can also be classified according to individual evaluation and subset evaluation approaches [52]; the former also known as "feature ranking" assesses individual features by assigning them weights according to relevance, whereas the latter produces candidate feature subsets based on a specific search strategy, which are subsequently evaluated by some measure.

There are many feature selection methods available in the literature and several have become particularly popular among researchers. In [3], there is a review of the behavior of different feature selection methods over different problems studied, such as noisy input, ratio of features to samples, etc. In our previous study [43], we have tested six different methods, all of them based on different metrics, returning different subsets of features. None of them was remarkably superior to others, although the ranker embedded method Recursive Feature Elimination for Support Vector Machines (SVM-RFE) [24] obtained quite good results. This embedded method performs feature selection by iteratively training a SVM classifier with the current set of features and removing the least important feature indicated by the SVM. For the sake of comparison, we have also used a completely different selection method, a subset filter method: Correlation-based Feature Selection (CFS) [26] that evaluates the worth of a subset of attributes by considering the individual predictive ability of each feature along with the degree of redundancy between them. This method was used in our previous analysis of the UDC [42] because of its simplicity and independence from any learning algorithm, returning good performance results. Besides these two methods, other subset feature selection methods were tested for the aggregate case study covering data from the three countries, such as INTERACT [54] and consistency [12], but similarly to the situation described in [43], the selected feature subsets were extremely small (1 feature in some case), and in consequence we have discard their use in this study. In relation to wrappers, although some were initially tested, their computational time was too long and not compensated with an improvement in the decision trees performance, and therefore they have not been included in this study.

Both feature selection methods finally selected, CFS and SVM-RFE, have been applied for each one of the desired outputs (behaviors), considering as inputs the questions indicated in Tables 2 (values) and 1. Table 4 shows the features selected by these two methods for the four behaviors. Note that SVM-RFE is a ranker method, so it returns the full set of features, but sorted in descending relevance. Therefore a threshold is required to make a proper selection. The problem of selecting a threshold for rankers has been and is still one of the focus of research in the field

Table 4 Features selected by CFS and SVM-RFE for the considered behaviors

FS method	*Country*	1. *Norms—subordinates*	2. *Mobility—commute by car*
CFS	All	3,5,**6**,8,20,46,50,55	**2**,5,7,23,25,36,55
	Spain	5,**6**,11,35,51,57,58,59,60	**2**,7,22,39,40,52,58,66
	Romania	3,5,**6**,46,50,58	**2**,5,37
	Italy	2,4,**6**,26,37,45,46,50,54,65	**2**,4,9,11,24,37,38
SVM-RFE	All	1,5,9,29,27,46,7,56,58,30,55,50	6,5,57,42,37,9,47,25,26,60,33,48,32
	Spain	19,60,35,22,12,40,15,29,33,59,30,34	40,43,22,46,59,47,41,13,15,11,7,60
	Romania	50,46,59,60,39,57,55,61,30,5,21,65	8,31,33,32,37,23,14,34,13,57,40,29
	Italy	9,27,42,64,10,29,39,20,26,55,13,65	59,62,2,64,23,55,33,13,57,4,65,2
FS method	*Country*	3. *Energy—lights off*	4. *Waste—recycled paper*
CFS	All	1,7,13,16,23,36,41,43,64	1,9,17,19,26,34,36,43,46,50,56,59,60
	Spain	5,11,17,22,43,51,54,62,64	2,4,6,9,14,30,34,38,49,50,62
	Romania	2,4,7,11,44,64	2,4,10,14,23,39,41,43,46,48,49,54,59
	Italy	2,4,7,9,11,12,19,47,51,63	2,4,5,7,10,15,21,24,25,31,32,38,43
SVM-RFE	All	16,50,46,26,25,34,47,59,56,7,51,28	25,21,5,25,36,49, 66, 58, 16, 50, 27, 1
	Spain	43,49,46,59,9,54,22,8,29,51,7,14	22,53,51,35,26,47,8,54,49,33,59,30
	Romania	24,64,43,16,42,20,12,41,54,38,44,13	23,49,57,5,41,53,13,12,35,14,58,46
	Italy	18,14,23,54,9,7,35,17,25,11,34,24	53,9,59,39,25,48,57,21,28,37,24,5

SVM-RFE results are in rank order. Marked in bold and underlined the features that appear in the four cases of the study using the same FS method of feature selection. At present, there is not yet an automatic and general method that allows researchers to devise a threshold for any dataset [3, 4]. For that reason, the thresholds used in many studies have to be tailored for each dataset. In our case, focusing on the goal of obtaining simple decision trees and bearing in mind the results already returned by the CFS method, we have set the threshold to include the twelve highest-ranked features.

The results shown in Table 4 indicate that CFS and SVM-RFE do not select the same features, as might be expected, because they are based on different metrics. An unexpected and perhaps surprising result is the low similarity between the features selected by a single algorithm for each of the case studies. If we focus in the first behavior—norm transmission to subordinates—it can be seen that only feature 6 (exemplary role) appears in all four case studies for CFS, and there are no features common to all four case studies for the SVM-RFE method. For mobility behavior, gender (feature 2) appears to be a determining variable, at least using CFS method. However, in the remaining behaviors, there is no match.

6 Clustering

The clustering process was carried out using only those 16 questions pertaining to values included in the questionnaire. Each question is associated with a profile, and there are four different profiles: altruist, biospheric, egoistic and hedonic (see Table 2). It is expected that individuals belonging to certain profiles behave more pro-environmentally than others, for example, biospheric *versus* egoistic. Thus, the aim of clustering is to identify these four profiles to then associate different behaviors, i.e., decision trees, to each one.

Each question in Table 2 has nine possible answers, one of them is entitled "Opposed to my values" , with a -1 value assigned. The other possible answers range between 0 ("Not important") and 7 ("Of supreme importance"). So, not all the ranges in the values part of the questionnaire have the same significance, as only the first column specifies opposing values, while the others specify a continuous rang. In order to take this fact into account, all -1 responses were multiplied by 10 to emphasize their significance.

To determine these four profiles, the well-known k-means algorithm was employed in this work. For this algorithm, the number of maximum iterations was established to 500 and the Euclidean distance was used as distance function to compare instances. This algorithm requires the specification of the number of clusters to generate. In our previous study on the University of A Coruña, UDC, [42], different initializations were tested for k-means trying to identify these four profiles. However, none of the partitions obtained using four clusters allowed for clearly distinguishing the profiles as indicated by the psychologist and sociologist experts working in the project. In discussion with them, and after testing different number of clusters, six clusters were identified that provide adequate separation of the samples and contain hybrid groups. Specifically four "almost-pure" profiles can be identified coinciding with the theoretical ones: egoistic, altruistic, biospheric and hedonic and two more hybrid groups, that mixed similar profiles (biospheric-altruist and egoistic-hedonic).

However, in this chapter we are considering several case studies: data from three countries separately and in aggregate. So, first for all the cases, again the number of clusters was established to 4 for the k-means algorithm, however the results achieved were not satisfactory as they did not allow to an adequate identification of the profiles. Then, as the data set is not large (387×16 when considering countries jointly) and the k-means algorithm takes around a second to provide a cluster, an automatic procedure was established trying to determine the optimal number of clusters for each case. So, consider the set of possible number of clusters formed by $K = 2, 3, 4, 5, 6, 7$. Then, for each $k \in K$, these steps were followed:

1. run the k-means algorithm 50 times with different initial points for the centroids
2. from these 50 runs, select the clustering that minimizes the within-cluster sums of point-to-centroid distances (C_k).

After executing these steps, a set formed by six possible optimal clusters $[C_2 - C_7]$ was obtained. Subsequently, the Calinski-Harabasz criterion [7] (sometimes called the variance ratio criterion (VRC) because it is based on the overall between-cluster and inner-cluster variance), and the Davies-Bouldin criterion [13], based on a ratio of within-cluster and between-cluster distances, were used to determine which was the optimum number of clusters of these six tentative values. The optimal number of clusters varied for each case study and the profiles identified were also different. As an example we will focus on the aggregate case study. For this case, two was the number of optimal clusters, independently of the criteria used. Graphically, the silhouettes in Fig. 2 show the same result. Note that these silhouettes show which objects lie well within their cluster (positive values), and which one are merely somewhere in between clusters (negative values) [39]. Clearly the fewest negative values are observed in the case of two clusters (top-left in Fig. 2), with slightly more in the case of three clusters (top-right in Fig. 2). Although the procedure assures that the number of clusters is optimal, it was unknown if it will be useful to distinguish the four profiles, so we studied the centroids of the clusters generated by the k-means algorithm. Since the number of clusters was less than the number of profiles, there must be some hybrid profile. In fact, two hybrid profiles were clearly identified (see Fig. 3): bio-altruistic and ego-hedonic. The differences between the answers to the questions in the egoistic profile are evident; in addition, individuals in the bio-altruistic cluster give slightly higher values in their answers to questions related to biospheric and altruist profiles, whereas the ego-hedonic individuals reply with higher values to the hedonic questions. If we increase the number of clusters to 3, the bio-altruist cluster is divided resulting in a cluster that would maintain the status of bio-altruist cluster and one that is not associated with any profile. Thus, for the purpose of this chapter, we retained the two-cluster model in Fig. 3.

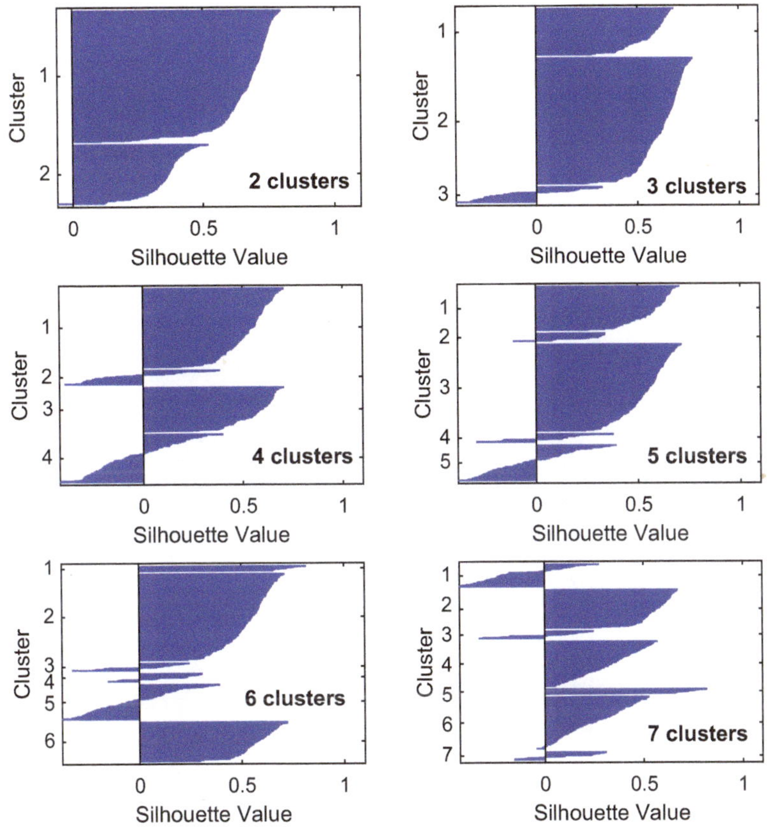

Fig. 2 Silhouttes for different number of clusters for the case study 'All'

For the other three cases addressed in this chapter, that is, each country separately, a summary of the results of the clusters using the same method as described for 'All' above is provided in Table 5. As it is fairly evident, the generated clusters are markedly different, in both Spain and Romania egoistic individuals predominate, while in Italy, the same hybrid profiles as those obtained in the three countries together prevail. Hence, in the top left of Fig. 3 ego-hedonic individuals are in the larger cluster. Another fact to highlight is that the clusters are clearly unbalanced, some with large numbers of samples and others with very low values. Remember that the number of samples was 185 for the UDC in Spain, and 106 for both Aquatim in Romania and ENEL in Italy. The smallest clusters have 12 samples for the Spanish case and 15 samples for the Romanian and Italian cases. In this scenario, even applying feature selection and discretization techniques, it became infeasible to derive decision trees with good generalization capabilities with such a small size of samples. In fact, when tested, the decision trees were not generated as all samples were assigned to the majority class.

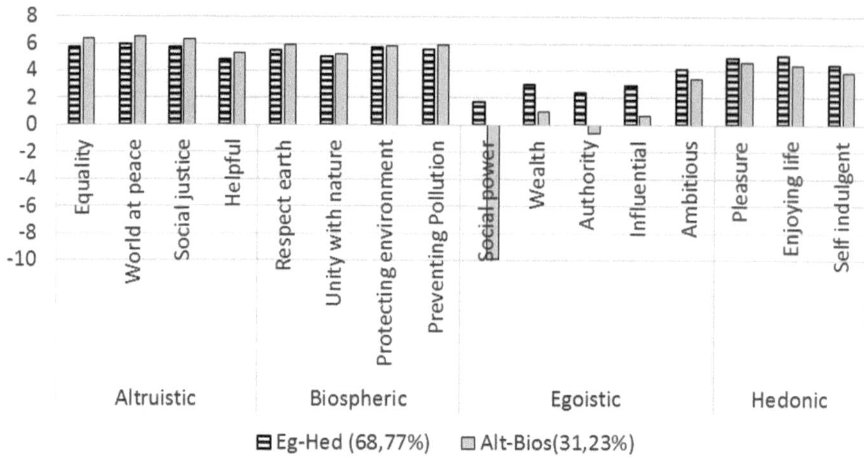

Fig. 3 Average answers to values questions for each cluster identified for the case study 'All'

Table 5 Summary of the clusters generated for the cases of study that consider countries individually

Country	Cluster	Profile	Percentage of samples
Spain	1	Altruistic	6.49
	2	Biospheric-hedonic	5.40
	3	Egoistic	60.0
	4	Indeterminate	28.11
Romania	1	Egostic	85.85
	2	Alt-Bios-Hed	14.15
Italy	1	Ego-hedonic	14.15
	2	Alt-biospheric	71.70
	3	Indeterminate	14.15

7 Experimental Results

The goal of this work is to obtain decision trees that exhibit good generalization capabilities, but at the same time are understandable, and have a simple structure for two reasons: (a) to be analysed by experts and (b) to facilitate a sensitivity analysis for any of the inputs within the agent-based model. In this section, we will show the results achieved generating the decision trees using various combinations of preprocessing techniques. In order to facilitate the readability of the subsequent tables, the different combinations of preprocessing techniques used and the acronyms given to each one is detailed below:

- DT: apply decision trees over the original data, without any preprocessing technique

- MDL-DT: apply discretization over the desired output, then build the decision tree
- Apply feature selection before training the decision tree
  - CFS-DT: CFS method as feature selector.
  - SVM-DT: SVM-RFE method as feature selector.
  - Apply both feature selection to determine the relevant inputs, and discretization to the desired output
    - CFS-MDL-DT: CFS method and MDL discretizer
    - SVM-MDL-DT: SVM-RFE method and MDL discretizer.

Table 6 shows the accuracy results for the test set for all the combinations above. The importance of discretization is clearly derived, since not all possible outcomes were covered by the available sample. However, feature selection plays also an important role because it helps to slightly improve the results in accuracy, but decreasing appreciably the number of variables involved. As it can been, most of the best results (marked using bold face) are in the columns entitled *CFS-MDL-DT* or *SVM-MDL-DT*. While this table clearly indicates that the generalization capabilities of the decision trees have been improved (compared to the DT column), it is necessary to check whether the decision trees are also smaller. Table 7 illustrates the same properties of the decision trees for CFS-MDL-DT as Table 3 does for DT alone to facilitate a comparative analysis. Besides the obvious differences in both

Table 6 Test accuracy (%) results for different behaviors using decision trees with and without preprocessing

Country	Beh.	DT	MDL-DT	CFS-MDL-DT	SVM-MDL-DT	CFS-DT	SVM-DT
All	1	42.1	68.8	65.2	**71.5**	39.8	38.8
	2	35.3	60.2	71.0	**72.3**	47.6	44.8
	3	33.2	38.5	**40.1**	39.8	39.0	37.5
	4	18.4	58.4	**70.5**	68.0	19.4	19.6
Spain	1	21.1	49.7	**56.8**	56.2	29.2	28.1
	2	40.5	69.2	77.3	**80.5**	55.1	58.4
	3	37.3	**42.2**	37.3	35.1	37.3	40.5
	4	20.5	35.7	**50.8**	44.9	27.0	23.8
Italy	1	31.1	82.1	83.0	**91.5**	31.6	32.1
	2	55.7	72.6	78.3	**83.0**	61.2	61.3
	3	23.6	29.2	26.4	21.7	**33.7**	25.5
	4	22.6	91.5	**92.5**	**92.5**	20.3	25.5
Romania	1	82.1	**99.1**	**99.1**	**99.1**	85.8	81.1
	2	25.5	34.9	**60.4**	39.6	28.3	21.7
	3	35.8	48.1	47.2	**50.9**	43.4	41.5
	4	16.0	60.4	**73.6**	71.7	16.0	27.4

Beh stands for behavior, see Table 3 for detailed information. Best result per row marked in bold font

Table 7 Results for the CFS-MDL-DT method

Country	Size	NumFeatures	AccTrain (%)	AccTest (%)
1. *Norm transmission (subordinates)*				
All	55	7	80.6	65.2
Spain	29	6	75.7	56.8
Italy	9	3	90.6	83.0
Romania	3	1	99.1	99.1
2. *Travel—How often do you commute by car*				
All	25	8	77.1	71.0
Spain	0	0	80.5	77.3
Italy	7	2	88.7	78.3
Romania	5	2	60.4	60.4
3. *Energy—How often do you have lights on at work when no-one is there*				
All	99	8	66.8	40.1
Spain	81	8	80.5	37.3
Italy	37	8	65.1	26.4
Romania	9	4	68.9	47.2
4. *Waste—How often do you use recycled paper*				
All	41	11	79.8	70.5
Spain	47	11	77.3	50.8
Italy	0	0	93.4	93.4
Romania	11	5	82.1	73.6

Size indicates the number of nodes and leaves of the tree, *NumFeatures* the number of features used to build the tree and *AccTrain* and *AccTest*, the accuracy in percentage for the training and test sets, respectively

accuracy and number of features, the decision tree size is lower in all cases, although the data sets with larger sample size (All and Spain) still have a large number of nodes in their decision trees. Note that there is no decision tree in some cases (for instance, behaviour 2 in Spain) because all samples were assigned to the majority class, there are two reasons for that: (a) the discretization step derives only one class or (b) no decision tree is constructed during the training phase. In the mobility case related to Spain, it is known that most of the employees commute by car (around 80 %), and that is in fact the assumption of the model.

As an example, Fig. 4 shows one of decision trees obtained, specifically, that associated with the behavior of commuting by car for the Romanian data, where we can see the strong influence of gender as might be anticipated from Table 4 (Behavior 2, Romania-CFS method).

In addition, some experiments were performed to check the relevance of using a clustering technique. For the reasons given earlier, we only analysed the 'All' country scenario (the aggregated case study), and so Table 8 contains the main results only for the case referred to as 'All' in Tables 6 and 7. The results shown in the table are the mean accuracy weighted by the percentage of samples in each cluster. Results improving or equaling those obtained by the same method but

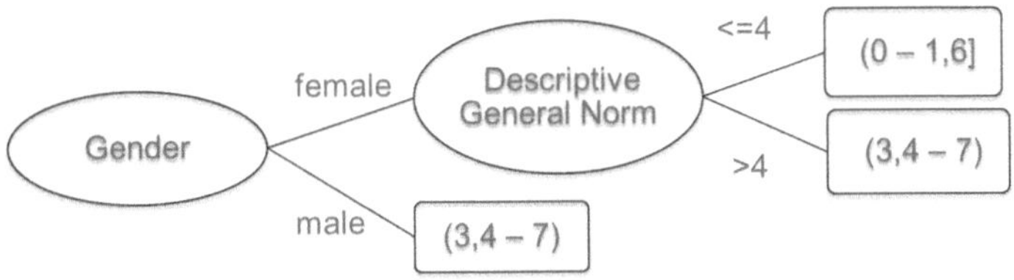

Fig. 4 Decision tree generated for frequency of using the car for commuting in the Romanian case study. Descriptive general norm question:*Most people who are important to me act pro-environmentally at work*

Table 8 Test accuracy (%) results for different behaviors using decision trees with and without preprocessing for the case study 'All' with weighted clustering

Beh	DT	MDL-DT	CFS-MDL-DT	SVM-MDL-DT	CFS-DT	SVM-DT
1	*42.8*	*68.8*	*68.5*	68.0	37.1	*39.6*
2	34.7	*65.7*	**72.3**	67.7	43.3	*46.3*
3	31.5	36.0	**45.6**	39.3	*43.1*	*38.8*
4	*21.4*	45.3	58.7	54.4	*19.9*	*20.4*

Beh stands for behavior, see Table 3 for detailed information. Results that outperform or equal the corresponding ones without clustering are marked in italics. Best values are marked in bold

Table 9 Results for the CFS-MDL-DT method for the 'All' case for each cluster

Beh	Cluster	Size	NumFeatures	AccTrain (%)	AccTest (%)
1	1	47	8	85.7	74
	2	37	7	83.9	56.5
2	1	0	0	72.2	70.0
	2	7	3	82.3	77.4
3	1	89	7	68.1	43.6
	2	9	3	58.1	50
4	1	61	11	75.5	56.4
	2	23	6	83.9	63.7

Size indicates the number of nodes and leaves of the tree, *NumFeatures* the number of features used to build the tree and *AccTrain* and *AccTest*, the accuracy for the training and test sets, respectively

without clustering are marked in italics. Also, the best values for each row are marked in boldface (only if the result is the best using clustering). As can be seen, in half of the cases, the use of a clustering technique allows improving the results obtained previously. It should be noted that in this study the number of instances used is relatively low. We expect that with a significant increase in the number of samples, using cluster would be beneficial in most cases.

Finally, Table 9 contains the detailed results for the best model (CFS-MDL-DT) applying it to each one of the clusters obtained previously. Remember that for the

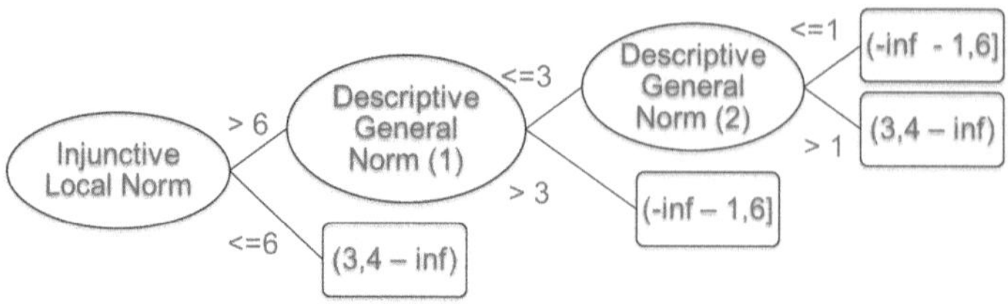

Fig. 5 Decision tree generated for cluster 2 (Alt-biospheric) of using the car for commuting for the 'All' case study. Injuctive local norm question: *Most of my supervisors think I should act pro-environmentally at work*. Descriptive general norm (1) question: *Most of my neighbors act pro-environmentally at work*. Descriptive general norm (2) question: *Most fellow countrymen act pro-environmentally at work*

'All' case the optimal number of cluster was two. Note that, although the results with clustering do not improve significantly those without clustering, these setups allow us to add variability to the agent model using agents of two types (cluster 1 and cluster 2), much as in [47], in which different actions are implemented depending on the type of agent. This can effectively be seen as treating the cluster as the root node in a decision tree combining those derived for clusters 1 and 2 independently. An alternative approach, not explored here, would be to add the cluster membership as a nominal explanatory variable to the feature selection (if used) and decision tree learning algorithms. Again, as an example, Fig. 5 shows a derived decision tree, in this case different questions related to norms predominate in the nodes, specially concerning the immediate environment of the agent (supervisors and neighbors).

8 Discussion and Conclusion

In this study we have examined the construction of empirically-derived decision trees for the purposes of implementing decision-making in an agent-based model. The decision trees were directly obtained from questionnaire data containing answers related to pro-environmental behavior at work in several contexts, in a scenario in which the number of samples available from each organization is limited. Decision trees were derived for each country separately as well as for an aggregate dataset. Although applying directly the C4.5 algorithm for obtaining the tree was feasible, it was demonstrated in our study that the resulting trees exhibit lower generalization capabilities and high number of nodes and branches, making them difficult to interpret. Although this situation was especially acute in the trees that were derived considering all three countries together (in which the size and number of features employed almost doubled those of each country separately), it was also inadmissible for each individual country. Regarding the difference in percentage of accuracy between train and test sets, it was similar for all the scenarios, with quantities above 80 % in most cases for training, while dropping to values around

20–40 % in most cases for test, and thus clearly pointing out the existence of overtraining in the method. Employing preprocessing techniques greatly improved the results, obtaining better percentages of accuracy in the trees, while at the same time making them simpler, and easier to interpret by the psychologists and sociologists in the project, who should review the automatically-generated trees to assure theoretical concordance. Several preprocessing techniques were employed, such as feature selection, discretization and clustering. Clustering was used in order to categorize individuals according to their values profiles (egoistic, hedonic, biospheric and altruist), and then feature selection (either SVM-RFE or CFS) techniques and discretization (using Fayyad and Irani's MDL method) were applied. In this way, simpler and smaller trees were obtained, diminishing in approximately three times the sizes of the trees for the case of all countries together. For this case, the reduction in the number of features is even more dramatic, ranging from 47 to 53 (depending on the behavior) for the decision trees without preprocessing to 7–11 using the best combination tested (*CFS-MDL-DT*); in any case, the training accuracy is maintained or even improved, but greatly increasing the test accuracy in more than 30 % in different cases, and thus paving the way for arriving at our goal in devising better and simpler decision trees that could be used by our agent-based model.

Agent-based models may have specific requirements of decision tree algorithms (and preprocessing) that do not apply in other contexts in which these machine learning techniques are applied. Feature selection, for example, is in the classical case concerned with using the minimum number of explanatory variables needed to develop the best decision tree. Choices made by the feature selection algorithm between subsets of explanatory variables with nearly equivalent decision tree performance could have a significant effect on the dynamics of the model. Ignoring the potential influence of a variable could limit the potential scenarios explored by the resulting model. At the same time, if the evidence in the data does not support the inclusion of a variable, there is little point in modelling it. As such, the instability in selection of features in Table 4 is not reassuring for the study at hand.

Social scientists with an interest in agent-based modelling are already concerned about the learning curve to be faced if they want to build an ABM themselves. If they are interested in using questionnaire data to develop the decision-making algorithms, they face an additional barrier in familiarising themselves with the machine learning literature needed to ensure that the decision trees are well-constructed. The potential for using questionnaire data in the way suggested here and in our previous work is an interesting prospect for future work. However, the diversity of machine learning literature, for each of the preprocessing techniques as well as the decision tree learning algorithms themselves, and especially the instability of results derived therefrom will be of concern for those considering attempting a similar exercise. That said, ABM is inherently an interdisciplinary exercise. With experts in ABM and machine learning working together, we have shown that it is possible to build algorithms to implement agents' behaviors that are good models of the data, open and transparent to colleagues in the conventional social sciences, and containing variables relevant to interaction (here, norms) in the decision nodes.

Acknowledgements This work was funded by the European Commission Framework Programme 7, grant agreement 265155 (Low Carbon at Work: Modelling Agents and Organisations to Achieve Transition to a Low-Carbon Europe) and by the Scottish Government Rural Affairs and the Environment Portfolio Strategic Research Theme 4 (Economic Adaptation).

References

1. Agar, M.: My kingdom for a function: modeling misadventures of the innumerate. J. Artif. Soc. Soc. Simul. **6**(3) (2003). http://jasss.soc.surrey.ac.uk/6/3/8.html
2. Ajzen, I.: Constructing a theory of planned behavior questionnaire (2006). http://people.umass.edu/aizen/pdf/tpb.measurement.pdf
3. Bolón-Canedo, V., Sánchez-Maroño, N., Alonso-Betanzos, A.: A review of feature selection methods on synthetic data. Knowl. Inf. Syst. **34**(3), 483–519 (2013)
4. Bolón-Canedo, V., Sánchez-Maroño, N., Alonso-Betanzos, A.: Recent advances and emerging challenges of feature selection in the context of big data. Knowl.-Based Syst. **86**, 33–45 (2015)
5. Breiman, L., Friedman, J., Stone, C.J., Olshen, R.A.: Classification and Regression Trees. CRC Press, New York (1984)
6. Brown, D.G., Robinson, D.T.: Effects of heterogeneity in preferences on an agent-based model of urban sprawl. Ecol. Soc. **11**(1), 46 (2006)
7. Caliński, T., Harabasz, J.: A dendrite method for cluster analysis. Commun. Stat.-Theory Methods **3**(1), 1–27 (1974)
8. Carrus, G., Bonnes, M., Corral-Verdugo, V., Moser, G., Sinha, J.: Socio-Psychological and Contextual Predictors of Sustainable Water Consumption, pp. 43–60. Nova Science Publishers, New York (2010)
9. Castella, J.C., Trung, T.N., Boissau, S.: Participatory simulation on land-use changes in the northern mountains of Vietnam: the combined use of an agent-based model, a role-playing game, and a geographic information system. Ecol. Soc. **10**(1), 27 (2005)
10. Cialdini, R.B., Reno, R.R., Kallgren, C.A.: A focus theory of normative conduct: recycling the concept of norms to reduce littering in public places. J. Personal. Soc. Psychol. **58**, 1015–1026 (1990)
11. Corral-Verdugo, V., Carrus, G., Bonnes, M., Moser, G., Sinha, J.B.: Environmental beliefs and endorsement of sustainable development principles in water conservation toward a new human interdependence paradigm scale. Environ. Behav. **40**(5), 703–725 (2008)
12. Dash, M., Liu, H.: Consistency-based search in feature selection. Artif. Intell. **151**(1), 155–176 (2003)
13. Davies, D.L., Bouldin, D.W.: A cluster separation measure. IEEE Trans. Pattern Anal. Mach. Intell. **1**(2), 224–227 (1979)
14. Dubois, E., Barreteau, O., Souchére, V.: An agent-based model to explore game setting effects on attitude change during a role playing game session. J. Artif. Soc. Soc. Simul. **16**(1), 2 (2013)
15. Evans, T.P., Sun, W., Kelley, H.: Spatially explicit experiments for the exploration of land-use decision-making dynamics. Int. J. Geogr. Inf. Sci. **20**(9), 1013–1037 (2006)
16. Fayyad, U.M., Irani, K.B.: Multi-interval discretization of continuous-valued attributes for classification learning. In: Thirteenth International Joint Conference on Artificial Intelligence, vol. 2, pp. 1022–1027. Morgan Kaufmann Publishers, San Mateo, CA (1993)
17. Fornara, F., Carrus, G., Passafaro, P., Bonnes, M.: Distinguishing the sources of normative influence on pro-environmental behaviours: the role of local norms in household waste recycling. Group Process. Intergroup Relat. **14**, 623–635 (2011)
18. García, S., Luengo, J., Herrera, F.: DAta Preprocessing in Data Mining. Springer, New York (2015)
19. Geller, A., Moss, S.: Growing Qawm: an evidence-driven declarative model of Afghan power structures. Adv. Complex Syst. **11**(2), 321–335 (2008)

20. Gigerenzer, G., Goldstein, D.G.: Reasoning the fast and frugal way: models of bounded rationality. Psychol. Rev. **103**(4), 650–669 (1996)
21. Gigerenzer, G., Hoffrage, U., Kleinbölting, H.: Probabilistic mental models: a Brunswikian theory of confidence. Psychol. Rev. **98**(4), 506 (1991)
22. Gostoli, U.: A cognitively founded model of the social emergence of lexicon. J. Artif. Soc. Soc. Simul. **11**(1), 2 (2008)
23. Groot, J.I.M.D.: Mean or green? Value orientations, morality and prosocial behaviour. Ph.D. thesis, University of Groningen (2008)
24. Guyon, I., Weston, J., Barnhill, S., Vapnik, V.: Gene selection for cancer classification using support vector machines. Mach. Learn. **46**(1–3), 389–422 (2002)
25. Guyon, I., Gunn, S., Nikravesh, M., Zadeh, L.A.: Feature Extraction: Foundations and Applications, vol. 207. Springer, Berlin, Heidelberg (2008)
26. Hall, M.A.: Correlation-based feature selection for machine learning. Ph.D. thesis, The University of Waikato (1999)
27. Hall, M., Frank, E., Holmes, G., Pfahringer, B., Reutemann, P., Witten, I.H.: The WEKA data mining software: an update. ACM SIGKDD Explor. Newslett. **11**(1), 10–18 (2009)
28. Holtz, G.: Generating social practices. J. Artif. Soc. Soc. Simul. **17**(1), 17 (2014)
29. Izquierdo, L.R., Gotts, N.M., Polhill, J.G.: Case-based reasoning, social dilemmas, and a new equilibrium concept. J. Artif. Soc. Soc. Simul. **7**(3), 1 (2004)
30. Janssen, M.A., Ostrom, E.: Empirically based, agent-based models. Ecol. Soc. **11**(2), 37 (2006)
31. Likert, R.: A technique for the measurement of attitudes. Arch. Psychol. **22**(140), 1–55 (1932)
32. Liu, H., Hussain, F., Tan, C.L., Dash, M.: Discretization: an enabling technique. Data Min. Knowl. Discov. **6**(4), 393–423 (2002)
33. MacQueen, J., et al.: Some methods for classification and analysis of multivariate observations. In: Proceedings of the Fifth Berkeley Symposium on Mathematical Statistics and Probability, Oakland, CA, vol. 1(14), pp. 281–297 (1967)
34. Murtagh, F., Legendre, P.: Ward's hierarchical agglomerative clustering method: which algorithms implement ward's criterion? J. Classif. **31**, 274–295 (2014)
35. Parker, D.C., Hessl, A., Davis, S.C.: Complexity, land-use modeling, and the human dimension: fundamental challenges for mapping unknown outcome spaces. Geoforum **39**, 789–804 (2008)
36. Perlich, C., Provost, F., Simonoff, J.S.: Tree induction vs. logistic regression: a learning-curve analysis. J. Mach. Learn. Res. **4**, 211–255 (2003)
37. Polhill, G., Craig, T., Alonso-Betanzos, A., Sanchez-Maroño, N., Fontenla-Romero, Ó., Dumitru, A., Bonnes, M., Bonaiuto, M., Carrus, G., Fornara, F., Maricchiolo, F., Steg, L., Ruepert, A., Kaizer, K., García-Mira, R.: Empirical agent-based modelling of everyday pro-environmental behaviours at work. In: Ames, D.P., Quinn, N.W.T., Rizzoli, A.E. (eds.) International Environmental Modelling and Software Society (iEMSs) Seventh International Congress on Environmental Modelling and Software, San Diego, CA (2014). http://www.iemss.org/sites/iemss2014/papers/iemss2014_submission_175.pdf
38. Quinlan, J.R.: C4. 5: Programs for Machine Learning. Elsevier, San Francisco (2014)
39. Rousseeuw, P.J.: Silhouettes: a graphical aid to the interpretation and validation of cluster analysis. J. Comput. Appl. Math. **20**, 53–65 (1987)
40. Ruepert, A., Keizer, K., Steg, L., García Mira, R., Dumitru, A., Vega Marcote, P., Muñoz, M., Ilin, C., Docea, A., Moza, D., Carrus, G., Maricchiolo, F., Fornara, F., Bonaiuto, M., Bonnes, M.: D4.3 final report on the role of individual factors in promoting or hindering GHG reduction behaviours and practices – four case studies: University of Corunna, the municipality of Groningen, Aquatim & Enel Green Power. Tech. rep., Department of Psychology, University of Groningen, Groningen (2013). http://www.locaw-fp7.com/userfiles/Deliverables/WP4_Deliverable_D4.3.pdf
41. Sánchez-Maroño, N., Alonso-Betanzos, A., Fontenla-Romero, O., Brinquis-Núñez, C., Polhill, J.G., Craig, T.: Influence of internal values and social networks for achieving sustainable organizations. In: ECAI, pp. 1179–1184 (2014)

42. Sánchez-Maroño, N., Alonso-Betanzos, A., Fontenla-Romero, O., Brinquis-Núñez, C., Polhill, J., Craig, T., Dumitru, A., García-Mira, R.: An agent-based model for simulating environmental behavior in an educational organization. Neural Process. Lett. **42**(1), 1–30 (2015)
43. Sánchez-Maroño, N., Alonso-Betanzos, A., Fontenla-Romero, O., Polhill, J., Craig, T.: Designing decision trees for representing sustainable behaviours in agents. In: Trends in Practical Applications of Agents, Multi-Agent Systems and Sustainability, pp. 169–176. Springer International Publishing, Switzerland (2015). doi: 10.1007/978-3-319-19629-9
44. Schreinemachers, P., Berger, T.: Land use decisions in developing countries and their representation in multi-agent systems. J. Land Use Sci. **1**(1), 29–44 (2006)
45. Schwartz, S.H.: Normative Influences on Altruism, vol. 10, pp. 221–279. Academic, New York (1977)
46. Schwartz, S.H.: Universals in the content and structures of values: theoretical advances and empirical tests in 20 countries. Adv. Exp. Psychol. **25**, 1–65 (1992)
47. Smajgl, A., Bohensky, E.: Behaviour and space in agent-based modelling: poverty patterns in east Kalimantan, Indonesia. Environ. Modell. Softw. **45**, 8–14 (2013)
48. Steg, L., De Groot, J.I.: Environmental values. In: The Oxford Handbook of Environmental and Conservation Psychology. Oxford University Press, Oxford (2012)
49. Steg, L., Perlaviciute, G., Van der Werff, E., Lurvink, J.: The significance of hedonic values for environmentally relevant attitudes, preferences, and actions. Environ. Behav. (2012). doi: 10.1177/0013916512454730
50. Stern, P.C., Dietz, T., Abel, T., Guagnano, G.A., Kalof, L.: A value-belief-norm theory of support for social movements: the case of environmental concern. Hum. Ecol. Rev. **6**, 81–97 (1999)
51. van der Werff, E., Steg, L., Keizer, K.: The value of environmental self-identity: the relationship between biospheric values, environmental self-identity and environmental preferences, intentions and behaviour. J. Environ. Psychol. **34**, 55–63 (2013)
52. Yu, L., Liu, H.: Efficient feature selection via analysis of relevance and redundancy. J. Mach. Learn. Res. **5**, 1205–1224 (2004)
53. Zeileis, A., Hothorn, T., Hornik, K.: Model-based recursive partitioning. J. Comput. Graph. Stat. **17**(2), 492–514 (2008)
54. Zhao, Z., Liu, H.: Searching for interacting features. In: International Joint Conference on Artificial Intelligence (IJCAI), vol. 7, pp. 1156–1161 (2007)

The Implementation of the Theory of Planned Behavior in an Agent-Based Model for Waste Recycling: A Review and a Proposal

Andrea Scalco, Andrea Ceschi, Itad Shiboub, Riccardo Sartori, Jean-Marc Frayret, and Stephan Dickert

Abstract In the near future, the waste management sector is expected to reduce substantially the adverse effects of garbage on the environment. However, the increasing complexity of the current waste management systems makes the optimization of the waste management strategies and policies challenging. For this reason, waste prevention is the most desirable goal to achieve. Despite this, low levels of household recycling represent the key factor that complicates the current scenario. Keeping this in mind, the present work investigates the determinants of recycling behavior through the development of an agent-based model. Particularly, we examined what would induce households to increase the probability to engage in recycling behaviors on the base of the individual attitude and sensitivity to social norms. The Theory of Planned Behavior (TPB) has been implemented as agents' cognitive model in environmental studies with the aim to predict recycling outcomes. Furthermore, in order to increase the realism of the simulation and the adherence of the model with the theory, we followed two strategies: firstly, we used real data to model a city district (Diong, Internship Report: Integrated Waste Management in Kaohsiung City, 2012). Secondly, we made use of the coefficients of the structural equation model presented in the work by Chu and Chiu (J Appl Soc Psychol 33(3):604–626, 2003) to build the agents' cognitive model. As a whole,

A. Scalco (✉) • A. Ceschi • R. Sartori
Department of Human Science, University of Verona, Verona, Italy
e-mail: andrea.scalco@univr.it; andrea.ceschi@univr.it; riccardo.sartori@univr.it

I. Shiboub
Higher Institute of Comprehensive Professions, Nalut, Libya
e-mail: i.s73300@yahoo.com

J.-M. Frayret
Department of Mathematics and Industrial Engineering, École Polytechnique de Montréal, Montréal, QC, Canada
e-mail: jean-marc.frayret@polymtl.ca

S. Dickert
School of Business and Management, Queen Mary University of London, London, UK

Linköping University, Linköping, Sweden
e-mail: s.dickert@qmul.ac.uk

A. Alonso-Betanzos et al. (eds.), *Agent-Based Modeling of Sustainable Behaviors*, Understanding Complex Systems, DOI 10.1007/978-3-319-46331-5_4

the results are in line with literature on descriptive social norms. Furthermore, the results indicate that the introduction of descriptive social norms represents a valuable strategy for public policies to improve household recycling: however, injunctive social norms are needed first.

1 The Problem with Waste

Environmental protection ranks very high on the global agenda. In 1987, the World Commission on Environment and Development (the Brundtland Commission) introduced a new term known as *sustainable development* [1]. This concept was later used to describe the international community's attitude regarding economic, social, and environmental development. So far, only some countries have taken advantage of the economic possibilities of waste management, exploiting the general need of countries to dispose of their waste and combining it with the equally widespread necessity to find sustainable means to generate energy. Currently, Sweden represents the best example: they have converted waste processes into a profitable sector, leading them, in the last few years, even to import waste from other countries [2].

Most of all, the waste management sector is expected to achieve significant results in the near future, with a substantial reduction of the adverse effects of garbage on the environment. However, the increasing complexity of the current waste management systems coupled with the demanding environmental protection targets makes the optimization of the waste management strategies and policies challenging. For this reason, waste prevention is the most desirable option, followed by the preparation of waste for reuse, recycling, upcycling and other recovery, with disposal (such as landfills) as the last resort.

With respect to recycling participation, ample evidence exists that the problem with household waste will continue to grow over time. This evidence includes sociological factors pertaining to overpopulation, the increasingly faster pace of resource exploitation, as well as the over-consumption made possible by higher incomes. In 2012, the United Nation (UN) made projections that the population of the earth may reach 8.3 and 10.9 billion by 2050 [1]: such a population increase would speed the rate of natural resource depletion and increase the production of wastes. Thus, the problem is twofold: we would be faced with the loss of both materials and energy; likewise, the problem of treating and disposing of the waste, which itself can cause environmental damage and additional costs to society. For instance, the European Commission has estimated that the per-year costs of municipal and hazardous waste disposal in Europe already exceeds €75 billion [3].

Given this annual cost, there is a great motivation to reduce expenses and, if possible, make the sector pay for itself or even turn it into a profit. For example, costs can be reduced by taking advantage of the possibilities of the waste-to-energy processes [4].

At any rate, in order to achieve a better future management of waste, governments need the cooperation of their citizens. Nowadays, low household participation represents a key factor able to complicate the waste-recycling scenario in most countries. In Sweden, recycling compliance significantly increased from 1975 to 2012 [2]. In fact, during 1975, landfills received almost 1,500,000 tons (62 % of municipal solid wastes; MSW), while, in 2012, this number was less than 33,000 tons (less than 1 % of MSW). While the municipal recycling rates only went from 6 % in 1975 to 32 % in 2012, other materials have been sorted and processed in beneficial ways with energy recovery going from 30 to 52 % and biological treatment going from 2 to 15 % in the same period. This means that consumer compliance to the environmental program is equal to, or at least near, 99–100 %, assuming that certain products may not feasibly be reprocessed into either energy or other goods.

If such a high rate of consumer compliance in recycling programs is not possible everywhere, what are the alternatives? There have been recycling programs that rely on sorting of household waste at a Material Recovery Facility (MRF) where commingled waste is processed. The problems associated with MRF waste separation is, first of all, a large investment in equipment such as "mills, cutters, screens, magnetic separators, float-sink separators, cyclones, drum separators" [5, p. 62]. In addition, there are "risks of contaminants for the workers" (*ibid*). Despite these obstacles, the crucial factor for most programs simply relies on the fact that the quality of the recovered materials is often substandard. Indeed, if recycled materials should replace raw materials inside production processes, the purity of the former becomes important, even from a financial perspective, and it is critical that valuable materials have not been mixed together with foodstuff and other contaminants [6]. In line with these considerations, if commingled collection with sorting at MRFs is problematic, we are left with the difficult task of creating citizen compliance with processes of waste separation at the source.

Consequently, a refinement of waste management strategies becomes urgent in order to implement policies able to go behind both preventing waste and creating a market for recycling. In such a framework, recycling household waste becomes crucial, as it would reduce waste while saving resources. Moreover, it is critical that the public sector examines incentives that would promote recycling in households. The degree and intensity to which people conform to these behaviors depend on several technical or sociological factors, as well as the demographic and economic facts about the households [7]. Overall, the success of a recycling programme is due to a mix of good public policy and efforts to increase public awareness and, thus, households' behavior. All of this must be take into account in order to achieve sustainable changes leading to new social norms.

Given these reasons, arising critical question is what would induce households to recycle their waste in a practicable way. One of the possible answers lies in a simple psychological phenomenon that is widely known but poorly understood: people's behavior is largely shaped by the behavior of those around them. In psychology, this phenomenon takes the name of *social norms*. These latter are in fact one of the most powerful customary rules that govern behavior in groups and societies.

However, traditional forms of market research (e.g. focus groups and surveys) are of limited use in a social norm campaign. When people are polled, they typically underestimate the effects of the campaign, because they are not usually aware that it had an effect on them. An issue that has received very little attention in the literature deals with the question of what is the most effective way to activate policy strategies in order to produce behavioral change. Therefore, to simulate possible scenarios for policy strategies, we created an agent-based model (ABM) representing a virtual society engaged in recycling behaviors. Indeed, agent-based modeling represents a promising alternative to traditional attempts to understand how social processes work over the time. Some authors even argue that "agent-based simulation (...) is the only feasible way of understanding the tangle of complex social phenomena, such as those that involve norms" [8, p. 47]. Indeed, modern computer simulations as a methodology of research within social sciences is a rather new idea, but it comes with great potential thanks to the fact that is «an excellent way of modelling and understanding social processes» [9, p. 1]. Overall, their major value lies in the ability to investigate how the macro-behavior of a system emerges as a result of micro-behaviors [10]. Within the current work, the micro-behavior is represented by virtual consumers and their propensity to recycle, whereas the macro-behavior is expressed by the virtual society and leads to promote or hinder pro-environmental behavior of agents.

In our work we chose to expand on the Theory of Planned Behavior (TPB), originally developed by Ajzen [11], as a valuable cognitive model of the virtual agents populating the simulation. An agent is here defined as a computational entity that we can use as the basis for simulating social processes, as though the entity were a human agent that could perceive, act, and interact within a virtual environment in a way that we can call autonomous [12]. Moreover, Ajzen's work was further developed by Chu and Chiu [13] into an integrated model on household waste recycling. Specifically, our work presents a model scaled from their original findings in order to assign probability distributions that satisfactorily simulate recycling behavior. In models such as this, the stochastic factor is important, given the fact that we can more realistically recreate the acts of agents that might not all act according to plan. This means that there is a strong possibility that different people will act differently even when provided the same instructions and given the same situation. By accounting for this in our model, we gain realism in our simulation [14].

2 Social Norms Theory and Recycling Behaviors

As suggested by Cialdini and Trost [15], norms are a widespread construct in social research because they indeed represent a worthwhile psychological phenomenon that can help explain human behaviors. Following their work, we chose to describe social norms as "rules and standards that are understood by members of a group, and that guide and/or constrain social behavior without the force of laws" (*ibid.*, p. 152). In other words, social norms can be easily conceived as unwritten rules: everyone experiences them daily, as they often guide our behavior without consciously asking or wondering about their validity. For instance, we know that it is a general rule to greet someone who we known when we hastily meet him/her on the stairs. We are not forced to do so, but we know that this can represent a violation of an accepted common rule.

Adherence to the norms of a social group allows members to avoid rejection and increase social approval [16]. In their work, Cialdini et al. [16] reported also the interesting study conducted by Aronson and O'Leary [17]. The research started from the notion that prompts and informational campaigns are not very effective most of the time to modify the behavior of people if they are asked to adopt an innovation or to change their habits. Instead, the adoption of new behaviors can be promoted if individuals observe others actively engaging in it. Following this consideration, they started monitoring the behaviors of several subjects when showering and the resulting usage of water. To reduce the consumption of this latter, they created two conditions with the aim of improving the awareness of the importance of avoiding water losses. In the first condition, the authors applied a sign outside of the shower room. This prompt explained in four consequential instructions that water must be opened under the shower just on a first time to wet down and after being soaped to rinse off. In this way, the prompt invited to turn off the water when soaping up. In a further condition, a confederate of the researchers was introduced into the shower room. In fact, research indicates that social norms are most compelling when people are shown evidence that the behavior they are being encouraged to adopt is already practiced by people similar to them (see Social Comparison Theory; [18]). When entered into the shower room, the confederate followed the instructions proposed by the prompt: thus, he modeled the proper behavior. Within this condition, the number of accidental participants who exhibit the right behavior increased up to 53 % (against the 6 % of people who followed the prompt in the previous condition). The authors concluded the study affirming that "having people model the appropriate behavior suggests to others that conserving water by turning off the shower is a reasonable and worthwhile thing to do" (*ibid*, p. 223). Therefore, the results demonstrate how powerful normative influence can be as social phenomena.

2.1 Understanding and Investigating Social Norms

An important distinction is usually made among studies regarding norms. In fact, within psychological and sociological literature it is rather common to find references about *descriptive* social norms and *injunctive* social norms. The former refer to informational influence and they are related to the observation of what most others do in a particular situation. In contrast, the latter type of norms can be seen as the source of normative influence, which is related to what other people consider as acceptable or unacceptable behavior [16]. Therefore, descriptive social norms simply consider how others behave, without a positive or negative evaluation of the behavior and without providing evidence of what is helpful behavior from the results of their actions [19]. As stated by Cialdini [20], descriptive social norms are able to transmit a simple but effective message: "If a lot of people behave in this way, this is probably the right thing that I should do". Besides, following the perspective proposed by Cialdini et al. [21], descriptive norms can represent a shortcut to make decisions in situations where there is a prevalence of ambiguity about the behavior that should be performed. Injunctive social norms, on the other hand, tend to be focused on social rewards (for instance, social approval) and punishment (in some cases, even the rejection one's own group) related to certain behaviors.

Moreover, there is an important aspect related to the psychological notion of saliency of norms. In fact, as reported in Cialdini et al. [21], norms do not have an equally powerful effect at all times and in all situations. Instead, norms must be made salient to elicit the proper response from people: that is to say, they have to be "activated" in the mind of individuals. For instance, Cialdini and Goldstein [22] experimentally demonstrated that an injunctive normative message can increase norm accessibility, and consequently promote the recall of the right behavior, when it is linked to a functional mnemonic cue that can easily be perceived in upcoming conditions. In addition, as shown by the work by Cialdini et al. [21], anti-littering norms can become salient by pointing out that littering constitutes a blameworthy action: in this way, they are injunctive norms as they bring with them a negative connotation. As expressed by Demarque et al. [23], "persons who are contextually focused on normative considerations are most likely to act in norm-consistent ways" (p. 167). Thus, it is when injunctive anti-littering norms are made salient, that people will tend to improve their pro-environmental behavior [21]. Finally, regarding salience, the previous authors specify that when only one (descriptive or injunctive norm) is made salient to an individual's mind, that norm will exercise the stronger influence on the subsequent individual's actions. Following the previous considerations, we can consider recycling behavior as a specific form of prosocial behavior, which is in turn related with social norms [21, 22]. Specifically, household recycling behaviors are motivated by social norms, whereas, instead, financial incentives may even reduce these actions, as they undermine the intrinsic motivations of people reducing the proneness towards recycling [24].

As an example, a rather interesting work about social norms has been provided by Savarimuthu et al. [25]. Following a bottom-up approach, they investigated the spread of a norm against littering inside a park within a virtual society. Particularly, they set up a (bi-dimensional) simulation environment with several agents that were able to interact in a social context. The agents interacted when they met on the same spot: in this situation, each agent was able to observe the behavior of the other one (littering/not littering). Furthermore, the authors developed a payoff matrix where pro-environmental behavior had a positive payoff (0.5), whereas littering had a negative payoff (−0.5). When an agent decided to pollute the park, the shared environment is ruined: this means negatively influencing the entire virtual society given that this action has an impact on the general productivity. Within the model by Savarimuthu et al., the term productivity is used to indicate the benefits that the agents receive when using the public park. Finally, the final payoffs are computed as the sum of the individual payoff and the park productivity. No central mechanism is present within the simulation; instead, each agent that considers littering as a blameworthy behavior has the ability to punish an agent engaged in an inappropriate behavior. Punished agents switch from littering to a pro-environmental behavior when the number of the received punishments exceeds their individual resistance to change. The main observable output of the simulation is constituted by the emergence of a norm (i.e. littering or not littering).

The results show that a norm against littering is established when the number of punishers is sufficiently high (at least 10 % of the initial population). Otherwise, the non-littering norm spreads across the population and the productivity drops gradually. As noted by the authors, this kind of process occurs commonly inside online-based encyclopedias: a norm of collaboration is established only when there is a sufficient number of reviewers that censor, or even ban, false contributors. Furthermore, the work highlights how social norms can be successfully being established among society if the costs related with enforcements are low.

3 Dealing with Social Norms from a Computational Approach

Jager and Jannsen [26] highlighted the importance to develop theoretical models of human decision processes starting from empirical research. Despite this, as pointed out by Ceschi et al. [27], currently there is still a lack of real integration between computational modeling and cognitive theories, both from a methodological and theoretical perspective. Indeed, cognitive psychological modeling can provide the means by which it becomes possible to identify the driving forces behind the recycling behavior and to determine the most likely successful factors for public policies. Literature indicates that environmental attitudes and situational and psychological variables are likely to be important predictors of the recycling behavior.

Interestingly, in their extensive work Elsenbroich and Gilbert discussed how to model norms [28]. Three fundamentals approaches can be useful to apply in agent-based modeling when dealing with social norms. One of these is represented by the well-known social network analysis. A social network is composed by two kind of elements: nodes (i.e. agents) and their ties (i.e. the relationships among agents). Social network analysis focuses primary on the latter. Given the fact that our model is aimed to investigating the spread of social norms without implying relationships among agents (at least, nothing more than closeness), we moved forward from this approach.

A second formalization invokes the social impact theory. This was firstly proposed by Latané [29] and it was aimed to turn the influence (the "impact") of one subject on another one into a mathematical formulation. Latané suggested considering three fundamental elements for his theory: social forces, the psychological law, and the number of targets. The first one, social forces, is composed of three main parts (the number of people that can exert influence, the strength of the influence—depending on the relationships established among the subjects and their individual features-, and the immediacy of the impact). Furthermore, the fundamental law states that the social impact experienced by an agent will increase with the number of agents who are exercising social pressure. This increment follows a logarithmic function, such that a new agent will exercise less influence than the previous one. Finally, the third component refers to the number of agents influencing a subject. The estimation of the final value of the social impact is promptly given by the sum of the previous three main components. However, as stated by Elsenbroich and Gilbert [28], even if social impact theory has the advantage to be generalizable, it is rather difficult to evaluate the social force and immediate component.

The last approach considered by the authors is the one that, more than the others, stems from a psychological background and that has been implemented inside the present work: the Theory of Planned Behavior (TPB; [11]), which provides a valuable theoretical and cognitive framework to understand and explain the influence of several psychological factors, including social norms.

3.1 The Psychological Bases of the Theory of Planned Behavior

Models of psychological cognitive functioning can be particularly useful to isolate the different aspects that may drive recycling behaviors, and, consequently, those successful factors of public policy that can enhance this kind of behavior. The Theory of Planned Behavior has been developed from the previous Theory of Reasoned Action [30]. They both assume that people have a basis for their behavior that is informed by reflection and deliberative thought such that they consider the implications of their actions. Particularly, the Theory of Planned Behavior represents a psychological theory that, more than other cognitive models, has been extensively used within environmental studies (see for instance: [31–42]).

According to the TPB, intentions to engage in recycling behavior stem from three main factors: subjective norms, individual attitudes and the perceived behavioral control. The concept of *subjective norms* refers to the individual's belief that people important to the decision maker see their behavior as the appropriate way to act. Aceti [43] argues that people are motivated to recycle by the actual pressure they receive from family and friends to do so. Furthermore, simply knowing that family, friends, and neighbors participate in recycling activities increases the likelihood of participation. In this spirit, Stern et al. [44] stressed the importance of considering the social structure within which individuals are embedded, based on the belief that social structures shape individuals' experiences and ultimately their personal values, beliefs and behaviors. Following Trafimov and Finlay [45], it may be suggested that subjective norms are relevant only for participants with higher accessibility of a collective self. However, according to Cialdini's Theory of Normative Behavior [21], it may be suggested that the actual impact of subjective social norms is underestimated when it is measured by means of anonymous questionnaires completed in private settings [46]. In fact, Cialdini et al. [21] showed that, in experimental settings, where an injunctive anti-littering norm was made salient, participants' littering behavior was significantly reduced. As indicated by Cialdini and Trost, those institutions that want "to activate socially beneficial behavior should use procedures that activate injunctive social norms, since these norms appeared to be more general and more cross-situational effective" [15, p. 161].

The concept of *attitude* refers to the individual's evaluation of the action. Boldero [47] found that intentions to recycle newspapers directly predicted actual recycling and that attitudes toward recycling predicted the recycling intentions. The expectations can reflect past experiences, anticipation of upcoming circumstances, and the cultural background. Davies et al. [48] argued that recycling attitudes should be separated into two components: an affective and a cognitive element. The former consists of the emotional approach to the recycling imperative, whereas the latter consists of the knowledge about the outcomes and consequences of performing the recycling behavior [40].

Finally, the concepts of *perceived control* and *moral obligation* refer to the individual's perception of their ability to perform behaviors. Taylor and Todd [49] found that both attitudes toward recycling and perceived behavioral control were positively related to individuals' recycling and composting intentions. According to TPB, perceived behavioral control will influence actual behavior only if the behavior is not completely under the person's volitional control.

4 Integrating an Empirical Model of Recycling Behavior

Agent simulations range from highly structured artificial worlds with few simple rules and constraints [50] to complex models where agent interactions constrain subsequent iterations of the simulation [51] and/or multiple structural layers are considered [52]. It is well known that the development of these algorithms is the most fragile aspect of the simulation analysis. Within the present work, in order to design a virtual society, a key activity is represented by the identification of an amount of the agent's attributes that are significant for recycling behavior. These attributes span from basic demographic attributes (i.e., age, education and income), to more specific features (i.e., environmental sensitivity, self-confidence and sense of social belonging; [53]). Most of the impact is due to these attributes and therefore it is important to consider them for the aims of the analysis. As a consequence, it is recommended to start from some empirical models, such as a structural equation model (SEM).

SEMs are a modeling technique rather widespread in social and psychological science [54]. They derive from the integration of three fundamental statistical techniques applied by social sciences: particularly, they combine path analysis, factor analysis, and multiple regression models. In this way, structural equation models are able to combine the methods usually applied by, respectively, sociology, psychology, and economy. Inside a structural equation model, the relationships among variables are expressed by regression coefficients: consequently, the entire model is developed following a cause-effect interpretation. The design of the model is firstly conducted following theoretical literature: that is to say, by connecting variables following findings provided by the current available research. Then, the model is tested statistically: starting from the covariance matrix of the examined variables, the fit of the model with the data is estimated by means of a maximum likelihood method. Usually, to obtain the parameters several iterations are needed until the "best fit" of the model with the data is achieved.

Among other social sciences, these models found a large usage within psychological research thanks to the fact that they are able to link latent variables to observable variables. In fact, as pointed out by Krishnakumar and Ballon [55], a remarkable benefit of this framework is that correlations of observed indicators are clearly made as arising out of subjacent factors that are accountable for the results. That is, SEMs are able to reveal and to quantify the relationship between a behavioral expression and its underlying psychological construct. For instance, they can corroborate the existence of latent factors, such as verbal and mathematical intelligences, starting from the observed responses of a psychological test.

Nevertheless, one downside of the structural equation modeling approach is represented by the difficulty to properly capture all crucial variables regarding a specific behavior during the beginning phase of a literature review and design of the theoretical model. In addition, given the complexity of human behavior, results extracted from literature sometimes can lead to confusing or overlapping variables. The model suggested by Ajzen [11] represents a fundamental schema of human behavior, as it is able to take into account three fundamental and distinct factors at the same time: the personal psychological attitude, the impact of the social sphere and the combination of perceived and actual factors that can hinder a certain behavior. Indeed, the schema proposed by the Theory of Planned Behavior represents a fundamental framework to properly design a structural equation model when dealing with pro-environmental behavior. In line with this, Zhang and Nuttall already stated how the TPB can summarize psychological, sociological and environmental elements related to decision-making processes and, at the same time, it still remains relatively easy to code: the authors concluded that these characteristics make the TPB "particularly suited to modelling consumer behavior in agent-based simulation" [56, p. 173].

A valuable example of the application of structural equation modeling designed following the Theory of Planned Behavior is given by the work by Chen and Tung [42]. They conducted research to develop an extension of the TPB aimed to explain and predict the consumer's intention to stay in green hotels. Following current literature, they started designing the research model, which should explain the antecedents of intention to visit green hotels, based on the individual attitude, subjective norms and perceived behavioral control. In addition, they extended the classical model of TPB by taking into account the perceived moral obligation of the studied subjects. By means of structural equation modeling, the authors were able to estimate path coefficients among the designed research model, uncovering the "force" of the causal relationships among variables. Furthermore, they were able to assess the indirect effect of consumer's environmental concern on the intention to visit green hotels. Finally, structurally equation modelling allowed revealing that the most indispensable factor of the model to predict intention to visit green hotel was the perceived behavioral control.

At any rate, as remarked by Hox and Bechger [54], it is important to note that a structural equation model (even when corroborated by the data) does not imply the truth of the model itself. There could be several other competing models able to achieve the same fit with the data.

In addition, a current limitation of structural equation models is related to the difficulty to take into account individual differences among people. Essentially, individual differences are characterized as a set that makes individuals particular, according to their inclinations, capabilities and outcomes. This set of characteristics can affect the result of the application of general psychological laws, making their results uncertain. For instance, the studies by Tversky and Kahneman [e.g., 57] within the framework of prospect theory revealed a general psychological law defined as "loss aversion" (also commonly known as risk aversion). Briefly, this law tries to explain why people are more prone to weight losses substantially more than objectively commensurate gains when evaluating economic prospects. However, this sensitivity to losses may differ among people [e.g., 53, 57–59]. That it is to say, people perceive losses more than their actual objective value, but individual differences modulate this perception. Starting from this, agent based modeling can help to dynamically represent, in a natural way, several scales of analysis and the importance of structures at different levels, none of which is easy to accomplish with other modeling techniques [60]. In this way, the limitation of SEMs regarding the modeling of individual differences may be seen conversely related with the advantage of ABMs to represents agents' heterogeneity (see for instance, [61]).

5 Specific Aim and Hypotheses of the Simulation

The aim of the current work is to present a model able to simulate a number of characteristics that have been scaled from the original work by Chu and Chiu [13], modeled, and assigned with probability distributions to simulate the recycling behavior. Usually, the purpose of this stochastic effort is to endow agents with a 'personality'. Contemplating the possibility of fuzzy logic implies greater simulation realism as different agents act differently in the same situation. Agents with personality lead to the modeling of more complex interactions where, for example, hypotheses may be tested more effectively by considering teams of agents with different personalities rather than single agents [14].

The built simulation tested two specific hypotheses related with the framework of the Theory of Planned Behavior. On the one hand, the first hypothesis is related to injunctive social norms. Specifically, we expect that those agents that are mostly sensitive to these types of norms will also be less susceptible with respect to the impact of external conditions on their intention to recycle. Assuming scenarios with extreme values of recycling rate, the intention of the householders to recycle will be stable over time. Instead, assuming a scarce recycling rate, only those agents that are most influence by injunctive social norms will engage in recycling behavior. We presume that the simulation will end with a stable equilibrium.

On the other hand, the second hypothesis is connected with descriptive social norms. Particularly, we think that those agents that most of all are sensitive to these types of norms will be influenced negatively by the impact of external conditions, reducing the probability to recycle. In low recycling rate scenarios, the intentions to recycle will be weak. This is due to the fact that descriptive social norms reduce the probability to recycle among the population. In contrast, in scenarios with a high recycling rate, the intention of householders to behave properly will be strong, thanks once again to the effects of descriptive social norms. We presume that the simulation will end with a self-reinforcing stable equilibrium.

6 The Planned Recycling Agent Behavior Model

Our analysis is based on a simulation model of the "Planned Recycling agent Behavior" (PRB_1.1)[1] that produces virtual neighborhoods with different agent types, waste generation and collection processes [Fig. 1; 62]. The scaling of the agents' features is based on the coefficients relating to the TPB and taken from a SEM on motivations to recycling behavior developed by Chu and Chiu [13], which represents an extension of Taylor and Todd's [49] efforts to suggest ways to influence recycling behavior. The application of scaling allows us to accelerate the simulation lowering hardware requirements to run the algorithm, leaving untouched the original ratio between agents' variables. Particularly, the model that has been presented in Chu and Chiu [13] included four basic coefficients expressing the recycling behavior, which include the force of subjective norms (SN_r), the individual environmental attitude (AT_r), the moral obligation perceived by the agent (PMO_r) and the perceived behavioral control (PBC_r). These factors reflect the traditional model proposed within the Theory of Planned Behavior [11], but the inclusion of moral obligation extends the original model. Thus, the mathematical expression of the model can be represented by the subsequent formula:

$$B_r \cong BI_r w_1 (AT_r) + w_2 (SN_r) + w_3 (PBC_r) + w_4 (PMO_r)$$

Where the term B_r refers to the actual expression of the behavior, and BI_r expresses the intention toward that behavior. As there are no components between these elements, the Theory of Planned Behavior [11] assumes that intention of behavior is itself a reliable measure of the probability to engage in that particular behavior. In line with the proposal by Ajzen [63], the four terms indicated with w are empirically determined regression coefficients used to weigh each element of the

[1]The full model, code, and documentation is available via the www.openabm.org website.

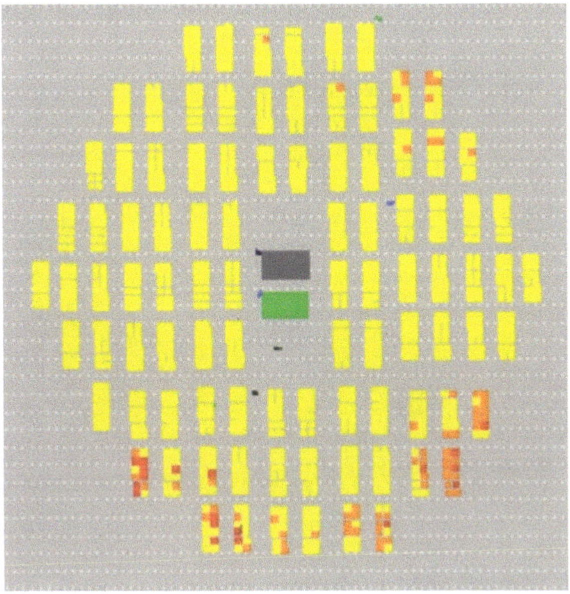

Fig. 1 Example of the PRB_1.1 simulation. The simulation presents three different types of agents. (1) Neighborhood agents turn their color from *yellow* to *red* assuming different shades. *Yellow color* indicates a stable situation, *orange* represents a state close to the critical level, *red* means instead that *R* achieved the critical level. Each one of the over 700 *yellow square* represents a household composed on the average of 1.5 agents. (2) Garbage transportation systems are represented in the model as *grey* and *green* small *rectangles* among neighborhoods. (3) Landfills are indicated by the *green* and the *grey rectangle* at the center of the world. They represent, respectively, the recycling and non–recycling landfills

formula. Moreover, the term AT_r refers to the personal attitude of a particular agent towards a certain behavior: thus, the agent computes the attitude on its expectations of the behavioral results. Boldero [47] suggested that the personal attitude could represent a good predictor of recycling behaviors. Again, the perceived behavioral control is comprised inside the equation by the term PBC_r. This refers to the actual difficulties that an agent might experience and the perceived control that it can potentially have on them. Taylor and Todd [49] reported how both behavioral control and attitude are positively related to the individual motivation towards recycle.

Finally, the subjective norms are included by the term SN_r. Taken together with moral obligation (PMO_r), they constitute the social determinants of the recycling behavior. While the subjective norms of the model are related with the behavior of the neighborhood, the moral obligation is connected with the injunctive norms shared by the society.

Table 1 Coefficients applied to the simulation PRB_1.1

Coefficient	Value
Population present in the virtual district	353,451
Total number of neighborhood agents	1100
Number of transportation systems	8
Landfields	2
Daily rubbish production for neighborhood (*R* and *Rre*)	427 kilo
Critical situation for a neighborhood	9 ton
Coefficient of the environment attitudes (*ATr*)	0.18
Coefficient of the subjective norms (*SNr*)	0.12
Coefficient of the perceived behavioral control (PBC_r)	0.33
Coefficient of the perceived moral obligation (PMO_r)	0.10

Values 1–6 are extracted from the work by Diong [64] and they refer to San-min district. Values 7–10 are taken from the standardized and normalized regression coefficients of the structural equation model presented within the work by Chu and Chiu [13]

6.1 The Formal Model

The values of the four previous constructs contained by the structural equation model illustrated by the work by Chu and Chiu [13] have been parameterized by a stochastics computation and used inside the simulation as probabilistic factors of behaving. In addition, in order to initialize the parameters (for instance, the number of households, trucks, waste production, etc.) we exploited the data contained inside the report about Kaohsiung City [64], used also by Chu and Chiu. Specifically, we referred to the values relative to the San-min district, the largest one of Kaohsiung with more than 353 thousand people and with a number of households equals to one-third of the population. All coefficients used to run the simulation are summarized inside Table 1.

The algorithm generates the virtual city and then, during the simulation, it manages three kind of agents: neighbored agents, garbage transporters, and landfills (see Fig. 1). More details about these agents are presented by the following subsections.

6.1.1 The Neighborhood Agent

All agents inside the simulation are able to generate recycled rubbish (*Rre*) and non-recycled rubbish (*R*). This is based on the probabilities of psychological constructs and other agent habits. Neighborhood agents recycle if they possess high levels of environment attitudes (AT_r), high subjective social norms (SN_r), and perceived behavioral control (PBC_r). This link is not mediated by other aspects (Fig. 2). Probabilities of these psychological constructs are normally distributed among agents.

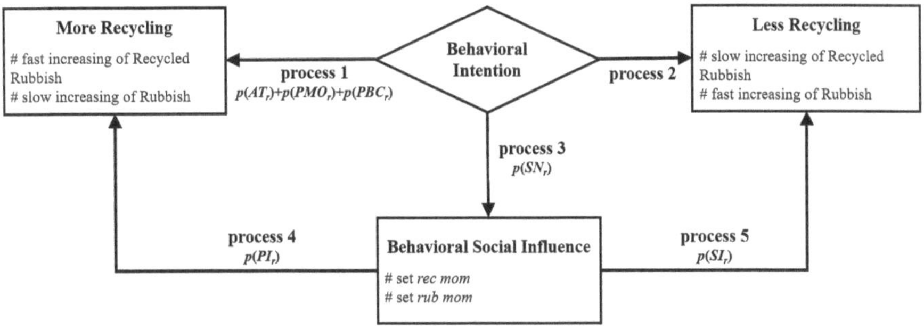

Fig. 2 Schema of reasoning of the agents inside PRB_1.1

If the level of subjective norms (SN_r) of an agent is sufficiently high, it can be socially influenced by other agents close to it. When this happens, neighbor agents close to the agent are observed and more recycled rubbish is produced if the neighbor observed is also recycling. We defined as "*peer influence*" (PI_r) the tendency of an agent to be influenced by others around it.

In addition, the general disposition of the agents to recycle is computed within the simulation by a decay (and an inverse decay) function aimed to resemble human psychophysical sensitivity (see, for instance, [65]). This function has been developed starting from the original model of motivation and satisfaction of needs over time proposed in the work by Jager and Janssen [66].

Furthermore, agents' recycling behavior is negatively influenced by the actual presence of rubbish around them. In fact, agents are endowed with the ability to observe the level of rubbish that is produced by others. When this exceeds the critical level, agents start to decrease the probability to recycle. We defined this phenomenon as "*surrounding influence in recycling*" (SI_r) and it is computed by means of another decay function related with the quantity of rubbish existing in the neighborhood at a certain instant of the simulation. Both peer influence and surrounding influence are determinants of the probability of an agent to recycle by being influenced by others.

The schema depicted in Fig. 2 explains agents' behaviors inside the simulation PRB_1.1. Considering a random agent *i*, at every cycle it compares the possible actions and then it executes one of them. The comparison is performed by assessing the probability levels: the value of p ranges between 0 and 1 and it is considered low when lower than 0.50, otherwise it is high. The strategy followed by an agent depends on five basic processes:

- *process 1:* agent *i* computes the value of probability related with its environmental attitude $p(AT_r)$, perceived moral obligation $p(PMO_r)$, and perceived behavioral control $p(PBC_r)$. If the sum of these values exceeds a probability threshold of 0.50, the agent follows this strategy and it will produce more recycled rubbish (*Rre*) than regular rubbish (*R*).

- *process 2:* if the sum of $p(AT_r)$, $p(PMO_r)$, and $p(PBC_r)$ does not reach an high level, process 1 is rejected, thus the agent applies process 2 and it recycles less. In this way, it increases the level of non-recycled rubbish (*Rre*).
- *process 3:* the agent computes the influence exercised by other agents (that is, *p*(*SNr*)). When it enters in this state, the agent will set the variables related with the recycle rate and not-recycle rate by observing another random agent close to it and the level of rubbish in the vicinity. Having this information, the agent estimates the peer influence (PI_r) and the surrounding influence (SI_r) in recycling and their probabilities. The PI_r level is computed by the agent each time using a specific function, which depends on whether the other agents are recycling (*1*) or not (*0*).
- *process 4:* the agent computes the level of the peer influence $p(PI_r)$: if it is high it decides to increase the probability to recycle.
- *process 5*: if the agent is scarcely influenced by the surrounding agents (i.e. there is a low level of $p(SI_r)$) the agent will recycle less.

6.1.2 The Garbage Transportation System and the Landfills

The model involves a transportation system, which takes away garbage from neighborhood agents and moves it to the collecting points. The pathways adopted by pick-up trucks are optimized considering distance and time. Pick-up trucks get to the closest neighborhood agents to collect *R* and *Rre*.

Two types of trucks have been designed and implemented inside the simulation: the first one is devoted to collect only recycled rubbish (the green truck in Fig. 1); conversely, the second one is dedicated to gather only non-recycled rubbish. In Fig. 1 they are indicate as, respectively, the green and grey truck. Both trucks assign priority to the neighborhood with the highest rubbish level. After a specific amount of *R* or *Rre* collected, garbage trucks move to the closest landfill.

Furthermore, there are two types of collecting points (i.e. landfills) in the simulation: one for unseparated garbage *R*, the other one for recycled garbage *Rre*. The landfill removes the garbage carried by pick-up trucks over time. Besides, the virtual environment reproduces in a two-dimensional space (specifically, a torus) a district composed of 1100 neighborhood agents. Agents are free to consume, recycle, and move within the boundaries of this virtual world.

7 Results and Conclusions

As stated, environmental protection ranks very high on the global agenda. However, the increasing complexity of the current waste management systems makes the optimization of the waste management strategies and policies challenging. For this reason, waste prevention is the most desirable result to achieve. Despite this, low household participation to recycling represents the key factor that complicates the current scenario. Recycling household wastes becomes crucial, as it would reduce waste while saving resources. The present work investigates the determinants of recycling behavior through the development of an agent-based model. Particularly, the programed simulation tries to answer to the following question: what would households induce to increase the probability to engage in recycling behaviors? In line with this, we chose to describe here social norms as "rules and standards that are understood by members of a group, and that guide and/or constrain social behavior without the force of laws" [15]. Moreover, we distinguished between the processes that lead the spreading of descriptive and injunctive social norms. While the former are related with the observation of others' behaviors, the latter are related to what other people consider as an acceptable behavior [16].

Besides the specific hypotheses and the results obtained by the research, the present contribution proposes a novel approach to agent-based modeling which includes integration of theories and quantitative methods commonly applied within psychological research. Specifically, we argue that the implementation of results obtained from statistical techniques such as structural equation models can add a significant validity with respect to the agents' behavior, due to the fact that a SEM is able to statistically express the link between a certain behavior and its psychological antecedents. In addition, due to the potential difficulties related with the design of a proper theoretical model of behavior, the scheme proposed by the Theory of Planned Behavior represents a valuable framework. In this way, in order to build a structural equation model regarding recycling behavior (as well as similar pro-environmental behaviors) it is recommended to take into account at least the three fundamental elements proposed by Ajzen's model [11]. The validity of the model should be successively tested by applying proper statistical procedures. In the end, the results can be smoothly implemented inside an agent-based model as exemplified by the current work: in fact, the values extracted from the SEM represent the basic coefficients of the agent's reasoning engine. Moreover, we argue that the limitation of SEMs regarding individual differences is overcome by the potential ability of computer simulations of generating heterogeneous agents.

In the current work, the TPB is applied as agents' cognitive model with the aim to predict the recycling outcomes on the base of the individual attitude and sensitivity to social norms. This approach may help to identify the factors of public policy that can enhance pro-environmental behaviors. We based the parameters in the simulation on the data contained inside the report about Kaohsiung City [64] to model a city district. We also made use of the coefficients contained by the structural equation model presented inside the work by Chu and Chiu [13] in order to build the agents' cognitive model. These values are parameterized by a stochastics computation and used inside the simulation as probabilistic factors of behaving. Undeniably, a potential limitation of the present study is based on using parameters and information provided by previous studies that might not fit perfectly for the proposed model: future research will have to corroborate the integration of the TPB and SEMs within an agent-based model by conducting the whole research process, from the design of the theoretical behavior model to the implementation into a virtual model. Agent-based models can simulate the efficacy of different

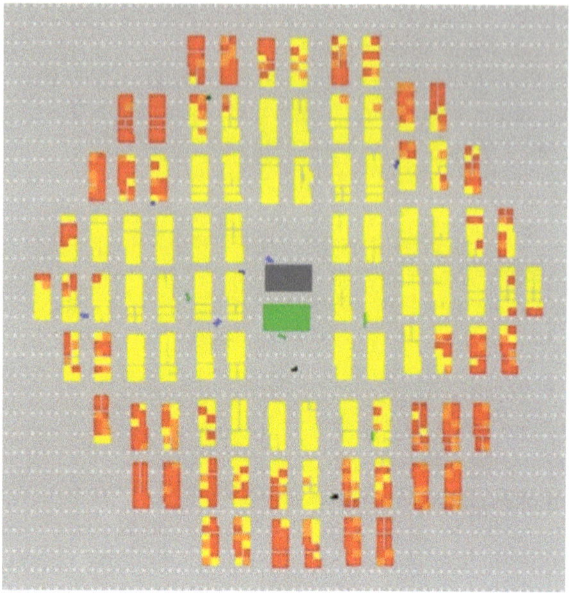

Fig. 3 A screenshot from one run of the ABM based on the PRB_1.1 with higher *R* levels than Fig. 1. Neighborhood agents turn color because of the *R* level. When *R* is equal to the critical level they turn *red*, *orange* if they are close to the critical level, *yellow* when the situation is stable

recycling campaigns under equal conditions and, at a subsequently stage, allow the simulation of specific policies under different conditions. Moreover, agent-based models are mostly structured on algorithms that illustrate the behaviors of agents, identify their causal effects, and specify critical parameter estimates. Therefore, stochastic simulation, while retaining its versatility, is also time-effective and cost-effective. However, it is important to state that the agent behavior is stochastic. As we suggested, factors of SEMs can be implemented inside ABMs, in contrast to equations of aggregation.

The preliminary results of the model available on the site owned by OpenABM Consortium show stability and reliability in relation to the outcomes of the simulation. The visual impact creates a virtual circle where household motivation to recycle is reinforced. This circle expresses the consequences of descriptive social norms. On the contrary, the failure in recycling when the environment is full of rubbish contaminates the neighbors' behavior (Fig. 3). As a whole, the results are in line with literature on descriptive social norms [16, 20, 31]. Findings in the literature about social norms and littering agree that in a 'dirty' environment individuals are inclined to litter more than those subjected to a 'clean' environment [e.g., 21], mainly because of the peer influence, due to the fact that agents continuously observe and mimic each other's behavior. Similarly, the surrounding has its own effect because the amount of garbage present in the system drives the trend away from its stable level. To conclude, the results obtained from several runs of the model indicate that the introduction of descriptive social norms represents a valuable strategy for public policies to improve household recycling. However, it is important to consider the sequence used to apply norms: injunctive social norms are needed in order to implement further policies based on descriptive social norms.

References

1. United Nations - Department of Economic and Social Affair: World Population Prospects: The 2012 Revision (2014)
2. Rousta, K., Richards, T., Taherzadeh, T.J.: An overview of solid waste management toward zero landfill: a swedish model. In: Taherzadeh, M.J., Richards, T. (eds.) Resource Recovery to Approach Zero Municipal Waste, pp. 1–22. CRC Press (by Taylor & Francis Group), Boca Raton (2016)
3. European Commission: Life and Waste Recycling: Innovative Waste Management Options in Europe (2007)
4. Psomopoulos, C.S., Bourka, A., Themelis, N.J.: Waste-to-energy: a review of the status and benefits in USA. Waste Manag. **29**(5), 1718–1724 (2009)
5. Rousta, K., Dahlén, L.: Source separation of household waste: technology and social aspects. In: Taherzadeh, M.J., Richards T. (eds.), Resource Recovery to Approach Zero Municipal Waste, pp. 64–75. CRC Press (by Taylor & Francis Group), Boca Raton (2016)
6. Sundqvist, J.: How Should Municipal Solid Waste Be Treated—A System Study of Incineration, Material Recycling, Anaerobic Digestion and Composting. IVL Swedish Environmental Research Institute, Sweden (2005)
7. Tobias, R., Brugger, A., Mosler, H.-J.: Developing strategies for waste reduction by means of tailored interventions in Santiago de Cuba. Environ. Behav. **41**(6), 836–865 (2009)
8. Edmonds, B.: Agent-based social simulation and its necessity for understanding socially embedded phenomena. In: Conte, R., Andrighetto, G., Campenni, M. (eds.) Minding Norms: Mechanisms and Dynamics of Social Order in Agent Societies (2013)
9. Gilbert, N., Troitzsch, K.G.: Simulation for the Social Scientist, 2nd edn. Open University Press, Maidenhead (2005)
10. Hughes, H.P.N., Clegg, C.W., Robinson, M.A., Crowder, R.M.: Agent-based modelling and simulation: the potential contribution to organizational psychology. J. Occup. Org. Psychol. **85**(3), 487–502 (2012)
11. Ajzen, I.: The theory of planned behavior. Organ. Behav. Hum. Decis. Process. **50**(2), 179–211 (1991)
12. Schwarz, N., Ernst, A.: Technological forecasting & social change agent-based modeling of the diffusion of environmental innovations - an empirical approach. Technol. Forecasting Soc. Change **76**(4), 497–511 (2009)
13. Chu, P.Y., Chiu, J.F.: Factors influencing household waste recycling behavior: test of an integrated model. J. Appl. Soc. Psychol. **33**(3), 604–626 (2003)
14. Garson, G.D.: Computerized simulation in the social sciences. A survey and evaluation. Simul. Gaming **40**(2), 267–279 (2009)
15. Cialdini, R.B., Trost, M.R.: Social influence: social norms, conformity, and compliance. In: The Handbook of Social Psychology, vol. 2, pp. 151–192. (1998)
16. Cialdini, R.B., Bator, R.J., Guadagno, R.E.: Normative influences in organizations. In: Thompson, L., Messick, D., Levine, J. (eds.) Shared Cognition in Organizations: The Management of Knowledge, pp. 195–211. Erlbaum, Mahwah (1999)
17. Aronson, E., O'Leary, M.: The relative effects of models and prompts on energy conservation. J. Environ. Syst. **12**, 219–224 (1983)
18. Festinger, L.: A theory of social comparison processes. Hum. Relat. **7**(2), 117–140 (1954)
19. Göckeritz, S., Schultz, P., Rendón, T., Cialdini, R.B., Goldstein, N.J., Griskevicius, V.: Descriptive normative beliefs and conservation behavior: the moderating roles of personal involvement and injunctive normative beliefs. Eur. J. Soc. Psychol. **40**(3), 514–523 (2010)
20. Cialdini, R.B.: Descriptive social norms as underappreciated sources of social control. Psychometrika **72**(2), 263–268 (2007)
21. Cialdini, R.B., Reno, R.R., Kallgren, C.A.: A focus theory of normative conduct: recycling the concept of norms to reduce littering in public places. J. Pers. Soc. Psychol. **58**(6), 1015 (1990)

22. Cialdini, R.B., Goldstein, N.J.: Social influence: compliance and conformity. Annu. Rev. Psychol. **55**, 591–621 (2004)
23. Demarque, C., Charalambides, L., Hilton, D.J., Waroquier, L.: Nudging sustainable consumption. The use of descriptive norms to promote a minority behavior in a realistic online shopping environment. J. Environ. Psychol. **43**, 166–174 (2015)
24. Brekke, K.A., Kverndokk, S., Nyborg, K.: An economic model of moral motivation. J. Public Econ. **87**(9), 1967–1983 (2003)
25. Savarimuthu, B.T.R., Purvis, M., Purvis, M., Cranefield, S.: Social norm emergence in virtual agent societies. In: Declarative Agent Languages and Technologies IV, vol. 5397. pp. 18–28 (2009)
26. Jager, W., Janssen, M.: The need for and development of behaviourally realistic agents. In: Proceedings of the 3rd International Conference on Multi-Agent-Based Simulation, vol. II, pp. 36–49 (2003)
27. Ceschi, A., Dorofeeva, K., Sartori, R., Dickert, S., Scalco, A.: A simulation of householders' recycling attitudes based on the theory of planned behavior. In: Bajo, J., Hernández, J.Z., Mathieu, P., Campbell, A., Fernández-Caballero, A., Moreno, M.N., Botti, V. (eds.) Advances in Intelligent Systems and Computing, vol. 372, pp. 177–184. Springer, Cham (2015)
28. Elsenbroich, C., Gilbert, N.: Modelling Norms, 1st edn. Springer, Berlin (2014)
29. Latané, B.: The psychology of social impact. Am. Psychol. **36**(4), 343–356 (1981)
30. Fishbein, M., Ajzen, I.: Attitudes and voting behavior: an application of the theory of reasoned action. Prog. Appl. Soc. Psychol. **1**, 253–313 (1981)
31. Botetzagias, I., Dima, A.F., Malesios, C.: Extending the theory of planned behavior in the context of recycling: the role of moral norms and of demographic predictors. Resour. Conserv. Recycl. **95**, 58–67 (2015)
32. Chan, L., Bishop, B.: A moral basis for recycling: extending the theory of planned behavior. J. Environ. Psychol. **36**, 96–102 (2013)
33. Cheung, S.F., Chan, D.K., Wong, Z.S.: Reexamining the theory of planned behaviour in understanding waste paper recycling. Environ. Behav. **31**(5), 587–617 (1999)
34. Do Valle, P.O., Rebelo, E., Reis, E., Menezes, J.: Combining behavioral theories to predict recycling involvement. Environ. Behav. **37**(3), 364–396 (2005)
35. Kaiser, F.G., Gutscher, H.: The proposition of a general version of the theory of planned behavior: predicting ecological behavior. J. Appl. Soc. Psychol. **33**(3), 586–603 (2003)
36. Mannetti, L., Pierro, A., Livi, S.: Recycling: planned and self-expressive behaviour. J. Environ. Psychol. **24**(2), 227–236 (2004)
37. Pakpour, A.H., Zeidi, I.M., Emamjomeh, M.M., Asefzadeh, S., Pearson, H.: Household waste behaviours among a community sample in Iran: an application of the theory of planned behaviour. Waste Manag. **34**(6), 980–986 (2014)
38. Ramayah, T., Lee, J.W.C., Lim, S.: Sustaining the environment through recycling: an empirical study. J. Environ. Manage. **102**, 141–147 (2012)
39. Tonglet, M., Phillips, P.S., Bates, M.P.: Determining the drivers for householder pro-environmental behaviour: waste minimisation compared to recycling. Resour. Conserv. Recycl. **42**(1), 27–48 (2004)
40. Tonglet, M., Phillips, P.S., Read, A.D.: Using the theory of planned behaviour to investigate the determinants of recycling behaviour: a case study from Brixworth, UK. Resour. Conserv. Recycl. **41**(3), 191–214 (2004)
41. Vicente, P., Reis, E.: Factors influencing households' participation in recycling. Waste Manag. Res. **26**(2), 140–146 (2008)
42. Chen, M.F., Tung, P.J.: Developing an extended theory of planned behavior model to predict consumers' intention to visit green hotels. Int. J. Hosp. Manag. **36**, 221–230 (2014)
43. Aceti, J.: Recycling: Why People Participate. Why They Don't. Massachusetts Department of Environmental Protection, Boston (2002)
44. Stern, P., Dietz, T., Kaloff, L., Guagnano, G.A.: Values, beliefs, and pro-environmental action: attitude formation toward emergent attitude objects. J. Appl. Soc. Psychol. **25**, 322–348 (1995)

45. Trafimow, D., Finlay, K.A.: The importance of subjective norms for a minority of people: between-subjects and within-subjects analyses. Pers. Soc. Psychol. Bull. **22**, 820–828 (1996)
46. Stiff, J., Mongeau, P.: Persuasive Communication. The Guilford Press, New York (1994)
47. Boldero, J.: The prediction of household recycling of newspapers: the role of attitudes intentions and situational factors. J. Appl. Soc. Psychol. **25**(5), 440–62 (1995)
48. Davies, J., Foxall, G.R., Pallister, J.: Beyond the intention–behaviour mythology an integrated model of recycling. Marketing Theory **2**(1), 29–113 (2002)
49. Taylor, S., Todd, P.: Decomposition and crossover effects in the theory of planned behavior: a study of consumer adoption intentions. Int. J. Res. Mark. **12**(2), 137–155 (1995)
50. Kohler, T.A., Gumerman, G.J.: Dynamics in Human and Primate Societies: Agent-Based Modeling of Social and Spatial Processes. Oxford University Press, New York (2000)
51. Sawyer, R.K.: Simulating emergence and downward causation in small groups. In: Multi Agent Based Simulation, vol. 1979, pp. 49–67. Springer, Berlin (2000)
52. Stinchcombe, A.L.: When Formality Works. Authority and Abstraction in Law and Organizations. University of Chicago Press, Chicago (2001)
53. Ceschi, A., Rubaltelli, E., Sartori, R.: Designing a homo psychologicus more psychologicus: empirical results on value perception in support to a new theoretical organizational-economic agent based model. In: Omatu, S., Bersini, H., Corchado, J.M., Rodríguez, S., Pawlewski, P., Bucciarelli E. (eds.) Advances in Intelligent Systems and Computing, vol. 290, pp. 71–78. Springer, Cham (2014)
54. Hox, J.J., Bechger, T.M.: Introduction to structural equation modeling. Fam. Sci. Rev. **11**, 354–373 (2009)
55. Krishnakumar, J., Ballon, P.: Estimating basic capabilities: a structural equation model applied to Bolivia. World Dev. **36**(6), 992–1010 (2008)
56. Zhang, T., Nuttall, W.J.: Evaluating government's policies on promoting smart metering in retail electricity markets via agent based simulation. J. Prod. Innov. Manag. **28**(2), 169–186 (2011)
57. Tversky, A., Kahneman, D.: Rational choice and the framing of decisions rational. J. Bus. **59**(4), S251–S278 (1986)
58. Kahneman, D., Knetsch, J.L., Thaler, R.H.: Anomalies the endowment effect, loss aversion, and status quo bias. J. Econ. Perspect. **5**(1), 193–206 (1991)
59. Mitchell, T.R., Mickel, A.E.: The meaning of money: an individual-difference perspective. Acad. Manage. Rev. **24**(3), 568–578 (1999)
60. Gilbert, N., Terna, P.: How to build and use agent-based models in social science. Mind Soc. **1**(1), 57–72 (2000)
61. Sartori, R., Ceschi, A., Scalco, A.: Differences between entrepreneurs and managers in large organizations: an implementation of a theoretical multi-agent model on overconfidence results. In: Omatu, S., Bersini, H., Corchado, J.M., Rodríguez, S., Pawlewski, P., Bucciarelli, E. (eds.) Advances in Intelligent Systems and Computing, vol. 290, pp. 79–83. Springer, Cham (2014)
62. Ceschi, A., Scalco, A., Dickert, S., Sartori, R.: Compassion and prosocial behavior. Is it possible to simulate them virtually? In: Bajo, J., Hernández, J.Z., Mathieu, P., Campbell, A., Fernández-Caballero, A., Moreno, M.N., Botti, V. (eds.) Advances in Intelligent Systems and Computing, vol. 372, pp. 207–214. Springer, Cham (2015)
63. Ajzen, I.: From intentions to actions: a theory of planned behavior. Action Control Cogn. Behav. 11–39 (1985)
64. Diong, K.: Internship Report: Integrated Waste Management in Kaohsiung City (2012)
65. Weber, E.H.: De Pulsu, Resorptione, Auditu et Tactu. Koehler, Leipzig (1834)
66. Jager, W., Janssen, M.: An updated conceptual framework for integrated modeling of human decision making: the consumat II. Paper presented at the paper for workshop complexity in the Real World@ ECCS, 2012

Social Simulations Through an Agent-Based Platform, Location Data and 3D Models

Alberto López Barriuso, Fernando De la Prieta, Sara Rodríguez-González, Javier Bajo, and Juan M. Corchado

Abstract This work presents an innovative agent based platform specially designed to simulate human activity at any urban area environment (buildings, apartments, houses, offices, gardens, parks, small residential areas, etc.) and manage data from sensors. The platform uses 3D models of the environment to perform accurate simulations, while simultaneously showing relevant and high quality data. This chapter also presents the platform case studies that have been conducted and that demonstrates their technical and conceptual validity, where, by merging data obtained by multiple sensors with information of the subject's activities and simulation data, the platform can extract and store information on typical situations in a real environment, and also observe possible technological or architectural barriers for people with disabilities.

1 Introduction

The simulation of human behavior is one of the approaches used to improve the daily life of people. It can be applied to some problems like disabled people integration, energy efficiency, emergency situations, among other. Concretely, the integration of disabled people is a major challenge in today's society. Not only can it result in the self-sufficiency of disabled people, but it can also facilitate their self-esteem improvement. According to studies carried out by the United Nations (UN) [1], approximately 15 % of the world population suffers some kind of disability, either physical or psychological, which impedes them from carrying out their regular daily routines (work, education, personal, etc.).

A.L. Barriuso (✉) • F. De la Prieta • S. Rodríguez-González • J.M. Corchado
Department of Computer Science and Automation Control, University of Salamanca, Plaza de la merced s/n, 37007 Salamanca, Spain
e-mail: albarriuso@usal.es; fer@usal.es; srg@usal.es; corchado@usal.es

J. Bajo
Department of Artificial Intelligent, Technical University of Madrid, Campus Montegancedo Bloque 2, Despacho 2101, Boadilla del Monte, Madrid 28660, Spain
e-mail: jbajo@fi.upm.es

A. Alonso-Betanzos et al. (eds.), *Agent-Based Modeling of Sustainable Behaviors*, Understanding Complex Systems, DOI 10.1007/978-3-319-46331-5_5

There is no doubt that our current society is concerned about this challenge. In this regard, since the publication of the *Standard Rules on the Equalization of Opportunities for Persons with Disabilities* by the UN [2], different governmental organizations have developed specific regulatory frameworks that have improved the everyday lives of people with disabilities, most notably, at the European level, the *Charter of Fundamental Rights of the European Union* [3], the *Treaty on the Functioning of the European Union* [4], the *Council Directive 2000/78/EC* [5] and the *European Disability Strategy 2010–2020* [6]. These initiatives are perfectly aligned with a growing academic attention, which has allowed the development of systems, techniques, models and methodologies oriented to improving accessibility. Beyond the development of these artifacts, one of the most promising study fields is social simulation, which in general terms, is applied to obtain a higher knowledge of a particular phenomenon being studied [7]. Within the context of the societal integration of disabled people, simulations permit discovering the problems encountered by people with disabilities before the problems occur.

Within the scope of this research work, this chapter proposes a new platform based on virtual organizations of agents (VO) to perform human behaviour simulations [8]. Additionally, the platform provides data gathering mechanisms in order to provide the simulation of data from real world sensors. This platform uses 3D models of the environment and locating data in order to perform accurate simulations, while at the same time showing relevant and high quality data. In order to test the platform, we have designed a different cases of study by fusing data obtained by multiple sensors, information of their activities, and simulation data. The high development of systems that extract and store information make it essential to improve the mechanisms required to deal with the avalanche of context data. In our case, the MAS approach results are appropriate because each agent can represent an autonomous entity (human beings, or environment elements) with different capabilities and offer different services while still collaborating among themselves.

Through the use of the proposed platform in two cases of study, it has been possible to investigate new mechanisms oriented to the decision making process related to the workflows present in the environment.

The first case of study is focused on improving accessibility to the workplace environment by detecting architectonical barriers through the performance of a simulation, while the second case of study is focused on testing how the platform behaves in a rapid and unpredictable scenario, so an evacuation scenario has been simulated. Several experiments have been performed to evaluate this platform and the preliminary results and the conclusions are presented in this chapter.

The chapter is organized as follows. Section 2 presents the state of the art on simulation, agents and similar current systems. Section 3 describes the created platform and its components (locating infrastructure, agent-based platform, 3D editor, 3D viewer). Section 4 shows the case studies carried out on the platform. Finally, Sects. 5 and 6 provide the discussion and conclusions obtained.

2 Background

ABSS (*Agent-based social simulation*) is one of the most representative techniques which are used in complex inquiries where a large number of active and heterogeneous objects are present. These objects can be humans [9], business units [10], functional or nonfunctional objects [11], animals [12], etc. The phenomenon to be simulated is a set of sequences or events in a system (natural or artificial), which can exist (or not) in the real (or artificial) world, and can be configured in the simulation model. These tasks are interrelated, since they are time or order dependent within the events which take place in the system. In this way, the simulation model allows the implementation of different and specific functionalities for each kind of target, applying different degrees of freedom. The complexity lies in the fact that this kind of model allows for the simulation of complex and changing events, so the use of intelligent agent based techniques is appropriate. The ABSS [13] technique is focused on the social phenomenon simulation, using MAS models. Therefore, ABSS is a combination of social sciences, agent-based computation and computational simulation. The use of this technique is specially indicated when it is necessary to capture different tasks, elements, objects, persons in dynamic complex environments, as long as they can be implemented without having a deep knowledge about the global interdependencies. Moreover, the technique also allows facilities on changing models, since it is not necessary to make local changes; global changes are made instead [14]. The benefits of agent based computing for computer simulation include various methods for evaluating MAS or for training future system users [15].

The contribution from agent based computing to the field of computer simulation mediated by ABSS provides benefits such as methods for evaluation and visualization of multi agent systems, or training future system users [15]. Many new technical systems are distributed systems that involve complex interactions between humans and machines, which notably reduces their usability. The idea is to model the behavior of human users in terms of software agents. However, it is necessary to define new middleware solutions that allow the connection of ABSS simulation and visualization software.

The use of this model and tools in ABSS makes it possible to model a great variety of tasks and environments. In fact, the model has been applied in previous works, obtaining good results [16, 17]. However, it has been observed that there are certain difficulties when modeling the physical environment which we intend to simulate. This feature is especially relevant since a high percentage of the problems to be solved are related to building structures or barriers to accessibility (steps, ramps, lifters, doors, corridors, etc.). To overcome these problems, the proposed platform, in addition to the simulated data obtained, uses environmental data sensors, which provide location data of people that enable to perform more realistic simulations.

Although outdoor locating is well covered [18] by systems such as the current GPS (*Global Positioning System*), indoor locating needs still more development, especially with respect to accuracy and low-cost and efficient infrastructures.

The use of optimized locating techniques makes it possible to obtain more accurate locations using even fewer sensors and with fewer computational requirements [19]. As presented in [20] the innovative smaller, portable and non-intrusive devices [21] are progressively more efficient when gathering context-information [22]. Thus, the new platforms as proposed should encourage the integration of such devices in order to create open, flexible and adaptable systems. For this reason we have concluded that MAS are an ideal option to create and develop open and heterogeneous systems such as those normally found in the ABSS process.

The emergence of new technologies has resulted in a number of projects that aim to improve interaction with an environment and that involve improving the quality of life and care of disabled people [23, 24]. It is possible to find projects involving wearable health devices [25–27] integrated with sensors providing the continuous monitoring of a person's health related issues and daily activities.

A survey [28] about wireless sensor networks for healthcare presents some interesting systems related to the field. Nevertheless, our system is technologically far removed from those mentioned to this point. Much more than a decision support system, this platform integrates other capabilities such as simulation for prediction of dangerous situations. Moreover, it is based on the multi-agent technology, which is highly compatible with our objective due to the need for fusing information from heterogeneous distributed resources and autonomous entities. The skeleton of the platform is a MAS, which provides the basic characteristics for the perfect functioning of the agents. The description of the system is shown below, including the main blocks, the technologies and features.

3 Agent-Based Simulation Platform

This section presents the proposed platform. One of the main aims to be achieved with its design and development has been the capability of modeling different environments in a flexible and dynamic way, allowing the simulation of human behavior in a urban area environment. For that reason, an agent based platform was used, as it can model the human organization, as well as gather location data and a 3D environment representation where simulations can be carried out. The combination of these strategies is a key factor in achieving the main objective which we intend to simulate. First, agents model the organization by implementing different tasks, objectives, purposes, etc. The 3D environment can then physically model the workplace environments, and finally, the indoor locating data improves the results of the simulations. Thereby, this approach can carry out simulations using the workplace environment itself, as the visualization of the results obtained. A representation of the proposed simulation architecture is illustrated in Fig. 1.

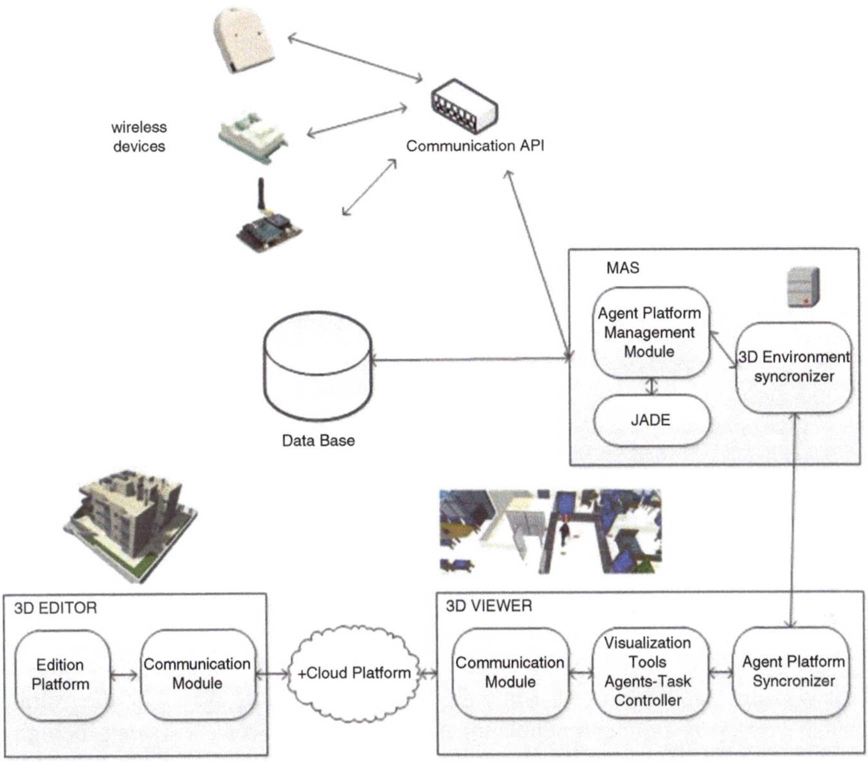

Fig. 1 Agent-based simulation platform

The basic infrastructure of the platform proposed in this article consists of the following elements:

- **Wireless sensor infrastructure**. The infrastructure consists of a set of physical wireless devices on which part of the low level middleware will be executed, and through which the rest of the system can access its functionalities (for example, obtaining readings from various sensors or calculating the location of the user).
- **Wireless devices**. They form part of the wireless infrastructure hardware. A set of low consumption and small wireless devices were deployed. Each one shares a common basic architecture composed of a microcontroller, a transceptor and a set of physical interfaces for the exchange of data between the device itself and the sensors and actuators to which the device is connected. According to the application, these devices can be either battery or externally powered.
- **Communication module**. It allows the different elements of the system to interact with the wireless devices in order to gather the information taken from the sensors they are connected to, and to send and receive data.
- **MAS**. This module, together with the wireless sensor infrastructure, is the other pillar of the system. It will be supported by an agent platform management

module, and will enable the transfer of tasks that are very difficult to implement in machines and devices. This infrastructure is essentially composed of the agents that provide the platform functionality and that can be found in a remote server. The agents execute the tasks (e.g., location, business task, movements simulations, etc.). MAS also provides the network communication data infrastructure, which makes it possible to connect with client machines and for agents to communicate with the services they offer. The agent module is composed of different agents that permit the exchange of data between the wireless devices and the other platform components. This MAS allows the exchange of information between the firmware included in the wireless devices, the API communication to access the functionalities, and the communication protocol.
- **3D Editor**. It is used to display environment on which the location data is shown and the simulations are carried out, so it requires having modeling properties. Therefore, it is necessary to add a subsystem around the graphic responsible for the process of creating and editing different virtual buildings on those wishing to deploy the services offered by the platform.
- **3D Viewer**. Last, but not least important, the platform includes a set of graphical 3D interfaces that can display an enriched form of all the information provided by the agents based on the data they have received. This makes it possible to access all the system information.

These modules are explained in the following subsections in detail. The two key concepts that come together in these modules are: using agents for the overall management, performing the simulations, and applying the control systems for high-level sensor data management. Furthermore, the sensors themselves will have to be managed and analyzed to extract information from them and apply it to the case study in question (assistance to disable people in their environments, barriers identification, etc.). The context information includes information not only about the environment, but about the people who live in these monitored environments.

3.1 Locating Infrastructure

A node is each element that is included in a sensor network. Each sensor node is usually formed by a microcontroller, a transceiver for radio or cable transmission, and a sensor or actuator mechanism [21]. Some nodes act as routers, allowing them to forward data that must be delivered to other nodes in the network. There are wireless technologies such as Wi-Fi, IEEE 802.15.4/ZigBee and Bluetooth that enable easier deployments than wired sensor networks [29]. At the sensor level, the basis of the infrastructure of the platform is made up of several ZigBee nodes. The ZigBee standard features make ZigBee an ideal supporting wireless technology for building indoor Real-Time Locating Systems. The possibility of working with low-power nodes that do not need large computational resources allows designers to reduce hardware costs when implementing the systems. In addition, these kinds of low-power nodes can reach a battery life of several years, with regards to the transmission range (transmitted power), the time resolution and the accuracy of the system. ZigBee-based Real-Time Locating Systems can use different locating techniques in order to estimate the positions of the tags in the environment. In the proposed platform, each ZigBee node includes an 8-bit RISC (Atmel ATmega 1281) microcontroller with 8 KB of RAM, 4 KB of EEPROM and 128 KB of Flash memory and an IEEE 802.15.4/ZigBee transceiver (Atmel AT86RF230). These devices are called n-Core Sirius B and Sirius D [30].

3.2 Agents Description

MAS architecture consists of agents which represent and can simulate the desired model. The MAS is based on VO, allowing the representation of a structure, roles, as well as a wide range of norms which will schedule the interactions among agents. In this simulation context, agents will represent the different actors involved in the simulation, which will be mainly human beings, although another additional agents will also intervene, representing different interactive environment elements, as telephones, lifts, sensors, etc.

The MAS has been developed by means of JADE agent platform [31]. However, JADE does not allow the implementation of organization characteristics, so it was necessary to include a higher layer, which is responsible for the MAS management from an organizational point of view. This layer provides additional self-organizational capabilities to the platform, including: (1) the capacity of grouping the MAS, enabling the agents to adopt different topologies [32]; (2) definition of the tasks attending to the role of the agent within the organization, as well as the definition of unique characteristics to each agent; and, finally, (3) the ability to define a set of rules to regulate the interactions among agents was implemented. Furthermore, this software layer which is included over JADE can also define scheduling and task-allocating oriented agents within the simulation. Finally, this layer also allows the communication within the 3D environment, so it can notify all the changes that take place bi-directionally between the agent platform and the tridimensional environment.

Figure 2 shows a detailed structure of each one of the agents, consisting of three main components: (1) the communication module, which will allow the agents to communicate with the other agents in the platform, (2) a reasoning module, which is the key component of the simulations based on the Belief Desires Intention model (BDI) [33]; and (3) the communication module between the agents and the 3D environment, that interacts turn made up of a sensor and an effector.

It is necessary to highlight that the sensor and effector of the agent-environment communication module are distributed: one component is present in the implementation of the agent itself (as a part of the agent platform).

The model of agent presented above is designed to represent different actors involved in the simulation, representing elements as human beings, or a similar kind

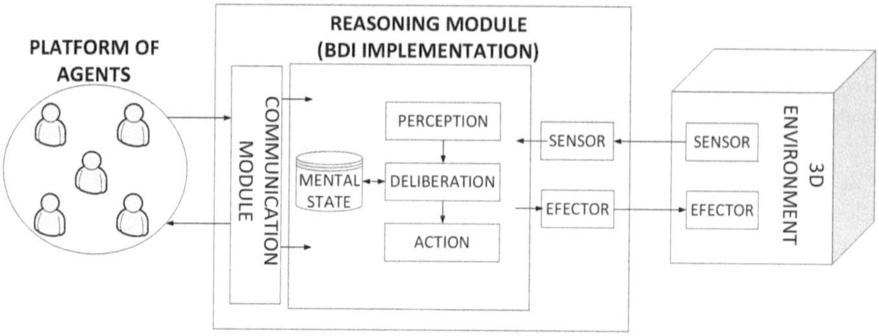

Fig. 2 Agent structure overview

of agent. However, we must also introduce another kind of agent, the environment agents. This agents will represent those elements of the environment with other agents can interact with. Examples of these agents may be telephones, lifts, copiers, etc. These agents are not aware of the environment around them, they will just be aware of their inner state, so they will lack of sensors. Any interaction with other agents cannot be initiated by an environment agent. When an agent requests an interaction with them, it will consult its inner state, and communicating relevant information to the other agent, which will finally establish if he can perform the task which pretends to carry out, according to this information.

It is also important to point that the wireless sensor infrastructure is distributed among the real environment. With this distribution of both the sensor and the effector, we allow the agent to be aware of the tridimensional environment state which was rendered for the current simulation, thus permitting the agent to monitor the changes which take place in the virtual world.

3.3 3D Environment Description

This is a 3D modeling tool that can model any building and monitoring it in real time through a three-dimensional experience. The graphical environment must support a wide range of services, and must always do so with a good performance, regardless of the size of the application environment. Consequently, it must include optimization techniques such as *culling*, lighting, shadows, modelling adapted to consume few resources, etc.

Figure 3 shows the different subsystems that compose the graphical environment. The most important subsystems match the two main features for which they are responsible: the Viewer, which allows 3D real-time monitoring; and the Map Editor, which allows modeling any building at runtime. In this way, we can also distinguish the assets repository that stores the 3D models.

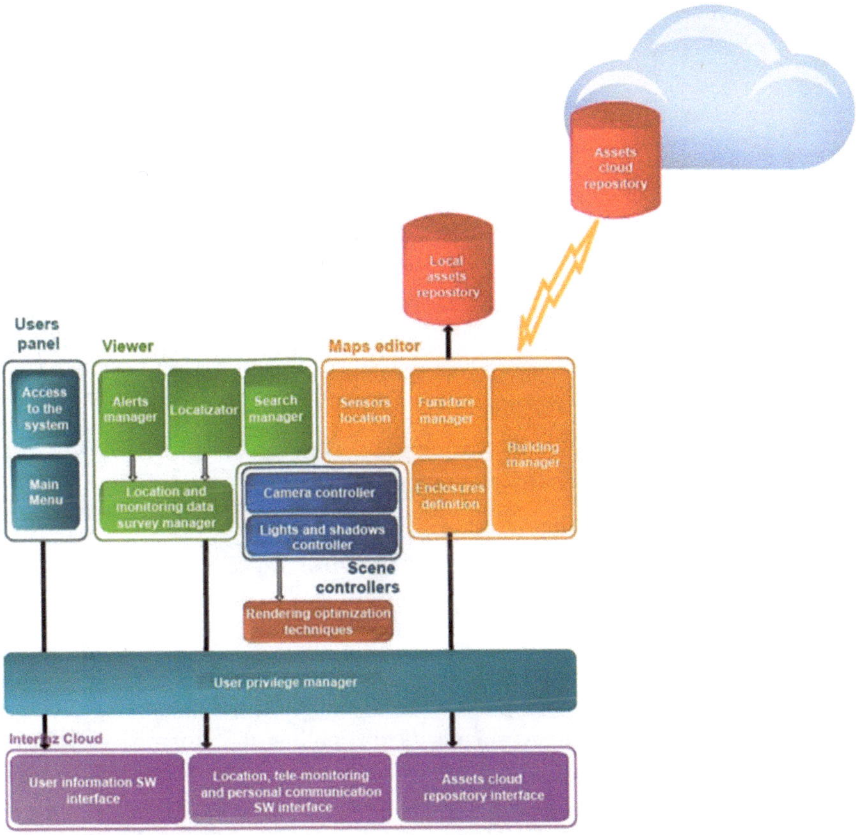

Fig. 3 3D tool components

3.3.1 3D Viewer

The Viewer is a subsystem whose main task is to display the environments in three-dimensions and interactively. There are three specific components of this subsystem: the location and remote monitoring data survey manager, the locator, and the alert manager.

The **location subsystem** is the core of the Viewer. In addition to including the main scene, it mainly manages the incoming data from the traceable objects. The scene generation must provide the proper lighting to the environment, as well as establish a ground on which to place the different models of the elements which we will work into the scene. Likewise, several cameras will be located on the proper angles, in order to provide the defined views. After creating this basic scene, the automatic modeling of the building on which the location and simulations are carried out—according the 3D map—can be performed.

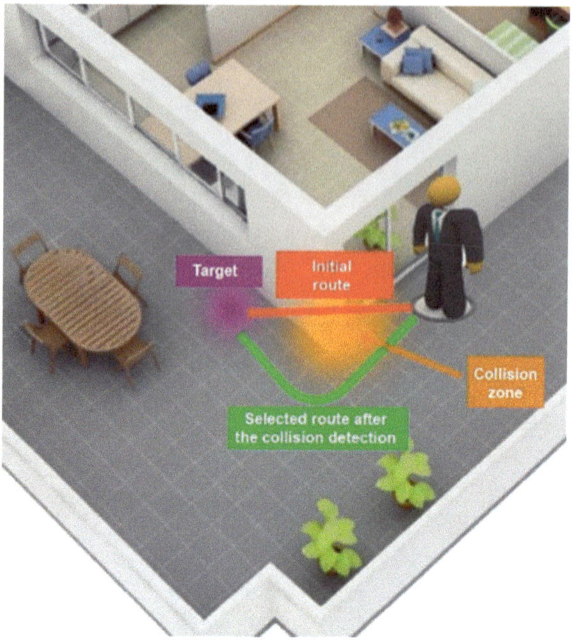

Fig. 4 Static collision avoidance

Once the process of developing the scene is complete, the monitoring process begins, in which the data survey monitor begins polling the progressive location data. The data from each node in the environment, which is associated with a user or a localizable object, are read individually. The node must be represented in the scene by using its model or avatar. Depending on its information (gender, age, etc.) the node will be associated with the model that better fits to its profile, then, the model is placed at the received coordinates.

If a node is already present in the scene, the process is different, since it is only necessary to update its position, simulating a movement that includes an animation. The movements of the models will be performed in a straight line, except when an obstacle is in the way of the trajectory, such as a corner (Fig. 4). Once a collision point is detected, a new trajectory is calculated in order to avoid it, while maintaining the original destination.

Finally, the location also includes components to allow the end-user to interact with the three-dimensional scene that is being generated. Therefore, the models that are representing the nodes are interactive. In this sense, by clicking on them, information of the represented user or object will be displayed to the end-user.

Other of the major components of the location data is the **monitoring survey manager**, which samples location data with a predefined frequency, using the polling technique. There is no loss of performance since the polling is always very large, and a single request is performed in order to obtain all the locatable user/object positions in the scenario. The location data polling module is mainly attributable to a thread, as shown in Fig. 5. It periodically invokes the web service that returns the last location data taken by the sensor network for a given building.

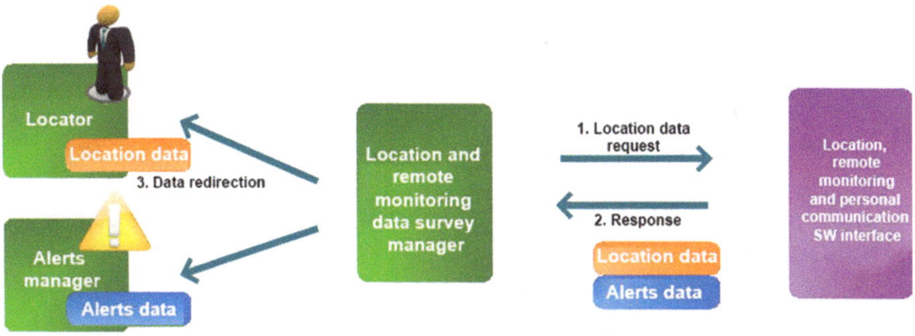

Fig. 5 Survey manager main functionality

The **alerts module** provides real-time information to the wireless sensor network. The three-dimensional representation to be rendered should be as appropriate as possible. Thus, the different types of alerts (movement, fire, light, etc.) that exist are related to the corresponding interpretations in the scene.

3.3.2 3D Editor

The visualization of the environment where the location and tele-monitoring are performed requires a model of each building so that it can represent the information on it. It is therefore necessary to add an extra component for creating and editing different virtual buildings. This component is divided into four modules that distinguish the four main functionalities that must be addressed: the management of the construction, the management of the furniture, the location module and the building enclosures definition module. These four modules come into line with the assets repository included in the graphical environment.

Figure 6 identifies the required steps for generating a building. First, a new map of the particular organization is required. Then, a user with the Editor role will be able to perform the construction, furniture and location elements (reader nodes) positioning. The order in which a user must perform is shown in the figure; however, the application will not restrict it, as the user might be interested in the performance of a previous modeling of the infrastructure, in which the *reader nodes* are positioned, but the furniture is not, allowing the building model to be completed in the future.

The **edification manager** is one of the most complex components because it must provide the necessary tools to design any building in perfect scale, regardless of its form.

The edification manager is divided into a total of six managers, each one responsible for the majority of the functionalities to be provided by this module.

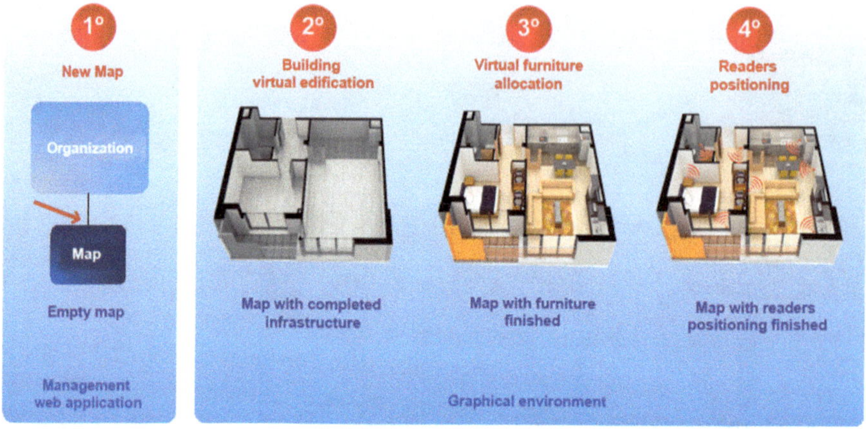

Fig. 6 Process of creation of the monitored environment

Fig. 7 Models placed on three different kinds of surfaces

These components are: walls manager, floor manager, door manager, window manager, stairs and ramps manager, and elevators manager. All of these components have one thing in common: the scene, which is made up of the progress made in the building, as well as the set of tools that allow different actions to be performed during the building creation process.

Meanwhile, the furniture manager provides a high level of realism to the visualization, as seen in Fig. 7. The building infrastructure is enough to correctly identify the room or point of the building in which each locatable item is found. However, it seeks to provide the system with versatile results, which should be attractive to a visual user, creating an innovative experience among the simulation systems.

Finally, the **location sensors module** can locate the reading beacons, thus ensuring that the location algorithms can properly calculate the position of the reachable entities. For this reason, the map Editor includes a module for pre-positioning these sensors, since it is the one that renders the building model. This process is performed by fixing the coordinate origin first, then locating each beacon from that point.

The enclosures module defines the enclosures, and is responsible for allowing the users to specify different rooms or dwellings inside a building. For proper management, the definition of enclosures, 2D views of each floor of the building, will be provided (see Fig. 8), allowing the user to define enclosures by two different methods: through the selection of a particular room, or through the definition of polygons. This allows a plan of each floor to be generated at runtime, according the building model which has previously been virtually defined by the user. The capacity of defining enclosures will allow to establish perimeters to be used by the different modules of the platform (e.g. establishing an enclosure to be used by the alerts module, in order to specify an area where certain users should not enter).

Fig. 8 Two methods of defining enclosures

4 Case Studies

In this section the proper functioning of the platform is evaluated. To do so, preliminary cases studies are proposed. The experiments that take place aim to validate the correct operation of the platform, allowing (1) verification of the proper integration between components; (2) the ability to represent a real 3D environment; (3) the ability of the MAS to model a human organization (roles, tasks, restrictions); (4) the correct operation of capturing data from location sensors; (5) simulations on simple tasks in the 3D modeling environment using the location data; and finally (6) verification that the 3D tool is capable of developing simulations in complex environments.

To meet these objectives, the case studies that are planned are:

- Simulations in an environment of dependency, such as a residence for elderly people.
- Simulations in an emergency situation, as is the case of a fire situation in a building.

Although both scenarios share the same philosophy, the contexts are, undoubtedly, very different, primarily because the individuals to locate are slow and programmed in the first case, and rapid and unpredictable in the second. Other factors include deployment time and construction of 3D simulation environment, with little time restriction in the first case study, and high time limitation in the second.

Within these case studies, the conditions that must be evaluated are:

- Performance of the simulation environment.
- Accuracy in locating individuals.
- Management notifications and alerts.
- Ease of building the 3D environment.
- Representing the accessibility barriers
- Modeling an organization through the MAS.
- Simulation of simple tasks

4.1 Case Study 1: Environment of Dependent People

The first case corresponds to "*Residencial La Vega*" (Fig. 9) located in Salamanca (Spain). The different facilities of this elderly residence include single and double rooms, deluxe double rooms, private garden and green areas suitable for taking walks and resting, a library, classroom, reading rooms, occupational therapy and rehabilitation, television and spaces for cinema, audiovisual activities, etc.

Facilities at this residence are divided into different zones according to the needs of the residents. Different areas are defined according to the degree of autonomy. One of these areas, the "Blue Zone" accommodates patients who have a high degree

Fig. 9 Dependence environment: Residencial La Vega

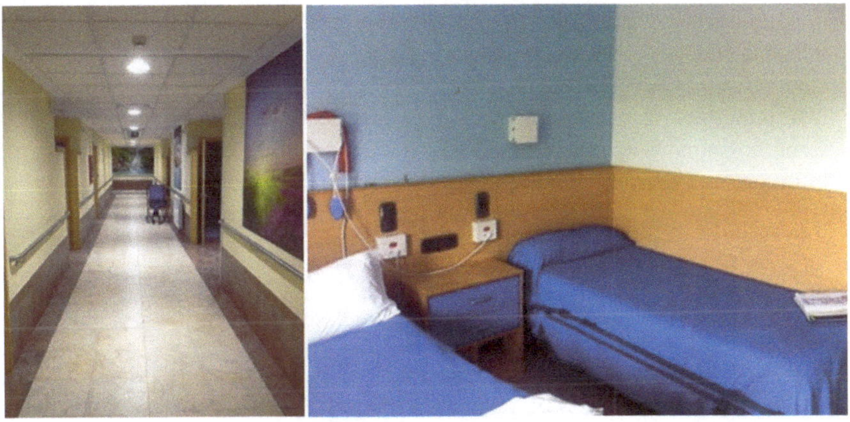

Fig. 10 Blue Zone at Residencial La Vega

of dependence because of their mental state. In this area it is essential to know the location of each patient continuously, in order to be able to resolve critical situations. Moreover, knowing the location of each patient is a key factor in the ability to foresee any risky situation before it happens. Therefore, this is the area that was used in the case study, as it is ideal for evaluating the platform.

The blue zone (Fig. 10) has forty-nine double rooms with bathrooms; all are connected to a central hallway with a main access point properly controlled and a secondary point as an emergency service.

During the case study, different experiments were performed, all of which focused on validating the following key features:

- *Location service in real time*. In the case study, checking the exact position of each of the patients and caregivers is a key factor. In this sense, the experiments show that it is necessary to increase the number of readers to ensure that the positioning system is adequate.
- *Modeling a human organization*. The experiments also validated the system's ability to represent all patients in the same scene, and that it can interact with

them independently. Despite being a rather large area composed of many models per room, including some fairly level of detail, an adequate performance was achieved, which makes it possible to navigating freely around the 3D space, allowing the avatars to simulate simple tasks of both patients and caregivers.

- *Validation of alerts*. It checks that the system is capable of launching automated alerts when certain users (patients) are not located within enclosures in which they are allowed. In addition, the case study also validated that actual users can launch alerts via a button that is included in the node location. All these alerts were tested and work perfectly, and can be seen in 3D environment with a delay of 1 or 2 s, making it possible to identify the position of the user instantly.
- *Representation of the environment in a 3D virtual space*. This requirement forces the virtual environment to be scaled correctly and both the property and the furniture to be positioned correctly as well. It is equally important that the obtained positions of the real world are mapped correctly in the virtual environment. The tools provided by the 3D editor have been a cornerstone, since it is important to pay attention to the construction and to make no mistakes when positioning architectural elements such as furniture, or when positioning information. Figure 11 shows the detail of the built environment.

4.2 Case Study 2: Emergencies

The second case study models an emergency situation. Figure 12 shows the fire station drill tower for the city of Salamanca (Spain), which allows the simulation of multiple risk situations with different requirements. The tower has four floors and a basement. Each floor has a different distribution in order to evaluate different realistic situations. Access between floors is by stairs with different sizes and designs. On each floor there are windows and exterior doors. The appearance of the building is simple, focusing exclusively on the practical and necessary features required for developing the tests. The unique peculiarity that the building presents is that it is slightly inclined to one side.

Experiments on this case study show the following results:

- *Construction of the building involved*. In an emergency situation, the construction of a 3D virtual environment must be rapid and effective, representing only spaces (walls, floor, etc.). In this case, the proposed tool is effective provided that the beginning is the blueprint (Fig. 13). The difficulty in this case study was the sloping wall, a design mode that was not foreseen a priori. However, the tool is able to remodel the 3D space according to the needs of the emergency itself, providing speed and flexibility to ensure the continuity of the emergency service.
- *Location service in real time*. The ease of deployment of the beacons was a priority in the tool; firefighters should simply "throw" the beacons, as they were previously configured with the exact coordinates where they should be deployed. Firefighters only had to place each beacon into place, without worrying about

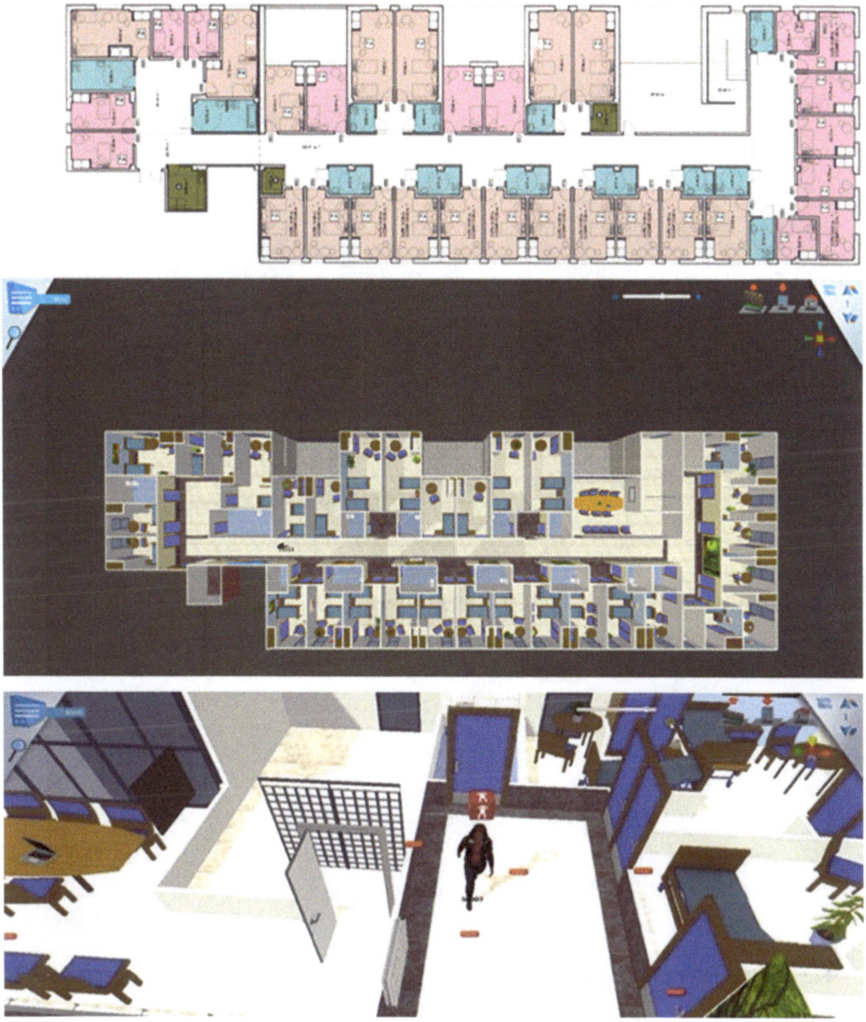

Fig. 11 Dependence environment represented in 3D

any other task. The correspondence between the positioning coordinates of the real and virtual environment must match and be as accurate as possible; however, negligible error was allowed for the location.

- *Alert management.* The system is able to filter and manage all types of notifications to the user, who is informed at all times of the possible risks and changing situations. Actuators can be included whose behavior is managed in the system in order to automate more services and achieve autonomous behavior.

Fig. 12 Emergency environment

- *Representation of evacuation systems.* A common situation may be to carry out a building evacuation after collapse. Using the module for managing alerts, the system can inform the user (fireman) of the need for evacuation. In this sense, the system also simulates possible evacuation routes in real-time, and assesses their advantages and disadvantages.

5 Discussion

The analysis carried out during the execution of the two case studies has helped to focus the application environments suitable for the proposed simulation tool. In both case studies, the services of tridimensional representation, modeling of organizations and localization functioned as expected. Specifically, Table 1 shows a summary of the results.

The platform seems to be a perfect solution for constant surveillance environments such as the dependence environment studied. With the location service, it is possible to know the location of an incident or risk situation in real time. The location service also complements the 3D visualization tool by helping to recognize the location immediately; a 3D re-creation is always able to provide great detail. Moreover, the ability to perform simulations on 3D environment can optimize the work of employees and reduce potential risks to dependent people. On the other hand, the results also show that the platform is not entirely appropriate in situations of extreme emergency. The experiments done in the drill tower at the station helped to determine that in a dangerous situation where the resilience of firefighting equipment is a critical aspect, the system may not behave in a desired manner due to errors of precision in its implementation, primarily because of the extremely short period of time that is available for deployment. However, on previously constructed

Fig. 13 Emergency environment represented in 3D

Table 1 Results of the cases of study

Condition to evaluate	Case study 1	Case study 2
Performance of the simulation environment	*MEDIUM*	*HIGH*
Accuracy in locating individuals	*HIGH*	*MEDIUM*
Notifications and alerts management	*HIGH*	*HIGH*
Ease of building the 3D environment	*HIGH*	*MEDIUM*
Representing accessibility barriers	–	*MEDIUM*
Modeling of an organization through the MAS	*HIGH*	*HIGH*
Simulation of simple tasks	*HIGH*	*HIGH*

buildings, the system continues to maintain of the ability to readapt virtual spaces, task assignments and simulation of cases of study that are useful in this kind of environment, such as evacuations.

In both case studies, the ability to generate alerts was adequate. In addition, the system has the ability to generate real-time alerts if it finds that an individual has accessed a restricted area, and to find patterns of user behavior after constant monitoring.

Similarly, the fact that the wireless sensor network is integrated into the platform allows any place to be suitable to perform real-time monitoring, whereby faults in a computer do not prevent monitoring from another computer.

Although the platform allows the quick deployment of the wireless sensor network, and the 3D editor provides all the necessary tools to perform accurate 3D modeling of the environment to monitor, it may respond too slowly in urgent situations. In addition, the platform needs to have a minimum amount of information about the place to simulate to work correctly this is, it requires the plans for the building to be simulated. When deploying a wireless sensor network, it is necessary to know to know the environment in which the beacons will be physically placed, and to be able to indicate their coordinates through the 3D tool. If there are errors during this process, the location algorithms will fail and, consequently, the system obtains erroneous positions of the tags to locate. In addition, 3D environments require a minimum time commitment to complete the modeling. In an emergency situation only the walls and access to the rooms were modelled; without having the blueprints or scale measures is a risk of accuracy.

Despite the limitations encountered, the platform stands out for its ability to represent human organizations, allowing simulations of simple tasks that can be performed in an environment and validating the difficulties of accessing these environments, so the initial objective of the platform is satisfied.

6 Conclusions

The presented platform and the collection of case studies are specifically oriented toward facilitating the integration of people with disabilities into the workplace to assist them in their daily lives and to help them to improve their quality of life as well. These systems were conceived after extensive research based on the needs of such users. The platform was being testing in independent case studies; however, because we believe that they are closely related, we chose to build an integral system and ultimately developed the new platform. In order to test the system, a MAS was developed to simulate an office environment and study the problems of accessibility experienced by people with disabilities in performing different jobs. The MAS is designed as an organization modeled based on reality. All workers, jobs, and interaction elements, such as architectural barriers, are modeled as agents. The platform makes it possible to deploy different agents, even those included in mobile devices, and to communicate with the agents embedded in the different sensors.

Acknowledgments This work has been partially supported by European Social Fund (Operational Programme 2014–2020 for Castilla y León, EDU/128/2015 BOCYL).

References

1. United Nations: Factsheet on persons with disabilities. In: http://www.un.org/disabilities/default.asp?id=18 [Last access: 27/01/2015]
2. United Nations: Standard Rules on the Equalization of Opportunities for Persons with Disabilities (1994)
3. Charter of Fundamental Rights of the European Union (2010/C 83/02)
4. Consolidated version of the Treaty on European Union and the Treaty on the Functioning of the European Union 2012/C 326/01
5. European Council Directive 2000/78/EC of 27 November 2000
6. Communication from the Commission to the European Parliament, the Council, the European Economic and Social Committee and the Committee of the Regions: European Disability Strategy 2010–2020: A Renewed Commitment to a Barrier-Free Eu-rope (COM/2010/0636)
7. Chung, C.A. (ed.): Simulation Modeling Handbook: A Practical Approach. CRC Press (2003)
8. Barriuso, A., De la Prieta, F., Li, T.: An agent-based social simulation platform with 3D representation for labor integration of disabled people. In Trends in Practical Applications of Agents, Multi-Agent Systems and Sustainability, pp. 55–64. Springer (2015)
9. Pepijn, S., Thomas, B.: An agent-based simulation model of human–environment interactions in agricultural systems. Environ. Model. Softw. **267**(7), 845–859 (2011) ISSN 1364-8152
10. Frans, P., Atle, F.: Relationship strength and network form: An agent-based simulation of interaction in a business network. Australas. Mark. J. (AMJ) **22**(1), 15–27 (2014) ISSN 1441-3582
11. Yu, Y., Kamel, A.E., Gong, G., Li, F.: Multi-agent based modeling and simulation of microscopic traffic in virtual reality system. Simul. Model. Pract. Theory **45**, 62–79 (2014), ISSN 1569-190X
12. Almeida, S.J.D., Ferreira, R.P.M., Eiras, Á.E., Ober-mayr, R.P., Geier, M: Multi-agent modeling and simulation of an Aedes aegypti mosquito population. Environ. Model. Softw. **25**(12), (2010)
13. Davidsson, P.: Agent based social simulation: a computer science view. J. Artif. Soc. Soc. Simul. **5**(1) (2002)
14. Borshchev, A., Filippov, A.: From system dynamics and discrete event to practical agent based modeling: reasons, techniques, tools. In: Proceedings of the 22nd International Conference of the System Dynamics Society (No. 22) (2004)
15. Davidsson, P.: Multi agent based simulation: beyond social simulation. In: Moss, S., Davidsson, P. (eds.) MABS 2000. LNCS (LNAI), vol. 1979, pp. 97–107. Springer, Heidelberg (2001)
16. De la Prieta, F., Rodríguez, S., Corchado, J.M., Bajo, J.: Infrastructure to simulate intelligent agents in cloud environments. J. Intell. Fuzzy Syst. **28**(1), 29–41 (2015)
17. Elena, G., Sara, R., Beatriz, M., Carolina, Z., Belén, P.: MISIA: middleware infrastructure to simulate intelligent agents. In: International Symposium on Distributed Computing and Artificial Intelligence Advances in Intelligent and Soft Computing. Advances in Intelligent and Soft Computing, vol. 91, pp. 107–116. Springer Science+Business Media (2011)
18. De Paz, J.F., Tapia, D.I., Alonso, R.S., Pinzón, C.I., Bajo, J., Corchado, J.M.: Mitigation of the ground reflection effect in real-time locating systems based on wireless sensor networks by using artificial neural networks. J. Knowl. Inf. Syst. **34**(1), 193–217 (2013)
19. Nerguizian, C., Despins, C., Affès, S.: Indoor geolocation with received signal strength fingerprinting technique and neural networks. In: 11th International Conference on Telecommunications, pp. 866–875. Fortaleza, Brazil (2004)

20. Rodríguez, S., De la Prieta, F., García, E., Zato, C., Bajo, J., Corchado, J.M.: Virtual organizations in information fusion. In: 9th International Conference In Practical Applications Of Agents And Multiagent Systems - Special Session on Adaptive Multiagent Systems, Salamanca, Spain (2011) pp. 195–202
21. Marin-Perianu, M., Meratnia, N., Havinga, P., de Souza, L., Muller, J., Spiess, P., Haller, S., Riedel, T., Decker, C., Stromberg, G.: Decentralized enterprise systems: a multiplatform wireless sensor network approach. IEEE Wireless Commun. **14**, 57–66 (2007)
22. Borrajo, M.L., Corchado, J.M., Corchado, E.S., Pellicer, M.A., Bajo, J.: Multi-agent neural business control system. Inf. Sci. **180**(6), 911–927 (2010)
23. DISCATEL: http://www.aeerc.com/teletrabajo.cfm. Last access date: 14/04/2016 (2013)
24. INREDIS: http://www.inredis.es/ Last access date: 14/04/2016 (2013)
25. Yu-Jin, H., Ig-Jae, K., Sang, C.A., Hyoung-Gon, K.: Activity recognition using wearable sensors for elder care. In: 2nd International Conference on Future Generation Communication and Networking, pp. 302–305. Hainan Island, China (2008)
26. George, P., George, X., George, P.: Monitoring and modeling simple everyday activities of the elderly at home. In: 7th IEEE Consumer Communications and Networking Conference, Las Vegas, USA, (2010) pp. 1–5
27. Seon, W.L., Yong-Joong, K., Gi-Sup, L., Byung, O.C., Nam-Ha, L.: A remote behavioral monitoring system for elders living alone. In: 7th International Conference on Control, Automation and Systems, Seoul, Korea, (2007), pp. 2725–2730
28. Alemdar, H., Ersoy, C.: Wireless sensor networks for healthcare: a survey. Comput. Netw. **54**(15) (28), 2688–2710 (2010)
29. Sarangapani, J.: Wireless Ad hoc and Sensor Networks: Protocols, Performance, and Control, 1st edn. CRC Press, Boca Raton (2007)
30. Tapia, D.I., García, O., Alonso, R.S., Guevara, F., Catalina, J., Bravo, R.A., Corchado, J.M.: The n-Core polaris real-time locating system at the EvAAL competition. In: Evaluating AAL Systems Through Competitive Benchmarking. Indoor Localization and Tracking Communications in Computer and Information Science, pp. 92–106 (2012)
31. Bellifemine, F., Poggi, A., Rimassa, G.: JADE–A FIPA-compliant agent framework. In: Proceedings of PAAM'99, vol. 33, pp. 97–108, (1999)
32. Dignum, V.: Handbook of Research on Multi-Agent Systems: Semantics and Dynamics of Organizational Models, p. 111. Information Science Reference, Hershey (2009)
33. Rao, A.S., Georgeff, M.P.: BDI agents: from theory to practice. In: ICMAS, vol. 95, pp. 312–319 (1995)

An Intersection-Centric Auction-Based Traffic Signal Control Framework

Jeffery Raphael, Elizabeth I. Sklar, and Simon Maskell

Abstract Vehicular traffic on urban road networks is of great interest to those who monitor air quality. Combustion emissions from transport vehicles are a major contributor of air pollution. More specifically, the release of fine particulate matter which has been linked to premature deaths. Travel and idle time are two factors that influence the amount of pollution generated by traffic. Reducing idle and travel times would have a positive impact on air quality. Thus, it is increasingly crucial to manage intersections effectively, particularly in congested cities and across a range of different types of traffic conditions. A variety of market-based multi-agent traffic management mechanisms have been proposed to improve traffic flow. In many of these systems drivers "pay" to gain access to favourable road ways (e.g., minimise travel time). A major obstacle in adopting many of these mechanisms is that the necessary communication infrastructure does not yet exist. They rely on vehicle-to-infrastructure and/or vehicle-to-vehicle communications. In this work, we propose a market-based mechanism which relies on existing technology (and in some places this technology is already in use). Experimental results show that our market-based approach is better at reducing idle and travel times as compared to fixed-time signal controllers.

1 Introduction

London's *Great Smog* occurred over 60 years ago. It is estimated that over 4000 people died prematurely due to health complications brought about by the dense mix of smoke and fog [4]. The smog was a result of coal burning (most notably

J. Raphael (✉)
Department of Computer Science, University of Liverpool, Liverpool, UK
e-mail: jeffery.raphael@liverpool.ac.uk

E.I. Sklar
Department of Informatics, King's College London, London, UK
e-mail: elizabeth.sklar@kcl.ac.uk

S. Maskell
Department of Electrical Engineering & Electronics, University of Liverpool, Liverpool, UK
e-mail: s.maskell@liverpool.ac.uk

A. Alonso-Betanzos et al. (eds.), *Agent-Based Modeling of Sustainable Behaviors*,
Understanding Complex Systems, DOI 10.1007/978-3-319-46331-5_6

for heating), industrial practices of the time, and atmospheric conditions. Even though London had experienced bouts of smog in the past, the *Great Smog* launched air pollution into the spotlight as a major health hazard. Decades later, despite a number of legislative acts to curb the presence of noxious gases and particulate matter, London continues to exceed the allowable limits defined for air pollution standards [12]. There is strong evidence linking air pollution to increases in mortality and morbidity (health issues such as low birth weight, strokes and heart diseases [12]) rates [4, 16]. Motor vehicles have been identified as a major source of air pollution (specifically nitrogen dioxide and particulate matter smaller than 2.5 μm) [12, 34]. The amount of combustion emissions are directly related to long idle and travel times which can be exacerbated in congested cities. In England and Wales, over 60 % of commuters drive to work [11]. In many cities, traffic congestion does not appear to deter travelling by car. For example, over a quarter of Londoners still choose to drive to work [11], despite having access to alternative modes of transportation. Traffic congestion does not just pose a health hazard, it also costs money. Across the UK, close to £3.76 billion (€4.94 billion) [27] are lost from fuel and increased cost of delivering goods. The UK is not alone in its struggle with traffic congestion and its effect on air quality. Other cities, such as Beijing, Los Angeles, and Delhi, have similar problems. Any effort to improve travel times (and traffic throughput) would have significant benefits to air quality and a financial impact as well.

Today, traffic managers (people) are responsible for the prevention and/or reduction of traffic congestion. Setting speed limits, installing road signs, speed humps, islands, implementing movement and parking restrictions are some of the tools at the disposal of traffic managers. This includes the authority to enact transportation policies aimed at improving traffic flow. Traffic managers must be proactive and identify potential network hazards and/or issues that may disrupt traffic flow. While most measures exist for the improvement of vehicular traffic, some are put in place to maximise the synergy between pedestrian and vehicle movements. In addition, some cities now include cyclists in their traffic management schemes, with designated cycle lanes and traffic signals. Overall, traffic managers utilise their power to control different components of the transportation infrastructure to ensure the safe and efficient use of road networks.

Traffic signals (or "lights") are one of the most common tools employed around the world to control traffic at intersections. Different municipalities have different criteria for the use of traffic signals. Traffic demand, number of conflicting vehicle manoeuvres and general delays are the most common reasons for installing traffic signals. Traffic signals manage conflicting movements by dictating which vehicle movement(s) are allowed at a given moment. A driver approaching an intersection may arrive only to discover her vehicle is not allowed through and must then wait. Although traffic signals are installed to improve the safety and flow of traffic, they nonetheless become a major source of delay in road networks.

Traffic signals are programmed to permit small groups of non-conflicting vehicle movements through an intersection for a short period of time, followed by another group of movements. Traditional traffic signals (and the most basic deployment method) repeat the same sequence of vehicle movements without changing their duration or order of execution—regardless of what is happening in real time at the intersections they control. Traffic signal timings specify the duration and order of vehicle manoeuvres at an intersection [22]. Appropriate traffic signal timings, or *phase plans*, are essential to the proper function of traffic signals. Poorly designed phase plans may lead to additional delays, traffic jams and even accidents. Although many traffic signals rely on simple fixed protocols, they are nonetheless a vital component of traffic management [1]. Historically, finding the best signal timings involved using mathematical models of traffic behaviour to determine ideal settings [22]. It later became possible to develop even better signal timings through the analysis of historical traffic data. Adaptive *Urban Traffic Controllers (UTCs)* employ information about current road conditions and determine, some in real time, the best signal settings. Adaptive UTCs attempt to harmonise the interplay between all aspects of traffic (private vehicles, public transportation, cyclist and pedestrians) in areas ranging in size from a few city blocks to entire cities. Adaptive centralised systems have been developed that apply optimisation algorithms, such as RHODES [17], OPAC [9] and SCOOT [18, 31].

In general, the traffic control problem can be stated as that of finding a policy for setting traffic signal states such that traffic flow improves while the safety of an intersection is maintained and conflicts amongst movements are resolved (including pedestrian and cyclist movements). Such a policy could take into consideration traffic conditions at the intersection and could also incorporate information from neighbouring intersections. The policy should determine which movement(s) are allowed at any given moment in time. An optimal policy could minimise travel time. Or, it could attempt to optimise other aspects of traffic, such as number of stops per vehicle or *queueing* time (i.e., how long a vehicle waits at a particular intersection before it is able to pass, which might mean waiting through multiple signal phases if traffic is heavy).

Finding the optimal traffic signal timings is a non-trivial operation for a number of reasons:

- Traffic control is geographically distributed, takes place in a dynamic environment and the interactions amongst its components are highly complex [6].
- Traffic signal timings function under rigid temporal constraints which may be represented as discrete variables. Therefore, traffic control behaves in many regards like a combinatorial optimisation problem (e.g., TSP) [19].
- Scale is always an issue with traffic control. Any reasonably sized road network will have dozens of intersections, compounding the problem of finding an optimal traffic signal timing [3, 19, 33].
- Adaptive traffic control systems that work in real time must find a solution within a very small time window in order to function properly.

Traffic consists of many independent components that are interconnected in a highly complex manner. There are vehicles, pedestrians, cyclists and traffic control devices, to name a few of the elemental components. Using mathematical models, it is difficult to capture the interacting behaviours of these individual components;

however, modelling them as a large collection of *autonomous agents* allows us to apply a wide range of methodologies designed to investigate the interplay between independent entities. For this reason, the *Multi-Agent Systems (MAS)* paradigm offers an ideal method for modelling the critical elements of traffic behaviour. The advantage of using a MAS approach over traditional mathematical models is two-fold. One, MAS does a better job of modelling the stochastic nature of traffic. Two, better models means better platforms to investigate novel solutions to traffic control. The MAS paradigm offers a flexible and inexpensive method for designing traffic control solutions [31]. There is a plethora of traffic control solutions that fall under the umbrella of MAS. Traffic control systems that are developed within a MAS framework are also easier to maintain and scale [31]. As well, the traffic domain offers many interesting challenges from a multi-agent systems perspective.

Our work focuses on solutions that utilise *market-based mechanisms.* Traffic control can be viewed as the management of a set of traffic signals in a road network in order to minimise, for example, the delay experienced by vehicles traversing the network. If traffic signals are considered agents, traffic control can be viewed as a *coordination* problem [2] where traffic signals work together to prevent congestion and keep traffic flowing. This perspective on traffic control is important as it drives our MAS design choices and sets us apart from other MAS solutions to traffic control thus far, as explained next. We propose using *auctions* to achieve coordination amongst traffic signal agents by providing a framework for the resolution of conflicts and enhancement of cooperation amongst traffic signal agents.

Our approach for controlling traffic signals has been greatly influenced by coordination efforts in *Multi-Robot Routing (MRR)* [7, 10, 13, 23]. Auctions are a form of market mechanism for resource allocation, and they can produce near optimal results in some MRR scenarios [15]. In MRR, auctions have been used to facilitate coordination amongst robots [26]; thus the same can be done with traffic signals. A common theme in the existing literature on auction-based traffic controllers is the need for a *vehicle agent*, which refers to a vehicle-borne software system responsible for tasks ranging from simple *vehicle-to-infrastructure (V2I)* communication to more demanding vehicle navigation and control. We believe that auctions can empower agents, acting either locally at a single intersection or in small groups of connected intersections, to find local solutions to traffic congestion that then emerge as global improvements in traffic performance.

There are a number of significant issues with regard to the widespread deployment of current market-based approaches to traffic management. The first issue is the development and distribution of vehicle agents. Car manufacturers will have to agree on international communication protocols, physical specifications and many other aspects of deploying vehicle agents to the millions of vehicles that are currently in use. Second, there is the current state of the transportation infrastructure worldwide. The communication systems necessary for *V2I* communication currently does not exist; and a range of issues, such as security and privacy, remain unaddressed in the traffic management domain. Lastly, there is the concept of drivers bidding for intersection usage, which introduces the issue of *fairness*. Fairness is a general term for such questions as: *Which drivers will have to pay?* and *How much will drivers have to pay?* Our overarching goal is to design a system that reaps the benefits of market mechanisms, but without its less appealing features such as driving fees and V2I communication requirements. Our approach not only does away with vehicle agents, but also does not have drivers bidding for intersection usage; hence, our approach allows us to utilise auctions without having to consider fairness at the level of each vehicle.

In this paper, we describe our theoretical and experimental work on multi-agent auction-based traffic control mechanisms. As above, our mechanisms utilise auctions without the need for vehicle agents. We demonstrate how such a system could be designed and implemented, and we ran a series of experiments to measure the effectiveness of our mechanisms. Three empirical evaluations found that our mechanisms perform better than fixed-time signal controllers. The remainder of this paper is organised as follows. Section 2 discusses other auction-based approaches to traffic control, focussing on the MAS literature. Section 3 presents our approach. Sections 4 and 5 describe our experiments and results. Finally, we close with some discussion (Sect. 6) and conclusions (Sect. 7).

2 Related Work

Dresner and Stone [8] designed a reservation-based traffic management system to reduce traffic congestion. In a reservation-based system, vehicles request time slots. The time slots are time spans when the vehicle is allowed to occupy the intersection. The reservation-based system functions on a *first-come, first-served* basis and relies on vehicle agents (autonomous cars) that have complete control of the vehicle. The authors measured the delay experienced by vehicles passing through the intersection. Dresner and Stone [8] compared their reservation-based approach to two other traffic control schemes: *overpass* and *traffic light.* "Overpass" simulates a road network with no signals; roads cross each other via bridges. "Traffic light" simulates how current fixed-time traffic signals function. Dresner and Stone [8] found that their reservation-based system did not just outperform normal traffic lights but under certain conditions eliminated delay due to intersection crossing.

Vasirani and Ossowski [30] expanded on Dresner and Stone's work and examined the performance changes to a reservation-based system where time slots were allocated using a market mechanism. The authors also proposed a market-based traffic assignment scheme using the same reservation-based system. In Dresner and Stone [8], reservations were allotted using a *first-come first-served* policy or FCFS. Vasirani and Ossowski [30] replaced FCFS with *a combinatorial auction (CA).* As drivers approach the intersection, reservations are "won" through the auction, instead of simply handed to the next arriving vehicle. In this way, a driver may express its true valuation for a contested reservation. For the market-based traffic assignment system, Vasirani and Ossowski [30] devised a protocol where route selection was accomplished through a *combinatorial traffic assignment (CTA).* The cost of passing through an intersection continually changed depending on demand. In turn, these costs caused vehicles to select alternative or cheaper routes.

In a network with a single intersection, Vasirani and Ossowski [30] looked at the delay experienced by drivers based on the amount they were willing to "pay" to use the intersection. They were interested in finding out if drivers willing to pay more would experience less delay. They also looked at the delay experienced as traffic volume increased across the intersection. Vasirani and Ossowski [30] found that initially having a willingness to pay does decrease delay, but eventually this levels off. However, CA was found to increase overall delay. As the intensity of traffic increases, CA experienced far more delays and rejected reservations than FCFS.

The performance of CTA was studied using a simulation of a simplified road network of Madrid, Spain. The authors examined the density (number of vehicles per kilometre of roadway) and travel time to measure its performance. Vasirani and Ossowski [30] found that CTA, which used FCFS, produced a more balanced network, i.e., vehicles were better distributed throughout the road network. As both CA and CTA are extensions of [8], they too rely on vehicle agents and thus are infeasible with current transportation infrastructure.

Although Dresner and Stone's work does not directly employ a market mechanism, it does represent the state-of-the-art in terms of futuristic visions of traffic control. The reservation-based traffic management system [8] (and Vasirani and Ossowski [30] market based derivative) requires the greatest advancements in current transportation infrastructure: V2I communications and autonomous cars. Our approach on the other hand, does not require neither V2I communications nor autonomous cars; although the former could be used to improve the performance of our mechanism.

Carlino et al. [5] described a traffic control system where auctions are run at intersections to determine use. Vehicles are embedded with a software agent (the *wallet agent*) which bids on behalf of the driver. A *system agent* also bids in a manner that facilitates traffic flow beneficial to the entire transportation system—while the *wallet agent* is solely concerned with getting its occupants to their destination in the least expensive (and quickest) way. The *wallet agent* is assigned a budget to pay for trips. Carlino et al. [5] used a second-price sealed bid auction mechanism. They tested four different modes: *FIFO* (this is how a typical intersection works), *Equal* (every driver submits a bid of one), *Auction* (drivers bid an amount equal to their account balance divided by the number of intersections remaining on their trip), and *Fixed* (drivers always bid the same amount based on the value they've assigned for the trip). The authors evaluated their traffic control mechanisms in four simulated urban cities. FIFO performed the worst in three of the four cities. *Auction* (with and without the *system agent*) had the best performance. There are two important distinctions between our work and [5]. First, Carlino et al. [5] assumes vehicles have specialised software that allow drivers to effortlessly participate in the auction; we do not need require any such software. Second, although we utilise auctions in our approach, in our work the auction provides a framework for coordination and is not monetised.

Schepperle and Böhm [24] describes an intersection controller called *Initial Time-Slot Auction* (ITSA) which is *valuation-aware* (meaning the controller considers the driver's value of reducing wait time). In ITSA, as vehicles approach an intersection, they *register* with the intersection. The *intersection agent* then executes a second-price sealed-bid auction for the most current time slot that's available for usage. Here a time-slot is a window in time where a vehicle may safely use the intersection. Only the vehicles at the very front of the traffic queue participate in the auction. Schepperle and Böhm [24] utilised the FIPA Contract Net Protocol to implement the auction. Schepperle and Böhm [24] also described two variants of ITSA. In the first variant, a mechanism is included to prevent *starvation* where auctions are suspended if vehicle waiting time has reached some fixed limit. Starvation is defined as the situation where traffic is prevented from flowing in a particular direction. The other variant, ITSA+Subsidies, considers subsidies where vehicles that have not participated in an auction yet can influence the auction of the vehicles in front of them. In the subsidised variant, vehicles boost the bid of a candidate vehicle (a vehicle in front of theirs). If the candidate vehicle wins the auction, then the vehicle that subsidised its auction would be able to participate (attain a time-slot) in an auction sooner.

Schepperle and Böhm [24] used *waiting time* to measure performance. The authors defined waiting time as the difference between actual travelling time and the minimum travel time. Schepperle and Böhm [24] also examined *average weighted waiting time* where the weighted waiting time is the product of the waiting time and the driver's valuation of a reduced waiting time. They compared their traffic controller to the reservation-based system in Dresner and Stone [8]. Both ITSA and ITSA+Subsidies were able to reduce average travel time while minimising average weighted waiting time compared to FIFO, although ITSA+Subsidies was better at reducing average weighted waiting time. Drivers that had the lowest valuations, that is those drivers that did not mind waiting, fared better under ITSA+Subsidies than ITSA.

In follow-on work, Schepperle and Böhm [25] created a *valuation-aware* traffic-control mechanism which allows concurrent use of the intersection through an auction mechanism. In a valuation-aware traffic controller, the intersection takes into account the driver's value of time; but many of these systems do not allow concurrent use of the intersection. Schepperle and Böhm [25] propose two auction-based mechanisms: *Free Choice* and *Clocked.* In Free Choice, the auction winner gets to select the time slot it wants from an interval; while in Clocked, time slots are auctioned off. Schepperle and Böhm [25] concluded that Free Choice reduced the average weighted wait time by up to 38.1 %. Clocked reduced the average weighted wait time for only lower degrees of concurrency and high traffic volume. Similar to [5], Schepperle and Böhm [24, 25] assumes that cars have a vehicle agent. Again, our approach, detailed in the next section, does not involve vehicle agents or other embedded software.

Bazzan [2] constructed a decentralised method of traffic control that utilises Evolutionary Game Theory. The traffic controller facilitates coordination among intersections while minimising communication overhead. *Intersection agents* coordinate by selecting the same action (phase plan). Different phase plans favour different sets of vehicle movements. Intersection agents function in two states: local and global. In the local state, intersection agents use local information and a mixed strategy to make action selections. Intersection agents also have a payoff function which is used to update its mixed strategy. Intersection agents experience a learning phase which allows them to update their mixed strategy taking into account most recent payoffs over past payoffs. In [2], there is an entity, "Nature", that has a global view of traffic and is able to see (and process) information from a macroscopic level. Nature recognises global traffic changes and initiates the change from local to global state in intersection agents. While in the global state, intersection agents use a payment function, given to them by Nature, to update their mixed strategy.

Bazzan [2] set up a traffic scenario with vehicles travelling through a roadway with several intersections. Bazzan [2] evaluated her traffic control mechanism under three different traffic conditions. The author compared her method to a centralised traffic controller where a central computer determined the best phase plan for the traffic signals. In the case of the centralised controller, the best phase plan is the one that produced the least delay for the traffic flow (going east or west) with the heaviest volume. The author used traffic density (discretised) to measure performance. The agent-based decentralised method performed better than the centralised method in two of the traffic scenarios. In the first, east and west bound traffic had medium to high volumes of traffic. While in the second, both directions had medium to low levels of traffic. In contrast, our approach is fully decentralised at the intersection level. It does not require a global perspective (i.e., Nature) of traffic flow. Finally, in [2], intersection agents select phase plans from a closed set of phase plans while in our approach we focus on fine tuning a single phase plan in lieu of replacing it.

3 Our Approach: Multi Agent Auction Based Traffic Signalling

For our multi-agent traffic controller, we decompose each intersection into two types of agents: *intersection agents* and *traffic signal agents*. At an intersection, there is a single intersection agent and multiple traffic signal agents. The intersection agent is responsible for making adjustments to traffic signal timings and ensuring that those changes do not violate any basic traffic regulations (e.g., minimum *green times*). Traffic signal agents, on the other hand, operate on behalf of a small set of legal vehicle movements that may occur at the intersection. That is, each traffic signal agent is assigned a number of movements to manage. The traffic signal agents compete against each other for control over traffic signal timing adjustments. An intersection agent and its associated traffic signal agents work together at the intersection level to adapt signal timings in real time. The adjustments are made to improve the efficiency of the intersection and maintain its safety. Figure 1 illustrates the key components of our multi-agent traffic controller and how they are used on a global scale (image on the right) to manipulate traffic flow.

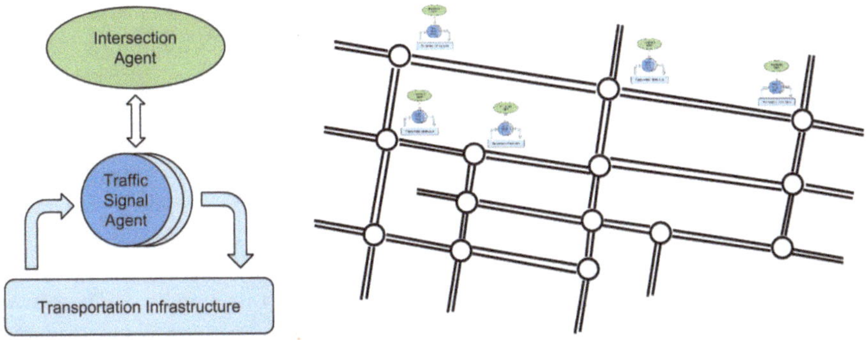

Fig. 1 Multi agent intersection controller

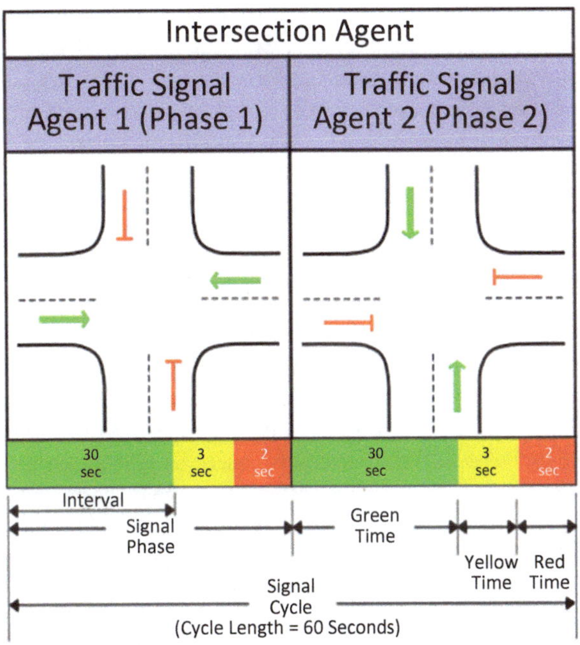

Fig. 2 Traffic signal agent and phase plan

The importance of defining our traffic signal agents in this manner is that it allows us to take advantage of in-practice methods of developing traffic signal timings. Traffic signals manage conflicting movements at an intersection by allowing and restricting movements during set time periods. A *phase plan* describes the sequence of lights a traffic signal will emit and for how long. Developing a phase plan is one of the most important first steps in the effective use of traffic signals. Without a good phase plan, a traffic signal may not be able to handle traffic demand and may even cause traffic accidents. A signal *phase* (illustrated in Fig. 2) is a portion of a traffic signal timing that is given to a set of vehicle movements [28]. A phase is a sequence of lights which includes a green *interval* followed by an amber (yellow) and then a red interval, all assigned to a single movement (or set of non-conflicting movements). The amber and red intervals are necessary so that vehicles have sufficient time to clear the intersection and come to a complete stop.

Our traffic signal agents are equivalent to traffic phases [28] in that they too represent a set of vehicle movements. Thus, for every phase in the phase plan, there is a traffic signal agent that functions on its behalf to tweak the time allotted to that phase. Together, all the phases form the signal timing for a traffic signal, while the traffic signal agents function as an intelligent counterpart to the phase. These two constructs, phase plan and traffic signal agents, address the needs of all legal vehicle movements as traffic demands change.

The design guidelines set by traffic engineers for phase plans (e.g., in the U.S., they use MUTCD [22]) provide a blueprint for determining which movements will be assigned to which traffic signal agent. Traffic engineers divide all the possible legal vehicle movements into subsets, to form phases. The most basic phase plan is the two-phase plan where each street in a standard cross junction (+) is given a phase. The two-phase signalisation plan was used in our initial work [21]. Figure 2 illustrates the relationship between our traffic signal agents and the traffic phases. As there are two phases, there are also two traffic signal agents.

There is a natural conflict that arises between traffic signal agents assigned to an intersection. Each traffic signal agent is designated to a single phase in the traffic signal timing. They compete for a slice of the limited amount of available green time in a *cycle* (see Fig. 2). Assuming the cycle length remains the same, giving more green time to one traffic signal agent means taking it away from another traffic signal agent. We needed a multi-agent interaction protocol [32] to determine an appropriate, adaptive allocation of green time to two competing entities.

As traffic flows through the intersection, auctions take place at fixed intervals which we call the *auction period*. The traffic signal agents participate in the auction and *bid* (explained below) against each other to increase the amount of green time in their respective phases. The winner is the traffic signal agent with the highest bid. The winning agent gains 5 additional seconds of green time, while the loser's green time decreases by the same amount. Although the cycle length remains the same, the amount of green time assigned to each phase changes. Note that the auction period does not (have to) match the cycle length. An auction may occur in the middle of a cycle or after a series of cycles have passed. Green time is only updated after the current traffic signal phase has completed. As a safeguard against starvation, traffic signal agents are prevented from having less than 10 s of green time. Using the taxonomy described by Parsons et al. [20], we could best categorise our auction as *single dimension*, *one-sided*, *sealed-bid*, *first-price* and *single-item*. Thus our implementation—the process that is executed for each auction—closely resembles a single-unit, seller-side English auction [20].

Traffic signal agents use *road sensors* to assess road conditions and generate an appropriate bid. Road sensors include, but are not limited to, inductive-loop vehicle detectors and cameras. The former is a loop of wire buried in the road with an

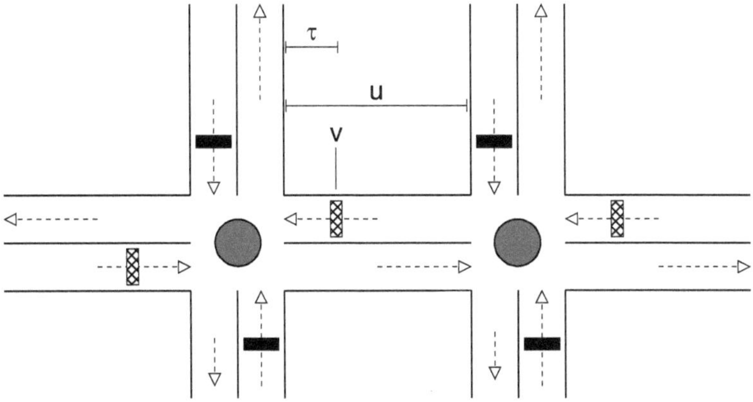

Fig. 3 Traffic Signalling Scheme. The hash-patterned *rectangles* represent the pre-existing *induction-loop* sensors for the west/east traffic signal agents; *black rectangles* for the north/south traffic signal agents. *Grey circles* indicate intersection agents (though they have no physical embodiment in the simulated system). In addition, the following parameters are indicated: **v** is the *volume* of traffic as measured by an induction-loop sensor; **u** is the occupation level between consecutive intersections; and τ is the occupation level between the sensor and the intersection

electric current running through it and is the primary sensor used in the SCOOT system (mentioned earlier). Vehicles are detected via disruptions in the magnetic field of the wire loop caused by the metal body of the vehicle. The induction-loop sensors are located 20 m from the intersection stop line (the hash-patterned and black rectangles illustrated in Fig. 3). The vehicle detectors provide data on traffic volume, measured in vehicles per hour (*vph*).

In our initial work, we defined *saturation* as the ratio of traffic volume on a road segment to its capacity and used this as a measurement of the level of use of a phase [21]. In general, a stream of traffic that is functioning closer to its capacity is more susceptible to traffic jams and delays [29]. Given a phase p, let d_p be the measure of its *saturation*:

$$d_p = \frac{v}{c}$$

where v is the traffic volume, measured by counting the number of vehicles N (reported by vehicle detectors) that pass a point on a road segment during time interval Δt [29], computed as $v = N/\Delta t$; and c is the capacity, representing the maximum possible traffic volume on a road segment, assuming the traffic signal was always green for that movement(s) [22], computed as $c = 3600/h$ vph. Headway, h, is the average amount of time that it takes vehicles in a queue to reach the intersection. For our simulated environment reported here, headway is set to $\sim$2.54 s, resulting in a capacity of $\sim$1417 vph.

We have implemented and compared two traffic signal agents which have different bidding rules: **Saturation (SAT)** and **Saturation with Queuing (SATQ)**. These are detailed next.

Saturation (SAT) In the SAT method, the traffic signal agents compute d_p for their road segment to use as their bidding rule. In the experiments conducted here, the traffic signal agents are only concerned with the single road segment preceding the junction they manage. For example, the west/east signal agent collects volume data one block west and one block east of its location. Equation (1) defines the SAT bidding rule:

$$bid = d_p \tag{1}$$

Saturation with Queuing (SATQ) The SATQ method extends the SAT method, by augmenting its bidding rule with road occupation, u, an indication of how "full" the road is. This provides a better picture of road conditions (e.g., whether there is a queue of vehicles leading up to the road sensor) than the saturation value alone. A traffic camera could be used to obtain this data. Equation (2) defines the SATQ bidding rule:

$$bid = d_p + u \tag{2}$$

4 Experiments

We evaluated our auction-based methods using the *Simulation of Urban MObility (SUMO)* traffic simulator [14]. SUMO is an open source microscopic traffic simulator and is often used in vehicle communication research [e.g., V2I or *vehicle-to-vehicle (V2V)*]; but it has also been used to study route choice and traffic control algorithms [14]. Although it has a GUI front-end, for our experiments we treated it as a back-end server in order to complete a statistically significant number of experimental runs across a range of traffic conditions. We developed a client application to control the simulation through a TCP socket in SUMO's Traffic Control Interface (TraCI).

As a benchmark for evaluating the effectiveness of our market-based methods, we also tested a **Fixed** method of controlling traffic signals. Fixed represents the traditional approach to tuning traffic signal timings. A fixed-time traffic signal maintains the same timings or light durations throughout the day. Fixed-time signals can be classified by their cycle length. So, we evaluated three types of fixed-time signals: *short*, *medium* and *long* cycle length (tested one at a time, i.e., one per experiment). Note that the starting signal timing (base timings) for our market-based approaches was initialised to the *medium* cycle length.

4.1 Simulation Environment

For the purpose of experimentation, we used the grid-style road network shown in Fig. 4. There are 25 intersections in a 5×5 grid layout. *Blocks* are square shaped and measure 200^2 m. The four traffic signals in the corners of the network are deactivated. These four traffic signals control streams that run without conflicts, meaning vehicles traversing these intersections will never have to yield to one another, therefore they are set to always show green. All the other intersections are on the two-phase signal plan. The signal plan does not include dedicated turning (right or left) phases, therefore left and right turns are given lower priority than *through* movements (going straight), i.e., vehicles turning left or right must wait until it is safe to do so. Induction-loop vehicle detectors are placed on roadways (as in Fig. 3) to collect traffic flow data. In Fig. 4, the vehicle detectors are represented by the black and hashed rectangles in the inset.

Vehicles in SUMO have a single goal: reach their destination as quickly and as safely as possible. Vehicles only perform legal manoeuvres including waiting to enter the intersection box until there is ample room to pass completely through it. Vehicles try to maintain a *safe* driving speed based on several pre-set parameters such as *maximum velocity*, *deceleration* and *acceleration*. The safe speed ensures that the vehicle will always be able to safely react to changes in the speed of the vehicle in front of it. Table 1 contains the SUMO driver model parameters that were used in our simulations. Also, drivers follow set routes which are determined before the simulation begins.

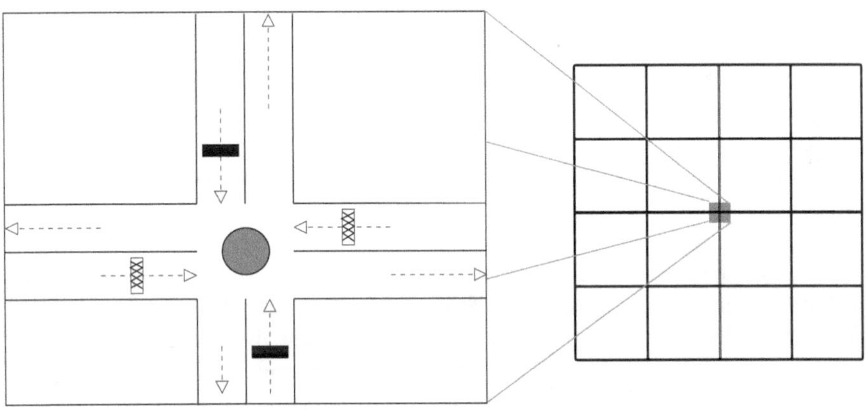

Fig. 4 Grid-based city plan with intersection layout

Table 1 SUMO driver model specifications

Driver model	
Parameter	Value
Acceleration	0.8 m/s^2
Deceleration	4.5 m/s^2
Sigma[a]	0.5
Tau[b]	1.0
Vehicle length	5 m
Minimum gap (between cars)	2.5 m
Maximum speed	16.67 m/s

[a] Used to perturb driver behaviour
[b] Driver reaction time

4.2 Experiment Setup

The aim of the experiments we present here is to clearly map the traffic "landscape" and assess the performance of our auction-based mechanism broadly across this landscape. Four different traffic conditions were simulated:

- **Structured** is traffic that flows through the network with an identifiable path with heavy flow;
- **Unstructured** is traffic flow with no identifiable path with heavy flow;
- **Regional** is identical to Structured, except that cross traffic is kept at minimal levels; and
- **Directional** is similar to Structured, but there is a shift in the direction of the heavy flow midway through each experiment.

In the scenarios, the level of cross traffic (east versus west) was varied. The rationale behind Structured, Regional and Directional is that these represent the ideal traffic conditions where an adaptive urban controller, such as SCOOT, would be used. We were interested in how our market-based approach performs under *normal* conditions, as well as in the face of *disruptions*. We produced two types of traffic flow disruptions: *intensity* and *direction*. "Intensity" simulated a sudden increase in overall traffic volume, while "direction" simulated a change in the direction of the flow of traffic with the heaviest volume. We raised the intensity of traffic at the 1 hour mark during Structured, Unstructured and Regional traffic conditions. With the Directional traffic condition, the disruption is the change in direction of the heaviest traffic stream, which occurs at the 1 hour mark as well. Traffic scenarios ran for 3 simulated hours in SUMO (simulations ran for a maximum of 7 simulated hours). Each set of experimental conditions was repeated 30 times to attain suitable statistics.

5 Results

In this section, we present the results of our experiments. We measured performance in terms of *average travel time* and *vehicle queue length*. Results for each metric are tabulated and analysed next.

5.1 Average Travel Time

Our market-based approach significantly reduced travel time for Unstructured traffic, which was the least predictable traffic flow (see Table 2). In fact, SATQ reduced the average travel time by over 25 % compared to the best **Fixed** traffic controller. For the other traffic flows—those that presented a patterned flow—SAT and SATQ performed second best. The best average travel times for Regional, Directional and Structured were attained by the fixed-time controller with the longest cycle length (FXL). The shortest traffic signal timing (FXS) had the worst travel times by a significant margin.

The *cumulative average travel time* of SAT and SATQ with Unstructured traffic, shown in Fig. 5a, remains fairly steady, with very little change, throughout the simulation. The cumulative average never reaches above 625 s for SAT and SATQ. Initially, with Unstructured traffic, FXS and FXM provided the best travel times, but halfway through the simulation, both controllers experience a sharp and steady rise in travel time. Meantime, SAT and SATQ remain relatively unperturbed. With the Regional condition, shown in Fig. 5b, the cumulative average travel time of SAT and SATQ is nearly identical to FXL—all relatively unperturbed. With the Directional and Structured conditions (Fig. 5c, d), the results are similar, though with more separation between SAT/SATQ and FXL.

We used the percent change between SAT, SATQ and FXM to measure improvements on average travel time (shown in Table 3). We compared our auction-based mechanisms only to FXM because SAT and SATQ began with the same signal timing as FXM, thus highlighting that any differences between our approach and FXM are due to the adaptive nature of SAT and SATQ. We found that SAT and

Table 2 Average travel time over all simulation runs—mean (and standard deviation) reported

	Average travel time			
Traffic signal control	Unstructured	Regional	Directional	Structured
SAT	*623.64*(42.31)	*140.26* (6.00)	159.42 (9.36)	160.22 (8.22)
SATQ	**604.78** (32.14)	143.14 (5.95)	*158.64* (7.16)	*150.31* (8.54)
FXS	1096.26 (169.99)	322.25 (10.96)	272.38 (6.42)	250.75 (8.46)
FXM	927.47 (107.39)	183.72 (1.44)	172.60 (0.94)	165.93 (1.38)
FXL	832.71 (52.46)	**139.13** (0.38)	**131.90** (0.43)	**129.04** (0.39)

Fastest times are highlighted in bold; second-fastest times in italics

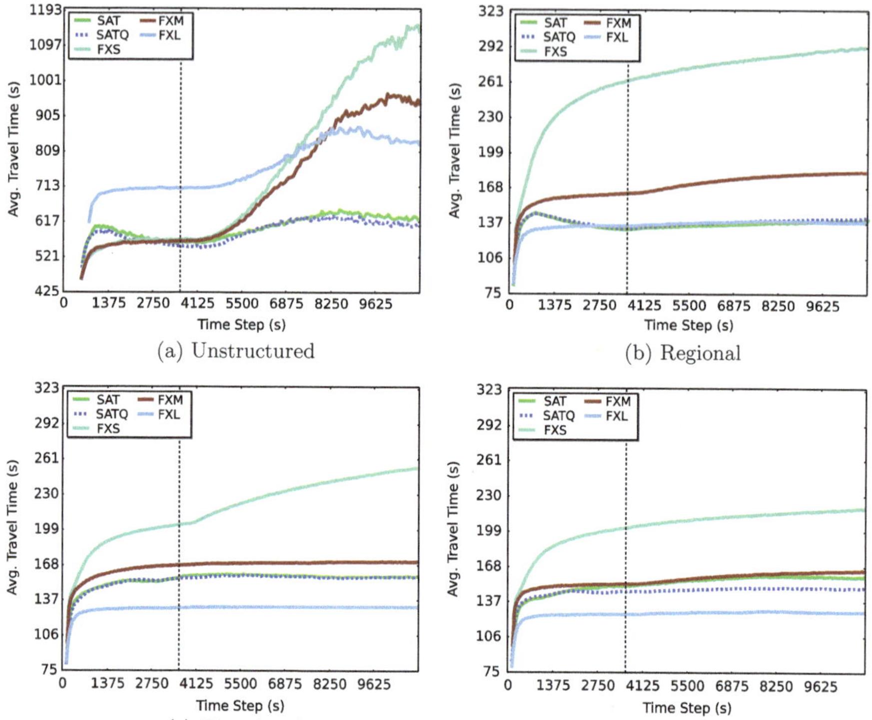

(a) Unstructured (b) Regional (c) Directional (d) Structured

Fig. 5 Cumulative average travel times. Disruptions occurred at the 3600 s mark. Note the y-axis range for (**a**) is larger and shifted, as compared to (**b–d**), further illustrating the distinctive flow exhibited by Unstructured versus the other traffic patterns

SATQ performed in a similar fashion under all four traffic conditions. SATQ's bidding rule, which utilises queue lengths, only provided a slight edge over SAT as Table 3 illustrates. SATQ, compared to SAT, reduced average travel time only slightly more than SAT under the Unstructured and Directional conditions. While SATQ did reduce travel time by nearly 10 % under the Structured condition, it increased travel time under the Regional condition. Under Unstructured conditions, both SAT and SATQ reduced average travel time by over 30 % compared to FXM. The Regional condition experienced similar reductions in travel time, over 20 %, using SAT and SATQ (see Table 3). Under Structured and Directional conditions, SATQ reduced average travel time by just under 10 %, compared to FXM. Overall, when compared to SAT, SATQ provided better travel times in three of the four traffic flows.

The average travel time of vehicles finishing their trip at each time step under the Unstructured condition was greatly reduced under the control of SATQ as compared to when they were using FXM (see Fig. 6). Although there is a slight rise in travel time around the 4500th second, SATQ quickly plateaus and eventually lowers

Table 3 Percent increase in average travel time

Difference in average travel time				
		% change		
Traffic pattern		SAT	SATQ	FXM
Unstructured	SAT	*N/A*	3.12	−32.76
	SATQ	−3.02	*N/A*	**−34.79**
	FXM	48.72	53.36	*N/A*
Regional	SAT	*N/A*	−2.01	**−23.65**
	SATQ	2.05	*N/A*	−22.09
	FXM	30.98	28.35	*N/A*
Directional	SAT	*N/A*	0.49	−7.63
	SATQ	−0.49	*N/A*	**−8.09**
	FXM	8.26	8.80	*N/A*
Structured	SAT	*N/A*	6.59	−3.44
	SATQ	−6.18	*N/A*	**−9.41**
	FXM	3.56	10.39	*N/A*

Each *(row, col)* entry in the table is computed as *(row −col)/col*

Bold values identify which of the two mechanisms (SAT or SATQ) had the best performance, highest decrease in average travel time, compared to fixed-time traffic signals (FXM).

travel time around the 8100th second. On the other hand, travel times increased dramatically under FXM after the 4500th second and remained elevated for the remainder of the simulation. Vehicles traveling under FXM during the Unstructured condition experienced a much broader range of travel times than under SATQ. We can see in Fig. 7 that with Directional, Regional and Structured conditions, average travel times fell within a very narrow band for both FXM and SATQ.

5.2 *Vehicle Queue Length*

In addition to analysing average travel time, we also measured the size of the *queue of vehicles* that formed at every time step as the simulations ran. The queue length was converted to a value, x, where $0 \leq x \leq 1$, representing the percentage of the road segment that was occupied with vehicles.[1] Figure 8 shows this occupancy measurement for the four incoming roadways at an intersection under the Unstructured condition. Around the 4000th second, FXM experiences an increase in its north-bound queue. During that same period, queues under SAT and SATQ control suffer only a slight increase in queue length. However, the opposite happens on the east-bound roadway, where there is an increase in queue

[1]This is the same as the u parameter included in the SATQ bid.

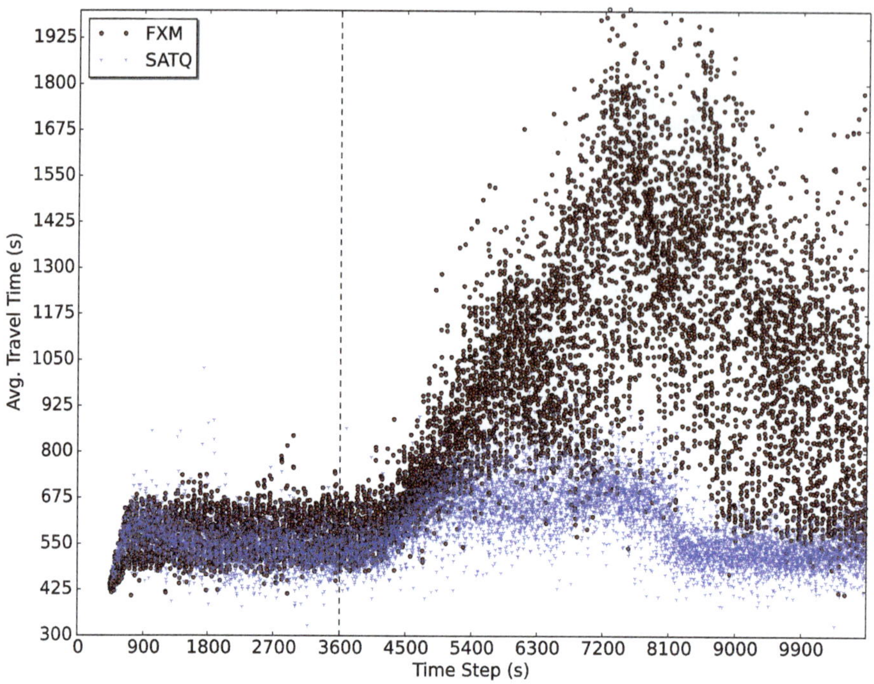

Fig. 6 A comparison of average travel times of vehicles that have completed their journey at each time step under the Unstructured condition. The first vehicles entering the simulation complete their journeys at around the 600 s time step; moving right along the *y*-axis, vehicles continue to complete journeys. The *darker dots* illustrate that the average travel times for vehicles passing through intersections managed by SATQ controllers is much more consistent than those managed by FXM. Figure 7 shows similar results under the other simulated traffic conditions

length for both SAT and SATQ. That increase in queue length on the east-bound roadway corresponds to the increase seen under FXM on the north-bound roadway, illustrating that the priorities for traffic flow vary with the different mechanisms.

6 Discussion

Our findings demonstrate that auction mechanisms can be used to manage traffic flow effectively, without the need for vehicle agents. Our results show that under certain traffic conditions, our auction-based approach to traffic control is superior to **Fixed** time traffic signals. SAT and SATQ reduced travel times the most with Unstructured traffic. This was an important finding because the Unstructured traffic represented the sort of unpredictable traffic flow that is often found in the real world. The other traffic conditions, Regional, Directional and Structured, displayed predictable traffic behaviours, e.g., heavy traffic flow in a northerly direction. Predictable traffic behaviours, assuming they are the cause of congestion, are

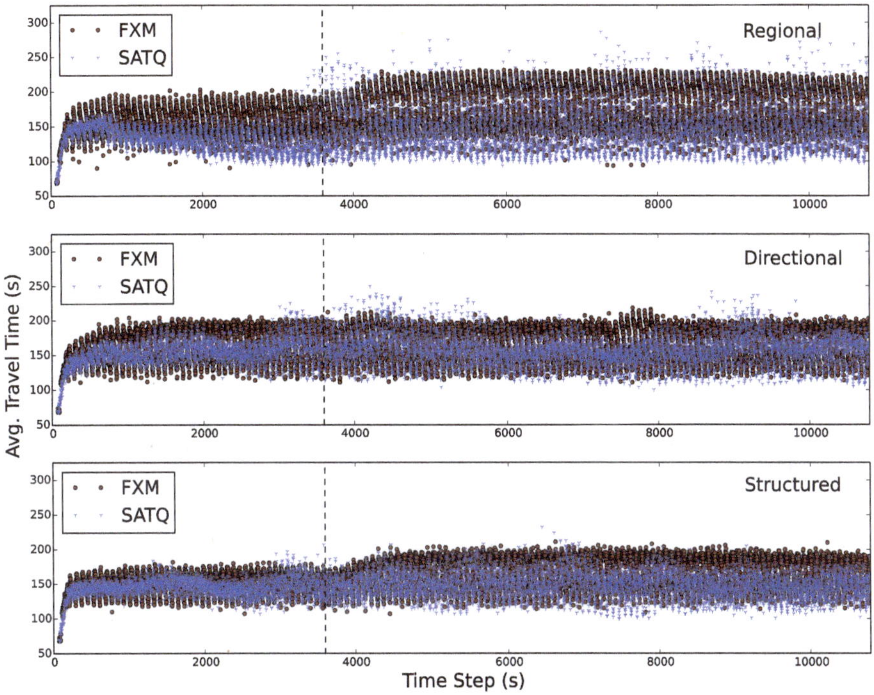

Fig. 7 A comparison of average travel times of vehicles that have completed their journey at each time step under Regional, Directional and Structured conditions. See Fig. 6 for an explanation of the plot format

much easier to manage, versus unpredictable traffic flows which are more difficult to regulate. FXL outperformed SAT and SATQ on the other traffic conditions most likely because longer signal timings fair better in heavy traffic than shorter cycles [8].

Recall that SAT and SATQ have the same cycle length as FXM. Yet in Unstructured traffic, compared to FXM, SAT and SATQ reduced average travel time by over 30 % (see Table 3). This strongly suggests that the shifts in green time, caused by the auction mechanism, resulted in a reduction in average travel time. Also, there is evidence that the savings in travel time is shared by all drivers. Figures 6 and 7 show that the average travel times for all the vehicles that completed their trips fell within a narrow band. In other words, throughout the timespan of each simulation, the vast majority of vehicles experienced a reduction in travel time.

Finally, we turn to the queue measurements shown in Fig. 8. Here we can use FXM to get an idea of what queue lengths would have been like without an auction mechanism. SAT (and SATQ) had increases in queue lengths on the east/west-bound roadways with a corresponding decrease in north/south-bound queues, suggesting that in order to improve travel time, green time was shifted to the north/south-bound lanes. This means that green time was given to the roadway that needed it the most—which is the intended goal of the auction mechanism.

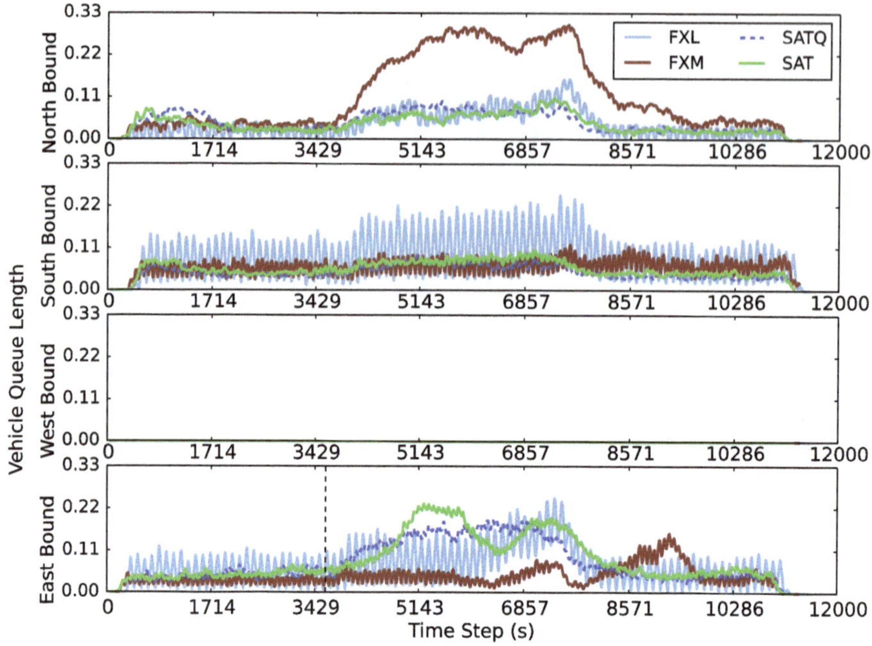

Fig. 8 Percentage of incoming roadway filled with vehicles at a single intersection within the city under the Unstructured condition. Note, for this particular intersection, the west-bound lane did not experience any traffic

Although FXL had the best average travel time under the Regional condition, a closer look revealed that SAT and SATQ behaved in a very similar manner to FXL (see Fig. 5b). This finding is interesting because the same behaviour is not seen under the Directional or Structured conditions. The difference in performance may be due to a lack of change in traffic demand. In Regional, there is very little cross traffic and the heavy flow is in a single direction. Most likely the majority of the green time was given to the heaviest flow; but with such little cross traffic, this did not cause an overall increase in travel time. Bazzan [2] utilised a similar traffic scenario in their work. However, they found the lack of demand in the opposite direction hampered their agent-based traffic controller. In contrast, our controller adapted well.

The Directional (and Structured) conditions probably experienced a similar shift in green time; but having greater cross traffic, this resulted in an overall increase in average travel time. The amount of cross traffic was not enough to influence the auction (as it remain constant), but enough to raise overall average travel times. The simulations under Directional and Structured conditions highlight the critical role the bidding rule plays in green time allocation and as a measure of traffic demand.

7 Conclusion

In conclusion, our work here and in [21] demonstrate the feasibility of a multi-agent auction-based traffic controller that does not require vehicle agents. SATQ reduced travel time by over 30 % under certain traffic conditions compared to a **Fixed** time traffic controller of initially identical cycle length. Traffic congestion costs urban areas billions of Euros in lost time, and vehicle emissions are a major source of air pollution. Auctions have been shown to be able to improve the management of intersection traffic, but thus far only when paired with vehicle agents. Our approach can be deployed in software, without the added cost of upgrading vehicles to include vehicle agents and working with existing transportation infrastructure hardware and control systems.

In this paper, we have outlined the framework for our auction-based traffic controller. Still, there is a need to further investigate the relationship between various bidding rules and traffic demand. Future work will focus on developing additional bidding rules and methods to measure a roadway's level of use, particularly under unpredictable and changeable conditions—the types of situations that currently stymie existing systems. We will also conduct experiments to compare our approach to other adaptive traffic controllers currently in use.

References

1. Bakker, B., Whiteson, S., Kester, L.J. H.M., Groen, F.C.A.: Traffic light control by multiagent reinforcement learning systems. In: Babuska, R., Groen, F.C.A. (eds.) Interactive Collaborative Information Systems. Studies in Computational Intelligence, vol. 281, pp. 475–510. Springer, Berlin/Heidelberg (2010)
2. Bazzan, A.L.C.: A distributed approach for coordination of traffic signal agents. Auton. Agent. Multi-Agent Syst. **10**(2), 131–164 (2005)
3. Bazzan, A.L.C., de Oliveira, D., da Silva, B.C.: Learning in groups of traffic signals. Eng. Appl. Artif. Intell. **23**(4), 560–568 (2010)
4. Cardiovascular Disease and Air Pollution. Technical report, Department of Health, Committee on the Medical Effects of Air Pollutants, February 2006
5. Carlino, D., Boyles, S.D., Stone, P.: Auction-based autonomous intersection management. In: Proceedings of the 16th IEEE Intelligent Transportation Systems Conference (ITSC) (2013)
6. Chen, B., Cheng, H.H.: A review of the applications of agent technology in traffic and transportation systems. IEEE Trans. Intell. Transp. Syst. **11**(2), 485–497 (2010)
7. Dias, M.B., Zlot, R., Kalra, N., Stentz, A.: Market-based multirobot coordination: a survey and analysis. Proc. IEEE **94**(7), 1257–1270 (2006)
8. Dresner, K.M., Stone, P.: Multiagent traffic management: a reservation-based intersection control mechanism. In: Proceedings of the Third International Joint Conference on AAMAS, pp. 530–537. IEEE Computer Society, Washington, DC (2004)
9. Gartner, N.H., Pooran, F.J., Andrews, C.M.: Implementation of the OPAC adaptive control strategy in a traffic signal network. In: IEEE Intelligent Transportation Systems. IEEE, Oakland (2001)
10. Gerkey, B.P., Mataric, M.J.: Sold!: auction methods for multirobot coordination. IEEE Trans. Robot. Autom. **18**(5), 758–768 (2002)

11. Gomm, P., Wengraf, I.: The Car and the Commute, The journey to work in England and Wales. Technical report, The Royal Automobile Club Foundation for Motoring, December 2013
12. Knight, S., Qureshi, M., Cleverly, J., Duvall, L., Gavron, N., Jones, J., Malthouse, K.: Driving Away from Diesel, Reducing Air Pollution from Diesel Vehicles. Technical report, London Assembly, Environment Committee, July 2015
13. Koenig, S., Tovey, C.A., Lagoudakis, M.G., Markakis, E., Kempe, D., Keskinocak, P., Kleywegt, A.J., Meyerson, A., Jain, S.: The power of sequential single-item auctions for agent coordination. In: Proceedings of the 21st National Conference on Artificial Intelligence. AAAI 2006, vol. 2, pp. 1625–1629. AAAI Press, Boston, MA (2006)
14. Krajzewicz, D., Erdmann, J., Behrisch, M., Bieker, L.: Recent development and applications of SUMO - Simulation of Urban MObility. Int. J. Adv. Syst. Meas. **5**(3 & 4), 128–138 (2012)
15. Lagoudakis, M.G., Berhault, M., Koenig, S., Keskinocak, P., Kleywegt, A.J.: Simple auctions with performance guarantees for multi-robot task allocation. In: Proceedings of the IEEE/RSJ International Conference on Intelligent Robots and Systems, pp. 698–705. IEEE, Sendai (2004)
16. Miller, B.G.: Report on Estimation of Mortality Impacts of Particulate Air Pollution in London. Technical report, Institute of Occupational Medicine, June 2010. Report P951-001
17. Mirchandani, P., Wang, F.-Y.: RHODES to intelligent transportation systems. IEEE Intell. Syst. **20**(1), 10–15 (2005)
18. Mladenovic, M., Abbas, M.: A survey of experiences with adaptive traffic control systems in North America. J. Road Traff. Eng. **59**(2), 5–11 (2013)
19. Papageorgiou, M., Diakaki, C., Dinopoulou, V., Kotsialos, A., Wang, Y.: Review of road traffic control strategies. Proc. IEEE **91**(12), 2043–2067 (2003)
20. Parsons, S., Rodriguez-Aguilar, J.A., Klein, M.: Auctions and bidding: a guide for computer scientists. ACM Comput. Surv. **43**(2), 10:1–10:59 (2011)
21. Raphael, J., Maskell, S., Sklar, E.: From goods to traffic: first steps toward an auction-based traffic signal controller. In: 13th International Conference on Practical Applications of Agents and Multi-Agent Systems (PAAMS'15). Springer, Cham (2015)
22. Roess, R.P., Prasas, E., McShane, W.R.: Traffic Engineering: International Edition, 4th edn. Pearson Education, Inc., Upper Saddle River (2009)
23. Sariel, S., Balch, T.R.: Efficient bids on task allocation for multi-robot exploration. In: Sutcliffe, G., Goebel, R. (eds.) FLAIRS Conference, pp. 116–121. AAAI Press, Menlo Park, CA (2006)
24. Schepperle, H., Böhm, K.: Agent-based traffic control using auctions. In: Klusch, M., Hindriks, K.V., Papazoglou, M.P., Sterling, L. (eds.) CIA. Lecture Notes in Computer Science, vol. 4676, pp. 119–133. Springer, Berlin (2007)
25. Schepperle, H., Böhm, K.: Auction-based traffic management: towards effective concurrent utilization of road intersections. In: CEC/EEE, pp. 105–112. IEEE, Washington (2008)
26. Schneider, E., Sklar, E.I., Parsons, S., Tuna Özgelen, A.: Auction-based task allocation for multi-robot teams in dynamic environments. In: Proceedings of the 16th Towards Autonomous Robotic Systems (TAROS) (2015)
27. The economic costs of gridlock. Technical report, Centre for Economic and Business Research, December 2012
28. Transportation Research Board. Highway Capacity Manual. Transportation Research Board, National Research Council, Washington, DC (2000)
29. Treiber, M., Kesting, A.: Traffic Flow Dynamics. Springer, Berlin/Heidelberg (2013)
30. Vasirani, M., Ossowski, S.: A market-inspired approach for intersection management in urban road traffic networks. J. Artif. Intell. Res. **43**, 621–659 (2012)
31. Wang, F.-Y.: Agent-based control for networked traffic management systems. IEEE Intell. Syst. **20**(5), 92–96 (2005)
32. Wellman, M.P., Wurman, P.R.: Market-aware agents for a multiagent world. Robot. Auton. Syst. **24**(3–4), 115–125 (1998)
33. Xu, Y., Zhang, Y., Liu, M.: Multiagent based decentralized traffic light control for large urban transportation system. Math. Probl. Eng. **2014** (2014)
34. Yim, S.H.L., Barrett, S.R.H.: Public health impacts of combustion emissions in the United Kingdom. Environ. Sci. Technol. **46**(8), 4291–4296 (2012)

AgentDrive: Agent-Based Simulator for Intelligent Cars and Its Application for Development of a Lane-Changing Assistant

Martin Schaefer, Michal Čáp, and Jiří Vokřínek

Abstract Intelligent cars represent a promising technology expected to drastically improve safety and efficiency of automobile transportation. In this paper, we introduce an agent-based simulation platform AgentDrive and argue that it can be used to speed up the development and evaluation of new coordination algorithms for intelligent cars. We present the high-level architecture of the simulator and characterize the class of tasks for which is the tool best suited. In addition, we present a case study of AgentDrive being used for development of a lane-changing assistant technology. We describe the developed solution in detail and present the benchmark result, which were obtained using AgentDrive simulator, that demonstrate that coordinated lane changing enables safer and swifter lane changing then the traditional non-coordinated approach.

1 Introduction

In recent years the research efforts in the field of intelligent and connected vehicles has greatly accelerated due to the advancements in the sensing and communication technology and the expectations over the impact of the new technology on the safety and efficiency of automotive transportation. Considering environmental sustainability of autonomous vehicles, a decrease of energy use is expected [5]. The benefits of automated coordination, shared vehicles or efficient driving should take over the expected increase in use of such transportation modes due to popularity.

Many of technologies for autonomous drive support or advanced driver assistance systems (ADAS) have been developed and deployed in production in recent decade. While the development of vehicle-to-vehicle (V2V) communication standards almost reached the point of wide-spread availability in production cars, the cooperative driving technologies enabled by the V2V technology are not quite following the trend. One of the reasons is that the research results in the

M. Schaefer (✉) • M. Čáp • J. Vokřínek
Department of Computer Science, Artificial Intelligence Center, Faculty of Electrical Engineering, Czech Technical University in Prague, Prague, Czech Republic
e-mail: martin.schaefer@fel.cvut.cz; michal.cap@fel.cvut.cz; jiri.vokrinek@fel.cvut.cz

A. Alonso-Betanzos et al. (eds.), *Agent-Based Modeling of Sustainable Behaviors*, Understanding Complex Systems, DOI 10.1007/978-3-319-46331-5_7

domain of coordinated vehicles are difficult to transfer to production due to the complicated testing that often asks for involvement of multiple vehicles and humans subjects. Indeed, the development and testing of coordination mechanisms for intelligent connected vehicles is a complex task. A simulator of connected cars can greatly simplify the development of wide range of technologies, since simulated experiments help to overcome expensive field setups and mitigate safety concern common during conventional on-road testing.

The simulation should model the relevant aspects of the real world. In the context of connected vehicles, the simulation should cover high-fidelity dynamic models of the vehicles, realistic communication model, and good sensor models. In addition, the simulator should be scalable in the number of cars and the size of the world, because some of the effects of the application of the studied algorithms will only emerge from interaction of large number of vehicles.

In this chapter, we introduce the AgentDrive platform, a novel tool aiding the development and evaluation of cooperative driving technologies. It combines a number of existing simulation tools into a single framework and provides a consistent interface for easy integration of the evaluated coordination algorithm. In the following section, we review related simulation tools. In Sect. 3, we describe the AgentDrive platform in more detail and in Sect. 4, we present a case study of using AgentDrive for the development of lane change assistant. Finally, we conclude the discussion in Sect. 5.

2 Related Work

Simulation platforms related to traffic, driving, sensor and network simulations are discussed in this section. We focus on platforms or projects that are open to modification, integration and research.

The most used traffic simulators are SUMO and MATSim. SUMO (Simulation of Urban MObility) [8] is an open-source simulation suite. It is a microscopic traffic simulator enriched by a set of tools related to a preparation of simulation scenarios. Import or generation of both road network and mobility model are the key features of the tool. Similar to SUMO, MATSim [1] is a relevant open-source framework for large-scale agent-based traffic simulations.

Agent-based modelling of traffic allows to model individual heterogeneous drivers, also to study the influence of local interactions to the complex system. The emergent effect of the social behaviour to the overall system is generally discussed in [6]. A review of usage of agent-based for traffic simulation providing motivation and examples of the agent-based approach is in [2].

The main applications of driving simulators are in research of driver behaviour, training of drivers and entertainment. There is a wide range of features addressed by driving simulators. One of the most important features is a realistic human driver experience.

There are high-end simulators in research laboratories of many manufacturers. These high-end simulators are equipped with large motion platforms providing realistic accelerations. The main disadvantage of these simulators is the cost of acquisition and operation.

Professional simulation companies like Vires,[1] CarSim[2] or Oktal[3] provide usually a scale of solutions. High-end simulator with a cabin on a motion platform is the most advanced, realistic and expensive solution. Simpler solution is a static cabin or only seats equipped with visual projection. The basic solution is the simulation software without any special peripheries.

OpenDS[4] is an open-source driving simulator open to modification and integration. OpenDS is based on a game engine jMonkeyEngine (jME).[5] The jME provides the simulator with a 3D visualization and game-like physics. OpenDS uses a possibility of importing various 3D models in jME. All scenes can be prepared in advance and loaded. Individual models can also be added using related XML format settings files or loaded directly in source code. OpenDS also contains so-called Drive Analyser, which allows to replay a drive and to perform further analysis after the simulation run itself. OpenDS is a driving simulator aiming mainly on human-machine interaction community. There are several prepared scenarios and drive tasks in OpenDS, nevertheless creation of a new experiment in OpenDS is possible.

Vehicular communication (V2X) is a big step towards transportation safety. V2X communication brings new possibilities and challenges into the related research areas, but the evaluation of concepts with higher number of real cars is expensive. It motivates the experimentation in simulations. The vehicles with communication units and infrastructure units form VANET—vehicular ad-hoc network. The network simulation can be performed in dedicated simulators (e.g. ns-2, ns-3, OMNeT++, JiST/SWANS), but there is a need of providing a mobility model in addition to network simulation. The common way to provide the mobility model is usage of traffic simulators.

The communicated information by network simulation can have impact on traffic simulation e.g. rerouting in case of a reported accident, so the integrated closed-loop simulation is necessary for realistic scenarios.

TraNS [9] is one of the first simulator with bidirectional binding of mobility model and network simulation. TraNS combines ns-2 with SUMO traffic simulator. It seems to be no longer developed since 2008. iTetris[6] project is a successor of TraNS as it combines SUMO with ns-3. The project aims to be compliant with European standard for V2X technologies.

[1]http://www.vires.com/products.html.

[2]https://www.carsim.com/products/ds/index.php.

[3]http://www.oktal.fr/en/automotive/simulators-operational-needs.

[4]http://opends.eu.

[5]http://jmonkeyengine.org.

[6]http://www.ict-itetris.eu/platform.htm.

Vehicles in Network Simulation (Veins) [13] is also using SUMO as a traffic simulator but for network simulation the OMNet++ is used. The V2X Simulation Runtime Infrastructure (VSimRTI) [14][7] is a comprehensive framework that allows to couple different simulators in simulation. The tests were done with traffic simulators SUMO, VISSIM and network simulators ns-3, OMNeT++, JiST/SWANS.

3 AgentDrive

The purpose of the AgentDrive platform is to support development and testing of coordination mechanisms for intelligent vehicles. It allows to deploy and to test coordination mechanisms in various levels of abstraction. The agent-based nature of the platform also allows to incorporate heterogeneous agents in the scenario. Besides that, the flexible level of simulation detail is essential for scalability in number of agents. Scalability is desired for development of coordination mechanisms for future (semi-)autonomous vehicles in the everyday traffic situations, where the high number of vehicles is needed to observe the emergent behaviours. The traffic simulation and realistic driving simulation features are combined in the AgentDrive platform.

AgentDrive simulation platform architecture is built on the top of two main components: a coordination module and a physics simulation. The architecture is presented in Fig. 1 to provide a reader with a the global picture. The architecture is described in detail in Sect. 3.1.

The heterogeneous multi-agent nature of the domain motivates usage of the agent-based modelling of car drivers. We consider drivers to be agents responsible for controlling cars' physics models. The coordination mechanism is to be designed among these agents. The development of coordination mechanisms is usually an iterative process. The methods are prototyped considering perfect execution in the first stage. Realistic execution with uncertainty and with more detailed models of vehicles is gradually incorporated into the development process. The agents—drivers are independent of the related car's physics model and the physics simulation detail can vary using various physics simulators—from perfect plan execution to realistic drive simulation.

The integration of realistic car physics models into the traffic simulation offers realistic interaction of cars with each other. Since driving simulators focus on believable realistic car physics, these can be integrated as a physics engine for traffic simulation. Various driving simulators exist and can provide the traffic simulation with realistic physics and visualization. The integration of human-in-the loop is natural in driving simulators, not in the traffic simulators. Integration of driving simulator into the traffic simulation brings the human factor into the traffic simulation. Therefore, mechanisms controlling vehicles in traffic simulator need to deal with human controlled vehicles.

[7] http://www.dcaiti.tu-berlin.de/research/simulation/download/.

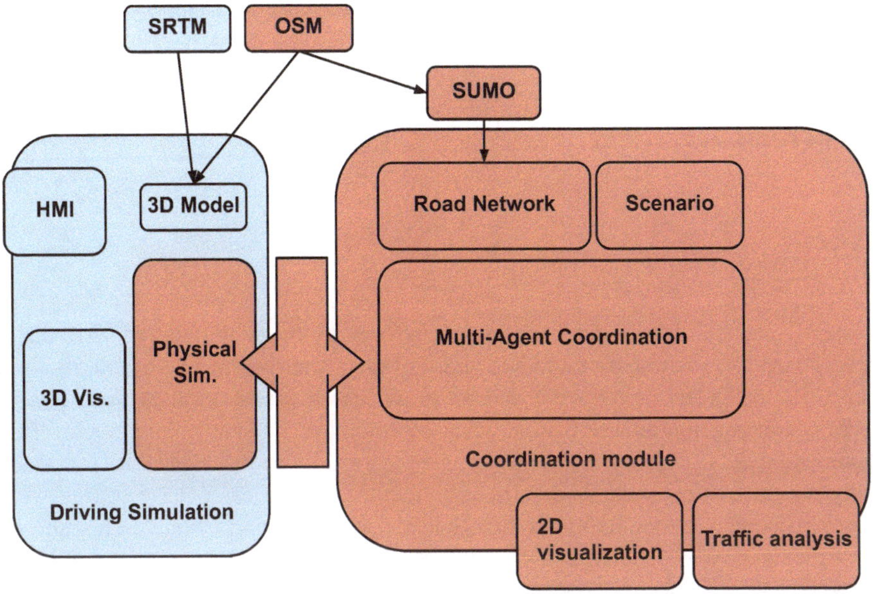

Fig. 1 AgentDrive architecture. Core of the AgentDrive platform is highlighted by *red color*. Coordination is to be implemented in multi-agent coordination module. The environment of the module—road network is generated using SUMO tools from OSM map data. Physics of the vehicles is simulated in external physics simulation. The physics simulation can be implemented by a driving simulator. In case the driving simulation is used, the *blue modules* are additionally available

From driving simulator perspective, the coordination mechanism among traffic can be seen as a module responsible for controlling traffic vehicles, thus it can provide realistic autonomous traffic in driving simulators. Such traffic autonomy is rare in driving simulators available on the market. Autonomous drive can be introduced into the driving simulation by implementing sophisticated sensors to sense and map the environment or by providing structured environment data. Structured road network is the base for most of the traffic simulations. The networks are commonly generated from real map data (e.g. OpenStreetMaps).

The integration of the traffic simulation with driving simulation is beneficial for users of both driving and traffic simulators. Traffic simulation is confronted by challenges related to realistic physics and human-in-the-loop tests. Driving simulation is extended by interaction with autonomous traffic.

Additional technologies are needed to integrate driving and traffic simulators. The driving and traffic simulation are based on the common world model. The world model is based on the real world—real map data. The driving simulation requires physical model of the world, the traffic simulation requires the structured world description—road network. The unified approach to the world model and it's compatible representation in driving and traffic simulation is vital for successful integration.

The domain of the traffic simulators usually focusses on traffic scenarios with statistical measures. On the other hand the driving simulators usually targets on realistic user single-car experience. AgentDrive fills the gap between these two approaches—it is focused on many-car realistic (cooperative) drive in the realistic traffic situations.

3.1 Architecture

AgentDrive simulation platform architecture is built on the top of two main components: a coordination module and a physics simulation. The decomposition of a car into a reasoning agent and a car physics model is to be seen in Fig. 1. The decomposition allows us to validate coordination methods on various level of simulation detail varying from simple mathematical movement model to a realistic physics drive simulation. The subset of modules needed for a basic setting with simplistic physics simulation is highlighted by red color in Fig. 1. This subset represents a core of the AgentDrive platform. The core consists of a coordination module, an interface to connect to physics simulation and a simplistic physics simulator. The coordination module—traffic simulation environment and multi-agent coordination module—is described in Sect. 3.2.

The simulator with idealistic plan execution is useful for development of coordination mechanisms. A developer can see what exactly the coordination mechanism is proposing thanks to the perfect execution of a plan. When the developer is satisfied with functionality in the simulator with perfect execution then the challenge of an imperfect execution in realistic physics is introduced.

The experiments with the realistic physics can be performed after replacement of the simple simulator by an advanced one. Advanced simulator can be used even for a single vehicle or a selected subset of vehicles. The specification of the interface between coordination module and physics simulation is crucial. Proper specification of the interface enables application of the coordination module in various simulators, while the development of the coordination methods within coordination module is independent from the specific physics simulation properties.

A realistic driving simulator can be used as the physics simulation module. The extensions of the platform by using a driving simulator are illustrated in Fig. 1 (blue modules). Driving simulator integration provides the platform with realistic simulation of physics, 3D visualization and possibility of the human-in-the-loop experiments. The integration of the coordination module and a realistic driving simulator in AgentDrive was proposed in [11]. In this work, we aim to introduce the coordination module in more detail to enlighten the scope of the cooperation techniques applicable.

3.2 Coordination Module

Purpose of the coordination module is to embrace individual agents that are responsible of controlling related individual cars. The module implementation is based on a simulation toolkit Alite [7]. The simulation environment is based on a road network model, that is generated from real-world data.

The coordination module is a multi-agent simulation environment itself where agents correspond to individual cars. The multi-agent simulation allows to use decentralized coordination methods as well as the centralized ones. An agent in this context is specified as an entity that can perceive and act on the environment. The sensors and actuators respectively are used to provide interactions with the environment, and eventually with each other. We use a new agent type for each coordination method.

Implementing an agent type in our context means to implement a mechanism to create a plan of the related vehicle(s) with respect to the sensed state of the environment. Our intention is to allow a user of the AgentDrive platform to use a wide spectrum of coordination mechanisms. Free-drive based methods for collision avoidance as well as methods based on structured road network can be implemented. Cooperative methods can be implemented by defining a communication protocol among agents.

The module provides environment representation that the instances of agents operate in. An agent senses the environment including the related car state. The module synchronizes the environment with a physics simulator. The environment is updated by data from physics simulation. The physics simulation (particularly the car models) executes the plans received from the coordination module (i.e., related agents).

3.2.1 Environment Representation

The environment of the coordination module consists of the car representation and structured model of the world. The world is considered to be 2-dimensional in the lowest level. The terrain is not considered, only the road network is taken into account. Obstacles are implicitly defined by border of the roads.

Scenario creation is based on the open real map data. Particularly, it is OpenStreetMaps.[8] The representation of the environment in the coordination module corresponds to representation used in traffic simulators. Actually the scenario specifications are built on the SUMO data model [8]. We use SUMO tools to import XML-based road definition from OSM map data.

[8] http://www.openstreetmap.org/.

The structured road network is used for routing on the level of a city. The path is obtained by graph-based algorithms (e.g., A^*). Also waypoint navigation is based on the road network, the waypoints trace the shape of the roads.

The data model of roads also contains information about junctions and allows the coordination module to handle junctions. The features obtainable from the data model of the road network are enumerated and described in the following list.

1. Shape representation—the data is describing the particular shape of the road infrastructure
2. Routing—data allows to find sequence of road segments to get from segment A to segment B
3. Navigating—data considers lanes in segments (e.g. only right lane allows to turn right in junction)
4. Smooth/drivable—lanes that are structurally connected are also geographically connected (without discontinuities)

Procedure of localization of a car in the road network is extendedly used. The reason is that a position of a car is obtained in coordinates and it is needed to be decided which of road network segments the position corresponds to. It is a consequence of our design where dynamics of a car is simulated in physics module (possibly with human controlled free ride) and an agent reasoning on the road network is in the coordination module. The complexity of the procedure is reduced by usage of kd-trees as a structure to store the road segments. Every vehicle is dynamically localized using the actual position mapped to the road network, so the error introduced by data, physics or human driver are a part of the simulation and provide a realistic noise for the coordination module (of course, it can be avoided by using the accurate data or simplified drive simulation with perfect execution of plans) (Fig. 2).

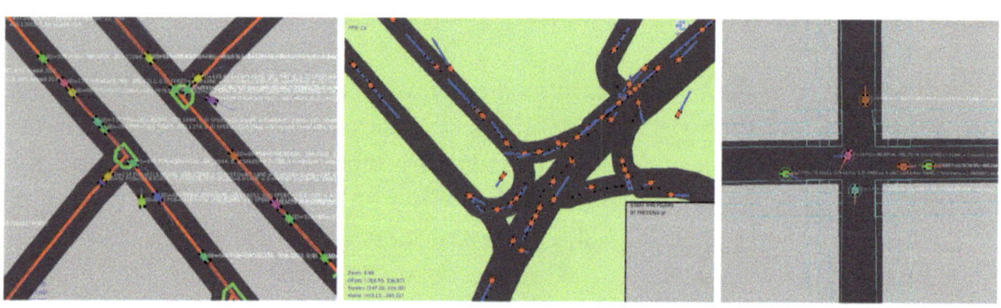

Fig. 2 Visualization layers in AgentDrive. The *left image* shows the graph visualization of the road network. Complex junction (city center of Prague) imported from OSM is in the *middle image*. The *right image* contains *blue areas* representing obstacle areas used by free-ride collision avoidance methods

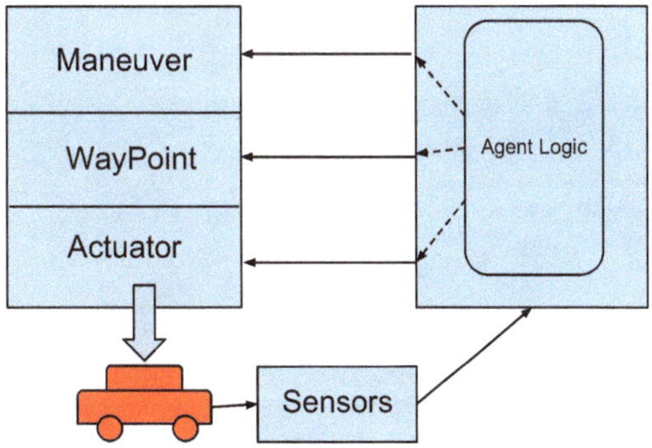

Fig. 3 Three layers architecture of physics simulation interface—maneuver, waypoint and actuator

3.2.2 Agent-Environment Interface Layers

AgentDrive allows to design a coordination mechanisms in various abstraction levels. It is enabled by a layered design of agent-environment interface. The interface is represented by sensors providing agents with environmental data and so-called actuators allowing agents to act on the environment. The sensors and actuators are divided into three layers. The layers allow to implement the simulation in desired level of detail (see Fig. 3). The actuator, waypoint and maneuver layers allow agents to interact on various levels of abstraction. The naming of the layers indicates their principle—the car physics model is controlled by sequence of maneuvers, waypoints or actuator adjustments. The maneuver layer is the highest level. A maneuver is to be performed by several waypoints, waypoint can be reached by a sequence of actuator adjustments. An agent can operate in various level of abstraction and choose the interface level accordingly.

The more abstract layers, i.e., waypoint or maneuver layers, can directly operate on the world model of the corresponding level of abstraction. This is often beneficial to use simplified model of the world that directly corresponds to the level of abstraction of the coordination mechanism. In such a case, there is no need to translate to the lower layers, because the lower level simulation is not to be performed. An example of such approach can be found in the case study in Sect. 4.

Of course, the usage of higher levels of the interface is also desirable if precise low level physics simulator is used. The translation from higher level plan to lower one (e.g. maneuver to waypoints) is usually better to be performed in the physics simulator. There is more information available about the controlled car physics model to implement effective control strategy. The choice of interface level depends on the level of abstraction of agent's reasoning and on the availability and quality of the controllers in the physics simulator.

The proposed architecture is reflected in an already implemented solution. Particular integration of OpenDS in the role of physics simulator was described in detail in [11]. OpenDS is a driving simulator that supports control of traffic cars on the level of waypoints. There is a controller of cars that navigates cars to follow so called follow box. The coordination module then feeds the trajectory of

Fig. 4 Driver's view in OpenDS with integrated HMI for collision avoidance support. The system provides the driver with the visual representation of a proposed plan. The proposed lane to follow is highlighted by *blue*, the speed adjustment is proposed via green or red signal of appropriate intensity. There is an example of an instruction to change lane to left to overtake (*left image*). Second case shows how the *red color* informs the driver of proposal to slow down (*right image*) because of speeding in this case

the follow box with waypoints. Note that OpenDS is a driving simulator, so there is a car controlled by a human-driver available. The manually driven car has a related agent in coordination module as the other traffic cars. In contrast to traffic cars the driven car's execution of plan is in responsibility of the human driver. The plan is presented to the driver by a human-machine interface (see Fig. 4). In fact, it can be seen by driver as a driver assistance system. The system using the coordination module as a back-end can exhibit autonomic features (i.e., self-configuration based on interaction with driver and other vehicles). More about autonomic properties of such car-to-car systems based on V2V cooperation is presented in [12]. Further, the human driver (human-in-the-loop) potentially introduces a huge source of uncertainty and error in the plan execution that need to be addressed in the development of car-to-car coordination schemes.

3.2.3 Visualization

The visualisation of coordination module results is a crucial component for the development. The visualization component is optional and can be easily switch off. The visualization is also easily expandable by new layers. New layers can be registered by for example an agent implementation. So any method used for coordination can provide a visualization specific for the method (e.g. planning graphs, velocity obstacles, potential fields, etc.). Examples can be seen in Fig. 2.

4 Case Study: Priority Based Lane Change Coordination for Road Vehicles

Emergence of vehicle-to-vehicle communication standards enables cooperation of vehicles on a road. One of the risky maneuvers that appears frequently especially on congested city highway is lane changing. It represents a major stress factor for many drivers as it requires good situation awareness and coordination with other drivers. Often, if a vehicle needs to switch to a lane that is congested, one must rely on gracefulness of the drivers at the congested lane to make space for the newly joining car. However, it is often difficult to distinguish whether the car in the congested lane is making space for the joining car or simply failed to follow the flow, leading to a possible misjudgement and consequently to a collision.

Such situations can be greatly simplified with the help of advanced driving assistants utilizing vehicle-to-vehicle communication. In this section we will propose a simple decentralized mechanism for coordination of lane change maneuvers with the neighboring cars. The presented mechanism borrows some of the underlying ideas from a more general multi-robot coordination algorithm called "asynchronous decentralized prioritized planning" [3] and employs them for coordination of lane change maneuvers at multi-lane roads.

In the rest of the section we present an abstract model of multi-lane driving, describe the proposed coordination mechanism, and discuss experimental result from a simulated deployment of the proposed mechanism.

4.1 Modelling Framework

To allow rigorous reasoning about the multi-lane driving scenario, we propose a simplified model of multi-lane driving. First, let us assume that there is an n-lane infinitely long highway. Second, let us assume that there is some nominal speed v_0 at which will the vehicles drive under normal circumstances. Since we are on a highway, the nominal speed is assumed to be $v_0 = 130\,\text{km/h} = 36.11\,\text{m/s}$. This will allow us to offset all speeds by $-v_0$ and create a velocity-offset projection of the highway, that is, for each speed v in the model we define $v' = v - v_0$, where v' will be the offset speed. In result, we get a velocity-offset projection of the highway, which is under normal circumstances (when all cars travel at speed v_0) steady.

To further simplify the reasoning about the problem, we imagine a regular grid over such a velocity-offset projection as illustrated in Fig. 5. The grid has infinite number of columns and n rows. Each cell of the grid has a predefined length and predefined width denoted by c_l and c_w respectively. Further, we assume that each cell can take at most one vehicle. If two vehicles get into the same cell, we consider such a situation a collision.

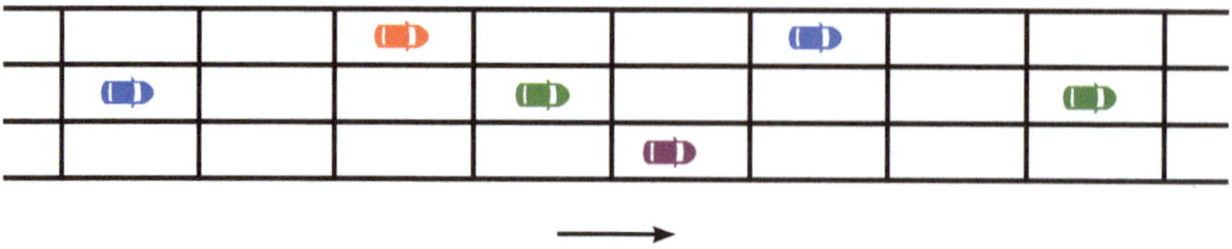

Fig. 5 Multi-lane drive simplified model in velocity-offset grid projection

4.1.1 Maneuvers

We assume that each vehicle is at every moment executing one of the following four maneuvers.

- **Maintain** (abbr. M) The cars maintains its cruise speed v_0 and keeps its current lane. In the grid, it stays in its current cell.
- **Left** (abbr. L) The car executes the lane change maneuver to the left at v_0. In the grid, it switches to the cell that is left from its current cell.
- **Right** (abbr. R) The car executes the lane change maneuver to the right at v_0. In the grid, it switches to the cell that is right from its current cell.
- **Retreat** (abbr. B as in Back) The vehicle slows down temporarily to speed v_B to create more free space to the first car ahead. In the grid, it switches to the cell that is behind its current cell. The velocity to which the car must slow down in order to create free space of one cell can be computed as $v_B = v_0 - \frac{c_l}{\Delta t}$, where c_l is the length of a cell in meters.

We assume that each of these maneuvers takes exactly Δt seconds to execute. However, note that in our model the vehicles need not to execute their maneuvers synchronously.

Now, we can assign concrete values to the parameters of the velocity-offset grid-based model. As we will see later, the vehicles try to maintain two free cells to the next vehicle in front of them. Real drivers are typically advised to keep two second safety gap from the car ahead of them. That is, an advisable safety distance on a highway is $2 \cdot 36.11 = 72.22\,\text{m}$. If we set the length of a cell $c_l = 30\,\text{m}$, then in the typical operation two cars will be travelling with 60–90 m spacing. The width of a cell can be set to a lane width, i.e. we can set the width of a cell $c_w = 3.5\,\text{m}$.

Next, we need to chose the maneuver duration parameter Δt. First consideration is that the maneuver length must be long enough to allow the driver to execute the lane change maneuver. Secondly, the longer the maneuver duration, the smaller the difference between v_0 and v_B. On the other hand, long maneuver durations slow down the reaction time of the system, i.e. one may wait too long for coordination to take place. As a compromise, we set $\Delta t = 5\,\text{s}$, which corresponds to $v_B \doteq 100\,\text{km/h}$.

4.2 Agent Deliberation Algorithms

The lane change assistant feature is realized by deploying a software *agent* on-board of each vehicle. The lane-change agent implicitly senses the lane-change desires of the driver by monitoring the state of the driver's direction indicator. If the driver triggers the left/right direction indicator, the lane-change agent assumes that the driver desires to change the lane to the right/left. The lane-change agents on each vehicle run a specific protocol (decentralized algorithm) that allows them to coordinate their movement and assist drivers when lane change is requested. The lane-change agent checks the safety of the maneuver and only if the maneuver is considered safe, the lane-change agent indicates to the driver that he/she should execute the right/left lane change maneuver. We also assume that the lane-change agent monitors the distance to the closest car ahead and can request the driver to execute a retreat maneuver if a car ahead gets closer than a specified safe distance.

We assume that the vehicle is able to provide to the lane-change agent its current position in a velocity-offset grid. We believe that this assumption is realistic since the row component of the position can be obtained by monitoring the lane changes, and the column position can be obtained either from a GPS device[9] or/and by monitoring the speed of a vehicle. The lane-change agent is able to communicate with the agents of other vehicles in the neighborhood using a short-range radio channel. In the following we will assume that all vehicles are equipped with the lane-change on-board agent and that the drivers will always execute the maneuvers requested by the agent. The interactions of a lane-change agent are depicted in Fig. 6. The internal architecture of lane-change agent follows the Belief-Desire-Intention (BDI) software model of agency [4].

4.2.1 Internal State and Deliberation

Every lane change agent is assigned a unique *id* that is used to determine the priorities of agents. Further, the internal state of the agent consists of beliefs and intentions and the desires come as an input from the driver:

- **Neighborhood view** (beliefs) is a data structure storing the agent's view over the positions and intentions of the lane-change agents in its neighborhood. These information are communicated using UPDATE messages broadcast through the radio channel.
- **Lane change desire** is a signal from the driver exhibiting his/her desire to change the lane to the right or to the left.
- **Lane-change intention** j is a variable that holds the position in the velocity-offset grid that represents the agent's intention to change the lane in near future towards the given target cell. If j = nil, then the agent has no intention to change lane.

[9]Recall that we set the length of a cell to 30 m and thus the accuracy of GPS-based positioning should be sufficient for this purpose.

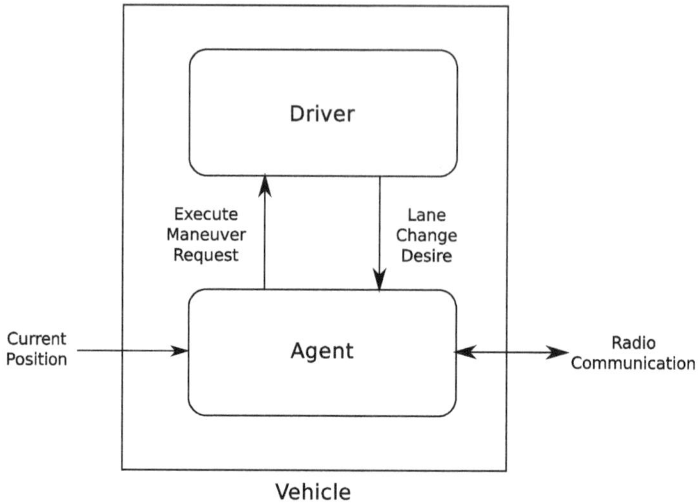

Fig. 6 Diagram of agent's interactions

Once the agent is started, the agent initiates an infinite repeated process that with the period of Δt performs a rule-based deliberation. The output of the deliberation is the maneuver to be performed in the next period. The deliberation is based on several behaviors that are executed in the order of their priorities. These behaviors are (1) Maintain safe distance, (2) Handle lane-change desire, (3) Allow lane-change, and (4) Cruise.

Further, at the beginning of each deliberation, the agent performs an *intention consistency check*, during which the agent checks whether his current lane-change intention is still sensible and at the end of each deliberation, the agent *broadcasts an update message* to the agents in the neighborhood containing the current position of the vehicle and the agent's current lane-change intention. In the following, we describe each of the behaviors in detail.

The highest-priority behavior is responsible for maintaining safe distance from the vehicle ahead. It attempts to maintain at least two empty cells in front of the vehicle. If the car ahead is less then two cells away, the retreat maneuver is returned as the result of this behavior. Otherwise, the deliberation continues with the lower priority behaviors. The communication link required for this behavior is indicated in Fig. 7a.

The handle lane-change desire behavior is the core behavior designed to assist the driver when changing the lanes. This behavior is considered only if the maintain safe distance behaviour generated no maneuver request. If the driver desires to change lane to the right (left), which he/she indicates by triggering the right (left) direction indicator, this behavior determines the target cell t, i.e. the cell that is right (left) from the current cell of the vehicle. Then it applies the following set of rules to assist with the lane change:

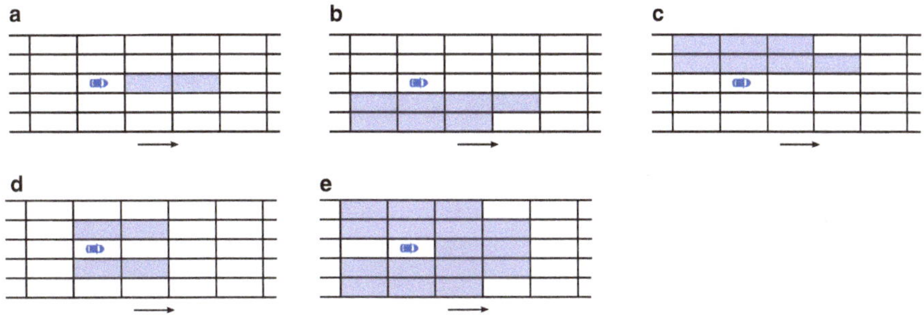

Fig. 7 Communication range requirements of individual behaviours. (**a**) Maintain safe distance; (**b**) handle right lane-change; (**c**) handle left lane-change; (**d**) allow lane-change; (**e**) all combined

1. If the target cell is too close to the closest car ahead (there should be at least two free cells), then the behavior returns retreat maneuver to create more free space between the target cell and the closest car ahead the target cell.
2. If the target cell has enough free space to the closest vehicle ahead, the agent sets the lane change intention variable to the target cell $j = t$ and returns the maintain maneuver. Consequently, the intention will be communicated to the vehicles in the neighborhood via an UPDATE message.
3. If the lane-change variable is already set to the target cell and the neighborhood of the target cell is safe, then the behavior returns the right (left) lane-change maneuver. The neighborhood of the target cell is considered safe only if the target cell is empty, there are two free cells in front of the target cell, one free cell behind the target cell and there is no higher priority agent that indicates an intention to change lanes having a target cell t' such that t' is the same as t,t' is two or less cells ahead of t, or t' is one cell behind t.

The communication link required for this behavior is indicated in Fig. 7b and c for right and lane change respectively.

The allow lane change behavior is a cooperative behavior of vehicle *id* designed to help vehicles from other lanes to change lane in front of the vehicle *id*. This behavior is considered only if the maintain safe distance behavior and handle lane-change desire behavior of vehicle *id* generated no maneuver requests. If a vehicle *id'* from another lane signals a lane-change intention with the target cell being either the cell of vehicle *id* or the cell that is one or two cells in front of the cell of vehicles *id*, then the this behavior returns retreat maneuver in order to make space for the other vehicle. The agent will perform such a cooperative action only if it currently does not have an intention to change lanes or if the other vehicle *id'* is a higher priority vehicle, i.e. if the $id' < id$. The communication link required for this behavior is indicated in Fig. 7d.

The cruise behavior is the default behavior that is applied if none of the higher-priority behaviors (i.e. maintain safe distance, handle lane change desire and allow lane change) returned a maneuver request. Then, the maintain maneuver is returned. The cruise behavior is not directly dependent on any communicated data.

4.2.2 Intention Consistency Check

In order to avoid deadlocks and other undesirable system behavior, the intentions of every BDI agent must be periodically re-considered. Typically, the intentions of a BDI agent must satisfy several rationality properties such as consistency with the agent's desires, the consistency with the agent's beliefs and the rational commitment principle according to which the intentions that cannot be achieved must be dropped [10].

Thus, the lane-changing agent at the beginning of each deliberation reconsiders the current lane-change intention according to the following rules:

1. (Consistency with beliefs) If the agent senses that the vehicle is already in the target cell, then the intention has been achieved and thus the lane-change intention should be dropped, i.e. agent sets $j = \text{nil}$.
2. (Consistency with desires) If the driver does not desire to change lanes anymore or his current desire leads to a different cell than to the target cell of the current lane-change intention, then the intention is not longer desired and as such it should be dropped, i.e. the agent sets $j = \text{nil}$.
3. (Rational commitment) If the target cell of the current intention is no longer accessible from the vehicle's current position (i.e. it is not directly left or right from the current cell), then the intention cannot be achieved and thus it should be dropped, i.e. agent sets $j = \text{nil}$.

4.2.3 Messaging

At the end of each deliberation, the agent broadcasts an $UPDATE(id, p, j)$ message that informs other vehicles in the communication range about its current position and its current lane-change intention. And vice-versa, if an agent receives an $UPDATE(id', p', j')$ messages from another agent, then it updates its neighborhood view data structure with the new position and new intention for agent id'.

The messaging between agents needs to be supported by the specific communication range and the bandwidth. The communication needs for specific maneuvers can be seen on Fig. 7.

If we combine the communication range requirements from all behaviors, we get a so-called *neighborhood of interest* shown in Fig. 7e. Each on-board agent must be able to reliably detect that another car moved away from the vehicle's neighborhood of interest in order to remove the car from the agent's neighborhood view. Therefore we define so-called *exit zone* which comprises the cells directly surrounding the neighborhood of interest. The exit zone is depicted as a red region in Fig. 8. The cars in the exit zone can be safely removed from the agent's neighborhood view. The neighborhood of interest combined with the exit zone constitute a necessary neighborhood, i.e. the region that must be covered by the radio link to ensure correct function of the on-board agent's behaviors.

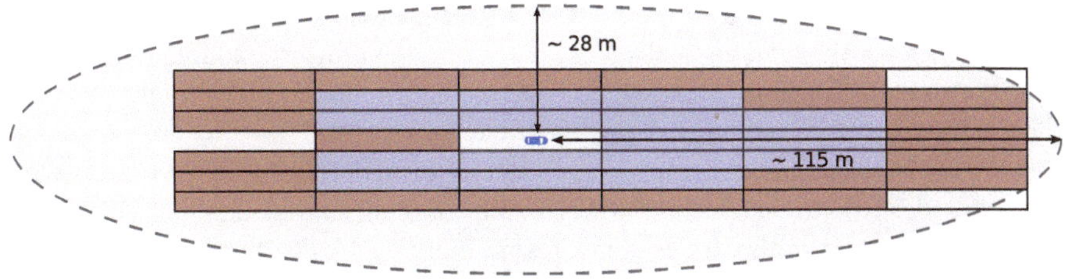

Fig. 8 Required communication range for a cell 30 × 3.5 m. The neighborhood of interest in *blue*, the exit zone in *red*. The *dashed ellipse* represents the needed radio range

The necessary neighborhood depicted in Fig. 8 must be covered by the communication channel of the vehicle. If we set the length of a cell $c_l = 30$ m and the width of a cell $c_w = 3.5$ m, then we can see that we require an ellipse-shaped communication range with radiuses $r_x \doteq 115$ m and $r_y \doteq 28$ m.

The $UPDATE(id, p, j)$ messages are sent by each agent with the period of Δt seconds. The *id* parameter can be represented as a 32-bit integer, the position and intention can be also represented as tuples of 32-bit integers. Thus the UPDATE message can be as short as 160 bits/20 bytes. Further, there is a limited number of vehicles that can fit into the limited communication reach of one vehicle. Suppose that the radio has a communication range radius *r* meters. Then, the radio can cover at most $n(2r/c_l)$ cells, where *n* is the number of lanes and c_l is the length of a cell. Thus, in the worst case situation, there can be $n(2r/c_l)$ cars in the communication reach of a single vehicle. For reasonable values $c_l = 30$ m and $n = 3$, we would require that the shared communication medium is able to transfer $3 \cdot (\frac{2 \cdot 110}{30}) \cdot 20 = 440$ bytes in a some reasonably short time (e.g. 100 ms).

4.3 Experiments

We performed experimental evaluation of the concept of cooperative multi-lane driving supported by an on-board lane-change agent using a simulation. The proposed high level coordination mechanism corresponds to the maneuver layer of AgentDrive (see Sect. 3.2.2). The model of the world is simplified and the simulation is performed on the level of the model. This is setting with the perfect plan execution and without the low-level simulation of physics.

We perform our experiments on a simulated artificial *n*-lane highway that is infinite to the both ends. There are *m* vehicles with their simulated drivers and lane-change agents placed randomly at the grid, such that each of the vehicles has at least two cell behind and two cells in front in its lane (safe-distance condition). The start positions of the vehicles are generated randomly from a 3 km (100 cells) long sub-region of the highway. The maximum numbers of vehicles m_{max} that fit to such a sub-region, i.e. *full saturation* situation, is $n \cdot \lceil l/3 \rceil$, where *l* is the length of the sub-region represented in the velocity-offset grid cells. The time at which the vehicles execute their first maneuver is taken from the uniform random distribution over the interval $(0, 5)$ seconds which ensures that the individual vehicles execute their maneuvers asynchronously.

For our experiments, we use two models of drivers: (i) an *active driver,* who generates desires to change lane towards the random lane at the start. Once the target lane is reached, the driver generates no more desires to change lane, and (ii) a *passive driver* that desires to stay in its current lane.

We have compared two modes of coordination—(i) **Non-cooperative** (NCA) mode, where the agents can communicate, but they do not attempt to coordinate their actions, and (ii) **Cooperative** (CA) mode, where each vehicle has an on-board agent installed and the individual agents can communicate with each other using short-range radio broadcasts.

4.3.1 Results

The results from the experiment are summarized in figures and measured quality metrics for CA mode are also presented in Table 1. The measured quality metrics are (a) the time to reach target lane *t*, i.e. the time between the lane change desire was generated and the time when it finished the lane change, and (b) the distance lost, i.e. the row-coordinate difference between the cell where a driver generates the lane-change desire and the cell where the vehicle reaches its target lane. These quantities were measured for varying values of the following variables: (a) ρ—density of vehicles relative to the full saturation (varied from 40 % to 80 %), (b) *adr*—active drivers ratio (varied from 20 % to 100 %), (c) number of lanes (varied from 3 to 11). The distance unit is a length of a cell and the time unit is a duration of a maneuver.

Each reported result is an average from five different random runs with different start positions of vehicles and intentions of simulated drivers. The collisions among the individual vehicles were monitored and all simulation runs were collision-free.

In Fig. 9 we can see that there is no difference in the distance measure between the two approaches in the heavy setting. However, Fig. 10 shows that CA performs better in more realistic lighter setting (60 % density and 60 % intended drivers).

As for the number of lanes, the CA mode beats the NCA mode in time to reach target line in the heavy setting (see Fig. 11). Even bigger difference can be seen in the lighter setting (see Fig. 12).

We have observed that the active driver ratio has bigger impact on the performance of the CA mode than in the NCA mode. It is due to the implementation of CA—agents prefer their own intentions before cooperation which means that in a region where there are only active drivers, the cooperation is harder to achieve. In the NCA mode, there is no cooperation, so the performance is not significantly affected (Figs. 13 and 14).

Table 1 Cooperative mode results table (ρ—density, *adr*—active drivers rate, 1 mvr = 5 s, 1 $cell$ = 30 m)

#lanes	3		5		7		9		11	
ρ/adr	$t(mvr)$	$d(cells)$	$t(mvr)$	$d(cells)$	$t(mvr)$	$d(cells)$	$t(mvr)$	$d(cells)$	$t(mvr)$	$d(cells)$
40/20	2.85	0.36	3.31	0.63	3.18	0.49	3.41	0.69	3.32	0.63
40/40	2.91	0.45	3.47	0.77	3.56	0.77	3.50	0.68	3.80	0.96
40/60	3.43	0.77	3.60	0.83	4.15	1.25	4.12	1.17	4.25	1.35
40/80	3.52	0.75	4.02	1.08	4.30	1.32	4.56	1.54	4.89	1.75
40/100	3.46	0.70	4.22	1.26	4.42	1.38	5.12	2.01	5.26	2.14
50/20	3.43	0.70	3.76	0.79	4.06	1.03	3.82	0.84	3.73	0.83
50/40	3.71	0.80	4.05	1.02	5.40	2.10	4.83	1.57	4.57	1.46
50/60	4.29	1.18	5.04	1.76	4.95	1.68	5.18	1.89	5.93	2.51
50/80	4.42	1.30	5.18	1.85	5.19	1.93	6.18	2.80	6.45	3.05
50/100	4.94	1.71	5.44	2.08	5.87	2.43	6.87	3.43	6.75	3.36
60/20	4.07	1.07	4.40	1.05	4.24	1.05	4.69	1.38	4.73	1.43
60/40	4.76	1.56	5.29	1.70	5.45	2.00	5.64	2.14	6.54	2.88
60/60	4.99	1.64	6.15	1.44	7.59	3.56	7.42	3.58	8.25	4.52
60/80	5.83	2.20	7.94	3.81	8.13	4.28	8.44	4.23	8.54	4.84
60/100	5.84	2.26	7.98	4.10	9.55	5.70	10.09	6.27	9.87	5.96
70/20	4.65	1.24	5.58	1.83	5.17	1.68	6.60	2.69	5.47	1.88
70/40	5.64	1.92	7.16	3.13	6.60	2.89	7.88	3.88	7.92	3.9
70/60	6.95	2.95	9.01	4.65	8.82	4.81	10.32	6.00	10.67	6.49
70/80	7.63	3.52	10.16	5.72	12.36	7.83	13.81	9.70	13.20	9.01
70/100	7.11	3.23	12.91	8.47	12.67	8.59	14.31	10.29	14.76	10.79
80/20	5.69	1.83	6.65	2.51	7.12	2.80	7.95	3.73	7.44	3.25
80/40	7.81	3.33	9.16	4.54	11.23	6.39	12.72	8.03	11.96	7.13
80/60	9.00	4.43	12.07	7.14	13.83	8.90	15.06	10.00	14.26	9.60
80/80	9.12	4.87	12.79	8.04	17.91	13.03	17.40	12.66	19.74	16.10
80/100	9.26	5.03	19.12	13.98	21.07	16.18	24.38	20.07	22.62	18.45

5 Conclusions

We introduced the AgentDrive platform, its implementation architecture and main components. We described the agent-based coordination module and why it can be used to implement coordination mechanism on various levels of abstraction. In the case study we present a decentralized protocol that allows cooperative lane changing between the vehicles on a simplified model of a highway based on a novel concept of a velocity-offset grid. The safety of the protocol is demonstrated experimentally by monitoring for possible collisions during our extensive experiments. The performance of the algorithm is evaluated and compared with the non-cooperative approach experimentally. We show that the cooperative mode allows the vehicles to change lanes faster compared to the case where the vehicles follow a non-cooperative mode. Presented cars modelling and simulation shows a

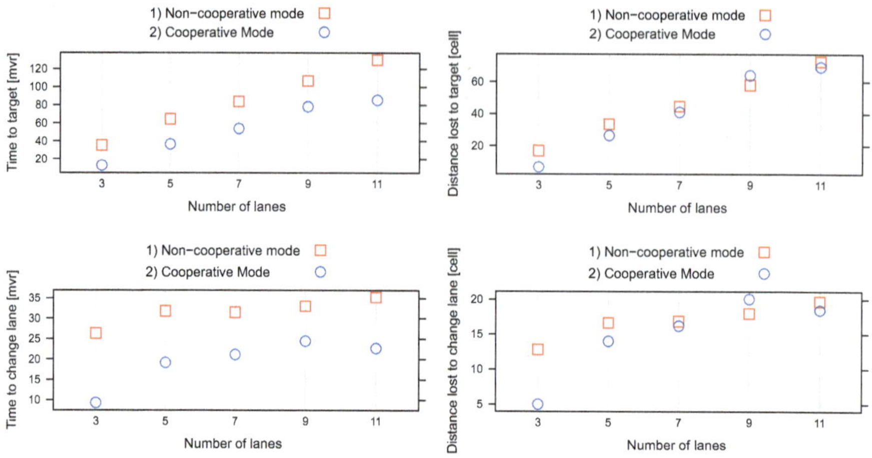

Fig. 9 Quality metrics as a function of *number of lanes* parameter for a highway with 80 % density and 100 % active drivers

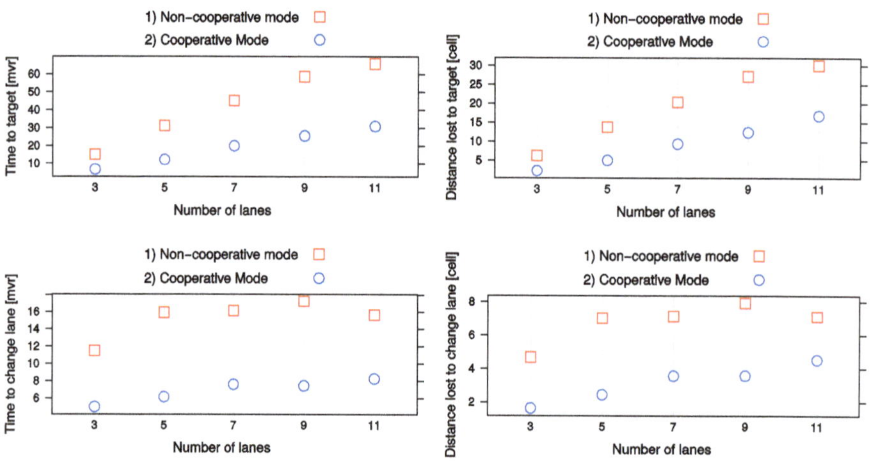

Fig. 10 Quality metrics as a function of *number of lanes* parameter for a highway with 60 % density and 60 % active drivers

promising potential not only for development of intelligent assistant systems, but also as an experimental platform for coordination algorithms and their validation via simulation. Examples of experiments with the AgentDrive platform including videos can be found at the project website.[10]

[10]http://agents.fel.cvut.cz/agentdrive/.

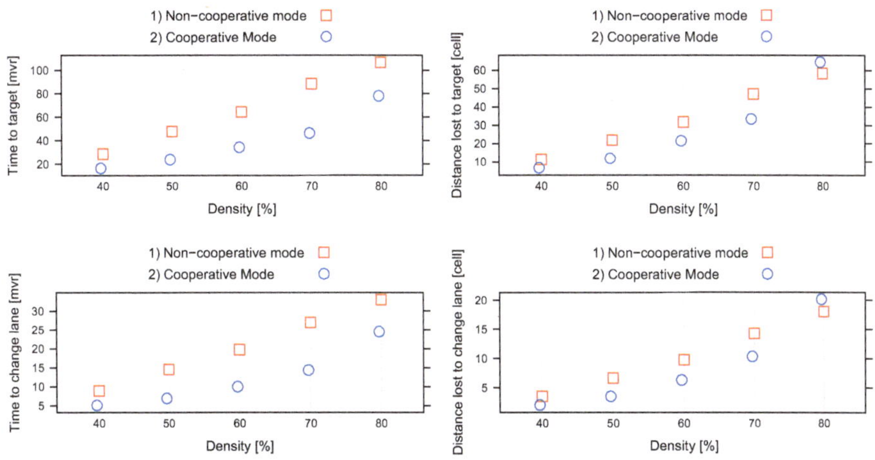

Fig. 11 Quality metrics as a function of *density* parameter for a highway with nine lanes and 100 % active drivers

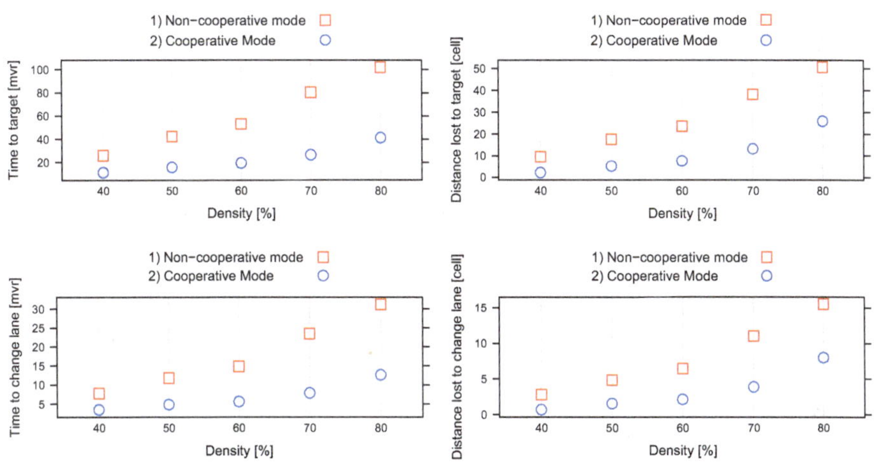

Fig. 12 Quality metrics as a function of *density* parameter for a highway with nine lanes and 40 % active drivers

Acknowledgements This work was supported by the Grant Agency of the Czech Technical University in Prague, grant No. SGS16/235/OHK3/3T/13.

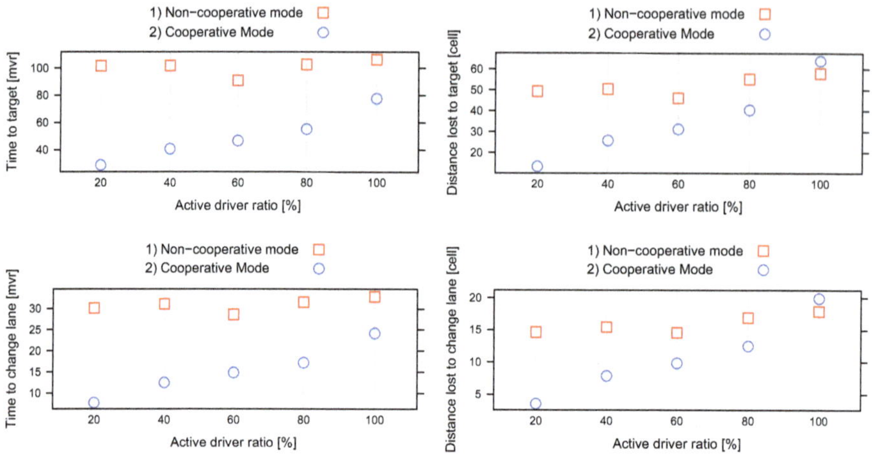

Fig. 13 Quality metrics as a function of *active drivers ratio* parameter for a highway with nine lanes and 80 % density

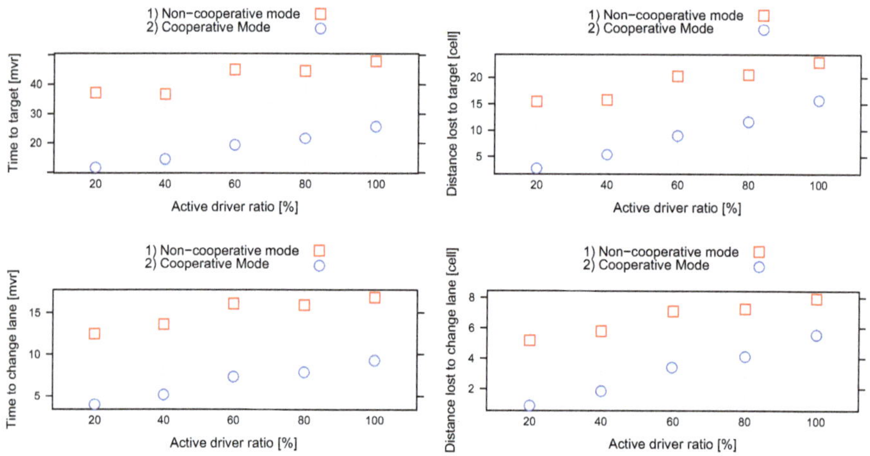

Fig. 14 Quality metrics as a function of *active drivers ratio* parameter for a highway with seven lanes and 60 % density

References

1. Balmer, M., Rieser, M., Meister, K., Charypar, D., Lefebvre, N., Nagel, K., Axhausen, K.: Matsim-t: architecture and simulation times. In: Multi-agent Systems for Traffic and Transportation Engineering, pp. 57–78. Information Science Reference, Hershey (2009)
2. Bazzan, A.L.C., Klügl, F.: A review on agent-based technology for traffic and transportation. Knowl. Eng. Rev. **29**(3), 375–403 (2013). doi:10.1017/S0269888913000118
3. Čáp, M., Novák, P., Selecký, M., Faigl, J., Vokřínek, J.: Asynchronous decentralized prioritized planning for coordination in multi-robot system. In: Intelligent Robots and Systems (IROS) (2013)
4. Georgeff, M.P., Pell, B., Pollack, M.E., Tambe, M., Wooldridge, M.: The belief-desire-intention model of agency. In: Proceedings of the 5th International Workshop on Intelligent Agents V, Agent Theories, Architectures, and Languages, ATAL '98, pp. 1–10. Springer, London (1999). http://dl.acm.org/citation.cfm?id=648205.749450
5. Greenblatt, J.B., Shaheen, S.: Automated vehicles, on-demand mobility, and environmental impacts. Curr. Sustain. Renew. Energy Rep. **2**(3), 74–81 (2015). doi:10.1007/s40518-015-0038-5. http://dx.doi.org/10.1007/s40518-015-0038-5
6. Helbing, D.: Social Self-organization: Agent-Based Simulations and Experiments to Study Emergent Social Behavior. Understanding Complex Systems. Springer, Berlin/Heidelberg (2012)
7. Komenda, A., Vokrinek, J., Cap, M., Pechoucek, M.: Developing multiagent algorithms for tactical missions using simulation. IEEE Intell. Syst. **28**(1), 42–49 (2013). doi:10.1109/MIS.2012.90
8. Krajzewicz, D., Erdmann, J., Behrisch, M., Bieker, L.: Recent development and applications of SUMO - Simulation of Urban MObility. Int. J. Adv. Syst. Meas. **5**(3&4), 128–138 (2012)
9. Piórkowski, M., Raya, M., Lugo, A.L., Papadimitratos, P., Grossglauser, M., Hubaux, J.P.: Trans: realistic joint traffic and network simulator for vanets. SIGMOBILE Mob. Comput. Commun. Rev. **12**(1), 31–33 (2008). doi:10.1145/1374512.1374522
10. Rao, A.S., Georgeff, M.P.: Intentions and rational commitment. In: Proceedings of the First Pacific Rim Conference on Artificial Intelligence (PRICAI-90). Citeseer (1993)
11. Schaefer, M., Vokrinek, J.: Agentdrive: towards an agent-based coordination of intelligent cars. In: Advances in Practical Applications of Agents, Multi-Agent Systems, and Sustainability: The PAAMS Collection: 13th International Conference, PAAMS 2015, Salamanca, 3–4 June 2015, Proceedings, pp. 211–224. Springer International Publishing, Cham (2015)
12. Schaefer, M., Vokrinek, J., Pinotti, D., Tango, F.: Multi-agent traffic simulation for development and validation of autonomic car-to-car systems. In: Autonomic Road Transport Support Systems, pp. 165–180. Springer International Publishing, New York (2016). http://link.springer.com/chapter/10.1007/978-3-319-25808-9_10
13. Sommer, C., German, R., Dressler, F.: Bidirectionally coupled network and road traffic simulation for improved IVC analysis. IEEE Trans. Mob. Comput. **10**(1), 3–15 (2011). doi:10.1109/TMC.2010.133
14. Stanica, R., Chaput, E., Beylot, A.L.: Simulation of vehicular ad-hoc networks: challenges, review of tools and recommendations. Comput. Netw. **55**(14), 3179–3188 (2011). http://dx.doi.org/10.1016/j.comnet.2011.05.007. http://www.sciencedirect.com/science/article/pii/S1389128611001629. Deploying vehicle-2-x communication

City Parking Allocations as a Bundle of Society-Aware Deals

Francesco Barile, Luigi Bove, Claudia Di Napoli, and Silvia Rossi

Abstract It is well recognized that parking in wide and highly populated urban areas is one of the main causes of traffic congestion, air pollution, wasted time, and frustration. In this direction, several initiatives, both from industry and research, are addressing this problem to improve the quality of life for citizens. They usually aim at supporting drivers when selecting parking spaces according to their preferences among competitive alternatives, which are well known in advance to the decision maker, without considering the needs of a city that may impose constraints on the selection process. In this work, an automated software agent negotiation mechanism is used to allocate parking spaces upon drivers requests by trying to accommodate the sometimes conflicting needs coming from the different actors that are involved in a parking allocation process in an urban area. A simulator of the negotiation mechanism is used to globally evaluate the social benefit of the overall allocation problem for a set of parking requests, processed one after another. The obtained results show that negotiation leads in average to an efficient allocations and a better social welfare when compared to baseline cases without negotiation.

1 Introduction

Most parking applications proposed and developed so far are based on Parking Guidance and Information (PGI) systems that provide drivers with dynamic information on parking availability within controlled areas, so a driver selects that

F. Barile
Dipartimento di Matematica e Applicazioni, Università degli Studi di Napoli Federico II, Napoli, Italy
e-mail: francesco.barile@unina.it

C. Di Napoli
Istituto di Calcolo e Reti ad Alte Prestazioni, C.N.R., Napoli, Italy
e-mail: claudia.dinapoli@cnr.it

L. Bove • S. Rossi (✉)
Dipartimento di Ingegneria Elettrica e Tecnologie dell'Informazione, Università degli Studi di Napoli Federico II, Napoli, Italy
e-mail: silvia.rossi@unina.it

A. Alonso-Betanzos et al. (eds.), *Agent-Based Modeling of Sustainable Behaviors*, Understanding Complex Systems, DOI 10.1007/978-3-319-46331-5_8

parking solution more suitable for his/her needs. On the contrary, smart parking applications are designed to automatically make the selection on behalf of the driver, but considering the process only from the point of view of one of the involved actors. For example, they are designed to help drivers in finding a parking space that meets their requirements usually regarding cost, and location, or modeling the allocation as an optimization process from the parking sellers revenue point of view.

Nevertheless, the problem of finding a vacant parking space in densely populated urban areas is a more challenging problem involving different entities: drivers who want to find a vacant parking space that meets their requirements; car parks owners, both public and private, who want to maximize their economic income by selling as many parking spaces as possible; city managers who want to avoid car circulation in specific areas of the city and to have a fair distribution of parking spaces among requesting drivers to limit traffic congestion in the proximity of car parks. Hence, a software system helping drivers to select and reserve parking spaces, should be able to find optimal parking spaces, where the concept of optimality is related not only to the driver looking for a parking space, but more generally to the society as a whole, intended as composed of drivers, city park owners, and city administrators taking care of the public interest of the city. In fact, a city administration may take care of different needs coming from city regulations, in terms of permitted parking areas, traffic congestion, car emission limitations, special events, or breaking events, so it is one of the entities directly involved in the smart management of city parking.

At this purpose, in a previous work [5], we proposed a multi-agent system that uses an automated negotiation approach to accommodate the different needs that have to be fulfilled when selecting parking spaces. Negotiation allows finding an agreement that satisfies different and sometimes conflicting needs of the entities involved in the selection process and to manage the dynamic nature of these needs depending on changeable conditions affecting the decision mechanism of both drivers and city managers. The adopted negotiation mechanism relies on the possibility to selectively propose the different parking alternatives, one by one, in a temporal sequence. Clearly, this temporal sequence has a very strong influence on the driver's final decision about the parking space, allowing to boost the allocation of those parking spaces that represent a viable compromise among conflicting needs. This approach is different from many decision-making solutions in transportation where the competitive alternatives and their characteristics are reasonably well known in advance to the decision maker.

In the proposed approach, a single negotiation occurs for each driver request, independently from one other. In [2], the proposed system was extended to include the possibility to use different sequential negotiations to satisfy a set of parking requests. Here, such approach is modified to serve and evaluate a set of parking requests as a global allocation problem. In this way, we are able to investigate whether negotiation can be used to improve the overall satisfaction of the city as a society composed of drivers, and city administrators, by evaluating the overall distribution of available parking spaces as a factor influencing the total social welfare depending on the utilities of the involved negotiators, and on the number of satisfied requests.

2 Related Works

Most of the previous research on smart parking systems deals with the parking decision problem starting from PGI systems that provide drivers with dynamic information on parking availability [12]. In these cases the burden of making the parking decision is left on the drivers based on the information delivered to them [1]. Shortcomings of PGI systems are due to the competition for parking spaces leading to the possibility of not finding a vacant space. Current practice shows that parking guidance systems usually do not decrease the time spent in searching for a free parking space [1], and from a traffic city authority point of view, PGI systems sometimes may cause even more congestion in the monitored areas [7].

In addition, these systems are designed to increase the probability of finding a parking space, but without considering the possibility to find a better solution for the driver (e.g., a solution nearer to the destination) and a better utilization of parking spaces [7]. Hence, there is a need for a system which can take all the relevant information into consideration to find an optimal parking space automatically, and let the driver concentrate on the road.

The optimal allocation of cars in car parks was studied in [14], where the authors propose a semi-centralized approach for optimizing the parking space allocation and improving the fairness among parking zones by balancing their occupancy-load. In this approach, parking coordinators are used to distributing the optimization allocation problem that is not manageable in a centralized way. In [7, 8], the parking space allocation strategy, is implemented as a global optimization problem, through the use of a Mixed Integer Linear Program. It is based on a user's objective function that combines proximity to destination and parking cost, while ensuring that the overall parking capacity is efficiently utilized. A set of requests are collected in a given time window, and they are processed by a software module producing an overall allocation that tries to optimize an ad hoc function describing both driver-specific requirements, and system-wide objectives. Moreover, in [8], an on-line reservation process has been adopted in the parking management system to improve parking guidance. Also in [19], the authors propose a real-time reservation system based on a reservation function depending on different parking tariff classes. It is used to make decisions regarding the acceptance or rejection of driver requests using simulation, optimization technique, and fuzzy logic with the overall objective of maximizing the revenue of a parking operator.

Some parking management systems adopt a dynamic pricing scheme to generate prices for parking spaces located in different car parks. The parking price reflects the real time parking availability and may encourage the selection of different parking spaces. In [20] the authors presented, as in our case, a smart parking application that tries to find a trade-off between benefits of both drivers and parking providers. To balance the needs of involved parties, they use a dynamic parking price mechanism as an incentive, as also used in [11], for the drivers to balance the convenience and cost in terms of parking price and the convenience in terms of parking distance from the user's destination. Differently from our approach, in [20] all the information is available and the parking selection is obtained as a maximization of drivers' utilities. Dynamic price mechanisms were also explored in [13], where the objective was to set up prices for available parking spaces in such a way to propose the most efficient parking allocation, in terms of social welfare, intended as the total utility value of all agents for which a parking space is allocated.

3 Problem Definition

It is widely recognized that searching for an available parking space in wide urban areas leads to a waste of time and fuel, so increasing traffic congestion and air pollution [16]. The fragmentation of public and private parking providers, each one adopting their own technology to collect occupancy data, makes it difficult to advise drivers of available parking in multiple zones, but more importantly, to help them in making decisions on where to park. Hence, smart parking applications should aim at coordinating individual parking solutions, both private and public, without involving end-users in the fragmentation of parking owners. Individual parking owners should be made aware of the benefits of such a global parking provision by showing them that the coordinated provision of parking solutions still guarantees their individual income and fair competition by better exploiting the parking spaces offered in a city. In this context, the provision of smart parking applications can dramatically improve the city sustainability in terms of air pollution, traffic congestion, and more generally the quality of life of citizens.

In this work, it is assumed that car park owners (that can be both public and private) agree to subscribe to a Coordinated Parking System by making it available a given number of parking spaces managed by a Parking Manager Agent (PMA). The PMA is responsible for the coordinated reselling of parking spaces located in different car parks of a city by distributing vehicles in the managed car parks. Its objective is to sell parking spaces by taking into account the economic needs of car parks owners that try to fill their car parks as much as possible to improve their profit, and, at the same time, the social needs of a city manager that tries to limit traffic congestion mainly in the city centers, and to distribute drivers in different car parks to limit the concentrations of cars in specific or more required city areas. Drivers are modeled as Driver Agents (DAs) interacting with the PMA to submit requests for parking spaces specifying their own preferences on the location of the parking space, and the time they want to park for, trying to pay as little as possible.

The selection of a parking space upon a driver's request is modeled as the result of an agent automated negotiation process occurring among a set of DAs, and the PMA. The allocation of a required parking space occurs if an agreement between the PMA and the DA can be found as the result of an automated negotiation process.

4 Automated Negotiation for Parking Space Allocation

The adopted negotiation mechanism is based on an alternating offers protocol [18] occurring between the PMA and a set of DAs issuing parking space requests to allocate parking spaces trying to satisfy all requests.

A request (`park_req`) is characterized by a geographical location, representing the required destination for the driver, located in an urban area, an hourly cost the driver would prefer to pay for the space, and a time interval the parking space is required for. The urban area is split in concentric rings (named *city sectors*) starting from the city center that are used to localize the considered car parks with respect to the city center.

A car park is characterized by static and dynamic attributes. A static attribute is its location within a ring, i.e. with respect to the city center, while a dynamic attribute is the number of available parking spaces at the time a parking space request is issued. The hourly static price for a parking space is assigned according to the criteria that car parks far from city centers are cheaper, so the adopted metric is to discount the price of a factor depending on the quadratic car park distance from the city center. In fact, it is assumed that car parks located in city centers are more expensive since they are located in the most requested and hence most densely populated city areas.

A negotiation process consists of all negotiations taking place between the PMA and each DA that issued a request over a set time window. Requests are collected and processed by the PMA, one by one according to their arrival order. At the first negotiation iteration, a DA sends a `park_req` to the PMA that replies with an offer (x_j), if any, or with a `decline` message. An offer has the form $x_j = < j, p_{1j}, p_{2j} >$, where j is a selected car park, p_{1j} is the static hourly cost (`static_price`) of an available parking space in the car park, p_{2j} is the travel distance (`travel_dist`) between the car park location and the destination specified in the request. The travel distance is evaluated in terms of the time necessary to reach the destination from the car park location either by walking, for distances within 500 m, or by public or other alternative means of transportation for longer distances.

It should be noted that that the PMA uses a Google Map service to compute the travel distance, but it is assumed that additional city services providing information on specific events that may influence the time necessary to cover such a distance, are made available from a city administration.

The DA replies to an offer with either an `accept` or a `reject` message according to its evaluation of the offer, i.e. if the selected parking space satisfies the driver's requirements. If an agreement is reached with the offer sent at iteration t, the negotiation ends successfully at that iteration, otherwise the offer is rejected and, if $t + 1 \leq t_{MAX}$, the negotiation continues with the PMA proposing another offer until the negotiation deadline t_{MAX} is reached, where t_{MAX} is the number of allowed iterations in the negotiation. The maximum number of iterations is the same as the number of car parks selected by the PMA, and it is not known to the DA. Note that a parking space offered at round t is not considered available at round $t + 1$ to model the possibility to assign a rejected parking space to another driver. So, the negotiation occurs in an incomplete information configuration from the driver agent side, since the information on all the available car parks is known only to the PMA agent. In fact, car parks attribute values may vary in time, so their sharing would require computationally expensive updates. The incomplete information setting leads to the possibility of accepting a sub-optimal agreement.

In the proposed negotiation approach, only the PMA may actually negotiate the values of these parameters, since it may propose a new offer, i.e. a new parking space with different attribute values, at each negotiation iteration. On the contrary, the DA does not issue a counterproposal, since it can only accept or reject the received offer.

In automated agent negotiation, agents are assumed to have preferences, which represent (partial) orderings on outcomes. Agent preferences can be mapped into values of utility, using an utility function that is simply a mapping from a space of outcomes onto utility values, so providing a measure of the satisfaction level associated to a given offer for the agent.

Both the PMA and the DA have their own private multi-dimensional utility functions, allowing them to evaluate the offers in terms of their own preferences, where each dimension relates to an attribute of the specific parking space.

In general, the utility of an offer x_j at round t is evaluated as follows:

$$U_i(x_j) = \begin{cases} 0 & \text{if } t = t_{MAX} \text{ and not}(\textbf{agr}) \\ v_i(x_j) & \text{if } t \leq t_{MAX} \text{ and } \textbf{agr} \end{cases} \quad (1)$$

where, $v_i(x_j)$ is the agent's evaluation function. The evaluation function is a weighted sum of the parking attributes (normalized in the range [0, 1]), assuming the independence of each attribute. The attributes for the PMA and the DA are different, since they have different preferences regarding a parking solution. Of course, an agreement between them is possible if their respective acceptable regions have a not-empty intersection, i.e. a parking space with attribute values acceptable for both of them.

4.1 The PMA Strategy

Upon receiving a DA request, the PMA selects the set of car parks located in the city sectors within a given radius (named *tolerance*) and centered in the driver's specified location. The tolerance value is private to the PMA, and it can be dynamically set by the PMA according to both the location specified by the driver, and the city needs. In fact, if the destination is very close to the city center, or to an area that for the time specified by the driver should be avoided, the considered radius value may increase to allow for more car parks to be selected, so having more alternatives to provide to the driver. In the proposed approach, the PMA evaluates each selected car parks according to its own private evaluation function, and it orders them in a descending order of their utility values. The PMA strategy to issue a counterproposal, i.e. a new offer, is to concede in its utility at each negotiation iteration, by offering one parking space at each iteration, in the same descending evaluation order, so applying a monotonic concession strategy.

The adopted evaluation function models the main objectives of the PMA that are: to incentivize drivers to park outside the city center, in order to limit car circulation in most crowded city areas, and to fill the less occupied car parks to allow for a better distribution of the traffic, and profit.

Hence, the evaluation function is the weighted sum of terms modeling the PMA preferences that are: the car park occupancy ($q_{1,j}$), i.e., the number of free parking spaces at the time the request is processed, and the car park distance from the city center ($q_{2,j}$), calculated as distance of two GPS-located points.

$$v_{PMA}(x_j) = \sum_{k=1}^{2}(\alpha_k * \frac{q_{k,j} - \min(q_{k,j})}{\max(q_{k,j}) - \min(q_{k,j})}) \qquad j \in \{1, \ldots, n\} \tag{2}$$

where, α_k are weights associated to each parameter (with $\sum_{k=1}^{2} \alpha_k = 1$), and n is the number of car parks selected for the request. Both terms of the summation are normalized w.r.t. the minimum ($\min(q_{k,j})$), and the maximum ($\max(q_{k,j})$) values of each parameter among all the selected car parks. The weights are used to take into account the possibility for the PMA to privilege one parameter or the other in its evaluation according to the specific city needs at the moment the request is processed.

4.2 *The DA Model*

Some research works are interested in the driver's behavior in choosing parking space, which is called "parking choice behavior model". Most works have proposed the utility function composed of several factors that affect parking choice [4, 10]. In this work, we consider as relevant attributes of a parking space for the DA only its hourly cost ($p_{1,j}$), and its travel distance from the destination specified by the user ($p_{2,j}$). Upon receiving an offer, the DA evaluates it according to these attributes using an evaluation function given by:

$$v_{DA}(x_j) = 1 - \sum_{k=1}^{2} \beta_k * \frac{p_{k,j} - c_k}{h_k - c_k} \tag{3}$$

where, β_k are weights associated to each parameter (with $\sum_{i=1}^{2} \beta_k = 1$), c_k and h_k are used for normalizing each term of the formula into the set $[0, 1]$. They are respectively the minimum and the maximum values for each k-th parameter that are set at the beginning of the negotiation by the PMA according to the strategy it adopts to select the car parks to offer during the negotiation, in terms of distance and price (i.e. they represent an estimated range of the attribute values used for selecting the set of car parks, and sent to the DA upon a parking request).

The weights are used to model different types of drivers according to their own preferences:

- **business**, i.e. drivers that consider very important the location of the parking space w.r.t. the location they need to reach, also being available to spend more money to get it ($\beta_1 < \beta_2$),
- **tourist**, i.e. drivers that are available to choose a parking space not so close to their preferred destination, provided that they can save money ($\beta_1 > \beta_2$).

The DA strategy is to accept an offer if its utility value is above a *threshold value* (DA_{att}) representing a measure of its attitude to be flexible on its preferred values for the considered parking space attributes. Since the utility function is normalized, its values may range in the interval [0, 1]. It should be noted that at each negotiation iteration, the DA utility varies according to the received offer, so it is not monotonic as the PMA one. This means that by keeping on negotiating does not guarantee the DA to find a better parking space in terms of its utility. In addition, the DA is not aware of the car parks available, so it could end up without reserving any parking space if he keeps on negotiating.

Also the threshold value is used to further characterize types of drivers according to their attitudes:

- **strict**, i.e. drivers who are quite strict on their preferences, i.e. they are characterized by a high threshold value,
- **flexible**, i.e. drivers who are more flexible on their preferences, i.e. they are characterized by a low threshold value.

5 Evaluating the Parking Social Welfare

As we said, we are interested in finding parking solutions for drivers that are the result of a compromise, where possible, among different and sometimes conflicting needs. In other words, we want to analyze how the distribution of parking spaces by means of negotiation affects the well-being of the agent society as a whole.

In order to evaluate the proposed multi-agent parking allocation mechanism, the economic efficiency can be used as a metric often adopted to evaluate multi-agent resource allocation mechanisms in the case of indivisible and not sharable resources, as in the case of parking spaces. This efficiency can be measured in terms of what is called the *social welfare* based on a more global perspective aimed at optimizing an allocation procedure not to increase the individual utility of a single agent, but rather at maximizing social welfare of the whole "society of agents".

5.1 Metrics for Social Welfare Evaluation

Both the DA and the PMA try to maximize their individual utility when negotiating with each other. The designed negotiation mechanism, proposed in [5], aims at finding an agreement between the conflicting needs of a DA and the PMA, leading to an outcome that is a viable compromise.

Here, a set of parking space requests to be globally satisfied are considered, each one processed through a negotiation process. The problem can be assimilated to a distributed indivisible resource allocation problem, where the selection of resources to be allocated for a specific request is carried out through a bilateral negotiation without considering the other requests. Differently from the typical resource allocation problem, here resources (parking spaces) once allocated (e.g., once that a parking reservation has been made), cannot be reallocated to optimize the global allocation results. Moreover, the problem is not to allocate all the resources, but to allocate as much resources as possible that respect the acceptance thresholds. Hence, a resource allocation may fail if the DA utility is low.

Given a set of available resources $\mathscr{R}$ (i.e., parking spaces), and a set of driver agents $\mathscr{DA}$, the overall process is to assign a single resource to each request (if available), in order to best match the DA request and, at the same time, to fulfill as many requests as possible. In resource allocation problems the social welfare is used as a metric to evaluate the efficient allocation of resources [6]. Hence, social welfare, computed for all requests, including the not fulfilled ones, can be used also as a metric to evaluate an efficient allocation of parking spaces.

Given a set $\mathscr{DA}$ of agents requesting a parking space, an optimal allocation of available spaces is the one that maximizes the social welfare of the driver agents, given by the sum of the individual outcomes (i.e. utilities) for all requests, fulfilled or not. Hence, $SW_{DA} = \sum_{i \in \mathscr{DA}} U_i(x_{agr})$, where U_i depends only on the agent i and on the selected parking space (agr). The overall utility of a set $\mathscr{DA}$ corresponds to the sum of the individual utilities. In order to get a global utility value that does not depend on the cardinality of $\mathscr{DA}$, a normalized version of the social welfare is used:

$$SW_{DA} = \frac{\sum_{i \in \mathscr{DA}} U_i(x_{agr})}{|\mathscr{DA}|} \tag{4}$$

Equation (4) accounts for the social welfare of driver agents and for the allocation problem in the sense that a high number of fulfilled requests with a high average utility will result in a high SW_{DA} value. However, in order to evaluate the social benefit of a global parking space allocation, the social welfare should include also the utility of the PMA for each allocated (or not) parking space. In fact, there could be two parking spaces that have the same utility for the DA, but one is more beneficial for the city welfare, i.e., it has a greater utility for the PMA, so being a Pareto optimal solution with respect to the other one. For this reason, a global social welfare (SW_{+}) is obtained, for each negotiation, as the sum of DA and PMA utilities, normalized in [0, 1].

$$SW_{+} = \frac{\sum_{i \in \mathscr{DA}} (U_i(x_{agr}) + U_{PMA}(x_{agr}))/2}{|\mathscr{DA}|} \tag{5}$$

While in multi-agent literature the definition of SW is taken for granted, the economic literature provides different definitions and interpretations. In particular, the adopted definition of social welfare does not account for situations with an imbalanced distribution of utilities among agents. In order to detect these situations the Nash Social Welfare [17] can be used, defined as follows:

$$SW_{*} = \frac{\sum_{i \in \mathscr{DA}} (U_i(x_{agr}) \cdot U_{PMA}(x_{agr}))}{|\mathscr{DA}|} \tag{6}$$

Because of its mathematical structure, increasing SW_{*} gives a balance between increasing the utilitarian welfare of the society, which is the sum of the utilities of agents, and fairness among agents (the balance between DA e PMA utilities).

Equations (4)–(6) are used to evaluate the outcome of negotiation for the parking spaces allocation problem.

6 Negotiation Simulations

We carried out an experimentation to evaluate if the adopted negotiation mechanism leads to a beneficial parking space allocation also when a set of requests is served within the same negotiation, i.e. to assess its outcome when considering the global allocation of parking spaces with respect to the social welfare of a city.

The use case considered for carrying out the experimentation is based on a real-word scenario referred to the city of Naples. Parking requests are issued, in a time window, by drivers that specify as their destination the city center that is a pedestrian area where car circulation is forbidden. As described earlier, the city is split in city sectors centered around the city center.

We carried out experiments with the following settings:

- the number of available parking spaces is 100 distributed in 20 car parks with different capacity;
- drivers' destinations are randomly generated in the city center, i.e. in the first city sector with a radius of 500 m;
- the selected car parks are distributed in all remaining city sectors except for the first one (i.e., the city center);
- the number of considered sets of requests is 6 composed respectively of 25, 50, 75, 100, 125, 150 parking requests;
- four classes of drivers are considered: *Flexible business*, *Strict business*, *Flexible tourist*, and *Strict tourist*, each one characterized by a threshold value (DA_{att}), and the weights of the utility function attributes (β_k), as reported in Table 1;
- the PMA weights on each utility function attribute α_k are the same, i.e. $\alpha_1 = \alpha_2 = 0.5$, since only one type of PMA is modeled.

Table 1 Classes of users

	DA_{att}	β_1	β_2
Flexible business	0.4	0.3	0.7
Strict business	0.6	0.3	0.7
Flexible tourist	0.4	0.7	0.3
Strict tourist	0.6	0.7	0.3

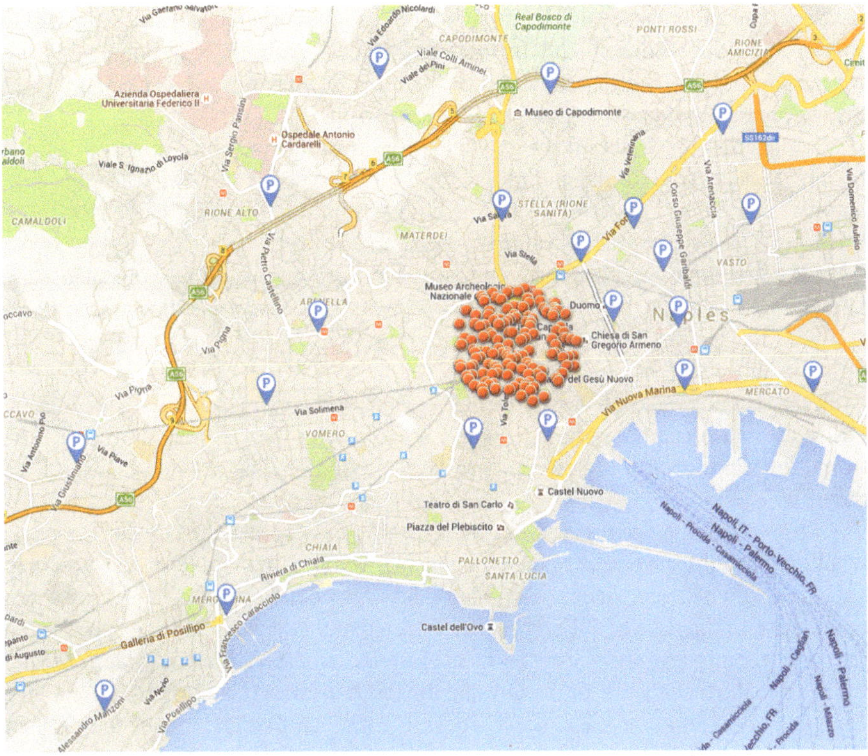

Fig. 1 A snapshot of the considered parking requests and car parks distributions

Positions of the parking requests (red dots) and the selected 20 car parks for the experiments are shown in Fig. 1.

The JADE framework [3] is used to implement the DAs and the PMA relying on its messaging primitives to simulate the adopted negotiation protocol. JADE is an open source software framework for developing applications that implements agent and multi-agent systems. It is a Java based agent development environment providing libraries designed to support communication between agents in compliance with Foundation for Intelligent Physical Agents (FIPA) specifications. The PMA is enveloped in an application server, more specifically Apache Web Server extended with Tomcat, and it is able to communicate with external services and information sources, such as:

- Google Map Server [15] to retrieve walking distance and travel time from a selected car park to the user's destination location,
- the Car Park Database to retrieve information on the available car parks,
- City Manager facilities to retrieve information regarding roads accessibility-related information.

The Car Park Database is implemented using PostgreSQL, an object-relational database management system, and PostGIS an open source software providing support for geographic objects to the PostgreSQL database. It is populated with car park information retrieved with the OpenStreetMap application [9], that is used also to obtain city maps for the user to interact with.

The negotiation outcome is evaluated for different settings in terms of the obtained social welfare depending on the utilities of both the PMA and the DA, and it is compared against five baseline cases without negotiation named *DA*-best, *PMA*-best, SW_{DA}-best, SW_{+}-best, SW_{*}-best.

In the first case, *DA*-best, the availability and locations of all parking spaces are known to the DA (i.e., there is a complete knowledge) that selects the parking space (x_i) with the highest utility ($x_i = argmax(U_{DA}(x_j)), \forall j$), and it reserves it if this utility is above its threshold ($U_{DA}(x_i) > DA_{att}$).

In the second case, *PMA*-best, the PMA selects the parking space with the highest utility ($x_i = argmax(U_{PMA}(x_j)), \forall j$) and it offers it to the DA that accepts it if its own utility for that offer is above its threshold ($U_{DA}(x_i) > DA_{att}$), otherwise it rejects the offer.

In the remaining three cases the social welfare is evaluated to detect the cases resulting in an improvement of the global social welfare. In these tests, the threshold value is not considered, since the allocation is assumed to occur once a parking space is selected. In the case of SW_{DA}-best the selected and allocated parking space is the one with the highest utility for the DA. In the case of SW_{+}-best it is the one that maximizes the mean value of the DA and the PMA utilities (i.e., $argmax((U_{DA}(x_j) + U_{PMA}(x_j)/2), \forall j$). In the case of SW_{*}-best it is the one that maximizes the product of the DA and the PMA utilities (i.e., $argmax(U_{DA}(x_j) * U_{PMA}(x_j)), \forall j$).

The requests are processed one by one, and if a request is satisfied the corresponding assigned parking space is reserved and it is not available for the other requests. If a request is not satisfied it is discarded and not processed anymore. We recall that the deadline of a negotiation (t_{MAX}) may vary for each requests according to the number car parks with available spaces for that request.

6.1 Experimental Results

The overall DAs and PMA average utility values (U_{DA} and U_{PMA}), and the percentage of successful allocations (*%all.*), normalized with respect to the number of requests, obtained by simulating 25, 50, 75, 100, 125 and 150 requests, are reported in Table 2. Such utilities are evaluated for the negotiation case (Negotiation), and

Table 2 DAs and PMA utilities and allocation percentage in different settings

	Negotiation	DA-best	PMA-best	SW_+-best	SW_{DA}-best	SW_*-best
25 req.						
U_{DA}	0.50±0.02	0.39±0.02	0.27±0.04	0.46±0.15	0.52±0.14	0.45±0.14
U_{PMA}	0.63±0.02	0.11±0.01	0.38±0.05	0.33±0.13	0.27±0.18	0.34±0.12
%all.	88±3	64±3	50±8	100±0	100±0	100±0
50 req.						
U_{DA}	0.45±0.01	0.33±0.01	0.25±0.03	0.43±0.16	0.48±0.16	0.40±0.15
U_{PMA}	0.54±0.01	0.06±0	0.31±0.03	0.26±0.13	0.20±0.17	0.28±0.11
%all.	82±2	55±2	45±6	100±0	100±0	100±0
75 req.						
U_{DA}	0.43±0.01	0.30±0.01	0.22±0.02	0.42±0.19	0.45±0.17	0.38±0.16
U_{PMA}	0.47±0.01	0.05±0	0.28±0.02	0.21±0.13	0.18±0.15	0.22±0.12
%all.	80±1	50±2	42±6	100±0	100±0	100±0
100 req.						
U_{DA}	0.41±0.01	0.26±0.01	0.21±0.02	0.40±0.19	0.42±0.17	0.38±0.17
U_{PMA}	0.55±0.15	0.048±0	0.25±0.02	0.18±0.13	0.18±0.13	0.18±0.13
%all.	75±1	44±1	40±4	100±0	100±0	100±0
125 req.						
U_{DA}	0.39±0	0.23±0.01	0.20±0.02	0.40±0.19	0.42±0.17	0.38±0.17
U_{PMA}	0.35±0.01	0.12±0.10	0.23±0.02	0.18±0.13	0.04±0	0.17±0.13
%all.	70±1	40±1	38±4	80±0	80±0	80±0
150 req.						
U_{DA}	0.35±0	0.21±0	0.19±0.01	0.40±0.19	0.42±0.17	0.37±0.17
U_{PMA}	0.30±0	0.12±0.10	0.21±0.01	0.18±0.13	0.04±0	0.17±0.13
%all.	63±0	37±0	37±3	67±0	67±0	67±0

for the five baseline cases without negotiation, i.e., when the best parking space respectively for the DA (DA-best), and the PMA (PMA-best), and according to the maximum value of SW_{DA} (SW_{DA}-best), SW_+ (SW_+-best), and SW_* (SW_*-best) are selected.

The results in Table 2 show that with negotiation a better parking space allocation is obtained (88 %), even though the percentage of allocation decreases while increasing the number of requests. In fact, since the number of available parking spaces is the same for all the configurations, it is easier to accommodate fewer requests. In the case of 100 requests (and 100 available parking spaces) the negotiation process is able to accommodate the 75 % of the requests.

The individual overall utilities U_{DA}, and U_{PMA}, in the negotiation case, are better with respect to parking spaces selected in the DA-best and PMA-best baseline cases respectively. However, note that in the last two cases there is a smaller allocation percentage. Also the cases SW_+-best and SW_*-best do not correspond to better individual utilities, even if they have an allocation percentage of 100 % (since there is no acceptance threshold). The only exception is given by the U_{DA} values of the

SW_{DA}-best that are higher than the ones obtained in the negotiation case. However, in such configurations the corresponding values for U_{PMA} are much lower.

In order to evaluate the aggregate utility values for all the actors involved in the negotiation, we have to consider the social welfare metrics as defined in Sect. 5. In Table 3, the social welfare values (SW_{DA}, SW_{+}, and SW_{*}), evaluated respectively with Eqs. (4)–(6), are reported along with their corresponding maximum values ($max(SW_{DA})$, $max(SW_{+})$, and $max(SW_{*})$) in the cases of different numbers of requests and different baseline cases. It should be noted that the definition of Equation (4) is exactly the overall DAs utility ($SW_{DA} = U_{DA}$).

With respect to the SW_{DA} evaluation, as already highlighted in Table 2, a better overall utility for the DAs is obtained in the SW_{DA}-best case with respect to the negotiation case. Trends of the SW_{DA} values, varying the number of requests, are shown in Fig. 2 for the negotiation, PMA-best, DA-best and SW_{DA}-best cases. For all other cases, the SW_{DA} values obtained with negotiation are always better than the others, while the ones obtained with the SW_{DA}-best can be considered comparable. In particular, in the case the number of requests is less than 100, the values obtained by SW_{DA}-best are slightly greater than the ones obtained with the negotiation since the SW_{DA}-best strategy satisfies every requests (e.g., there are no zero utilities). Instead, when the number of requests increases, the negotiation strategy has still room to accommodate more requests than the SW_{DA}-best strategy. This result is due to the fact that for the negotiation case, from the beginning, sub-optimal allocations of parking spaces are obtained (i.e., with lower U_{DA} values), leaving more chances to find, for the future requests, allocations with good DA utilities. So, when the number of requests exceeds the number of available parking spaces, it is better to have sub-optimal allocations for all the agents than trying to maximize their utilities, because, at the end the collective welfare is improved. In order to show this behavior, in Fig. 3 the SW_{DA} values for a single run with 150 requests in the case of negotiation and in the case of SW_{DA}-best are plotted.

When including the PMA utility in the social welfare (SW_{+}), the values obtained with the negotiation are greater than all the baseline cases (see Fig. 4). In this case the gap of the values obtained when negotiating with respect the SW_{+}-best case is more evident. In addition, these values are now closer to their respective optimal values ($max(SW_{+})$), i.e., the negotiation leads to near optimal global outcomes (see Table 3), with respect to the SW_{DA} cases.

Finally, negotiation allows for a better balancing of utilities among the involved agents, as showed by the values reported for SW_{*} in Fig. 5 and in Table 3 summarizing the average values for the evaluated welfare measures in all the considered cases. In particular, differently from the previous cases, only the negotiation and the *PMA*-best cases are able to achieve such balancing, while for the other cases the average social welfare is extremely low.

The last set of experiments evaluates the outcomes of negotiations for each class of drivers separately, in a configuration with 100 parking requests and 100 available parking spaces. The results in Table 4 report the percentage of allocated parking spaces *%all*, the DA and PMA utilities U_{DA} U_{PMA}, the normalized social welfare SW_{DA}, the global social welfare SW_{+}, and the Nash Social Welfare SW_{*}. In the

Table 3 SW_{DA}, SW_{+}, and SW_{*} values in different settings

	Negotiation	DA-best	PMA-best	SW_{+}-best	SW_{DA}-best	SW_{*}-best
25 req.						
SW_{DA}	0.50±0.02	0.39±0.02	0.28±0.05	0.46±0.01	0.52±0.01	0.45±0.01
$max(SW_{DA})$	0.56	0.43	0.40	0.48	0.53	0.47
SW_{+}	0.57±0.02	0.25±0.01	0.32±0.05	0.39±0.01	0.39±0.01	0.39±0.01
$max(SW_{+})$	0.63	0.27	0.45	0.41	0.40	0.40
SW_{*}	0.37±0.02	0.06±0.01	0.22±0.03	0.14±0.01	0.12±0.01	0.14±0.01
$max(SW_{*})$	0.41	0.07	0.30	0.15	0.13	0.15
50 req.						
SW_{DA}	0.45±0.01	0.33±0.01	0.25±0.03	0.43±0.01	0.48±0.01	0.40±0.01
$max(SW_{DA})$	0.48	0.36	0.30	0.44	0.49	0.41
SW_{+}	0.49±0.01	0.20±0.01	0.28±0.03	0.34±0.01	0.34±0.01	0.34±0.01
$max(SW_{+})$	0.52	0.22	0.33	0.35	0.35	0.34
SW_{*}	0.30±0.01	0.04±0.01	0.18±0.02	0.10±0.01	0.08±0.01	0.11±0.01
$max(SW_{*})$	0.31	0.04	0.21	0.11	0.09	0.11
75 req.						
SW_{DA}	0.43±0.01	0.30±0.01	0.22±0.02	0.42±0.01	0.45±0.01	0.38±0.01
$max(SW_{DA})$	0.45	0.33	0.29	0.43	0.46	0.39
SW_{+}	0.46±0.01	0.18±0.01	0.25±0.03	0.32±0.01	0.32±0.01	0.30±0.01
$max(SW_{+})$	0.47	0.20	0.32	0.32	0.32	0.31
SW_{*}	0.26±0.01	0.03±0.02	0.15±0.01	0.08±0.01	0.07±0.01	0.08±0.01
$max(SW_{*})$	0.27	0.03	0.19	0.08	0.07	0.09
100 req.						
SW_{DA}	0.41±0.01	0.26±0.01	0.21±0.01	0.40±0.01	0.42±0.01	0.38±0.01
$max(SW_{DA})$	0.42	0.28	0.28	0.41	0.43	0.39
SW_{+}	0.41±0.01	0.16±0.01	0.24±0.02	0.29±0.01	0.30±0.01	0.28±0.01
$max(SW_{+})$	0.42	0.17	0.29	0.29	0.30	0.28
SW_{*}	0.23±0.01	0.03±0.03	0.14±0.01	0.06±0.01	0.06±0.01	0.07±0.01
$max(SW_{*})$	0.23	0.03	0.17	0.06	0.06	0.07
125 req.						
SW_{DA}	0.39±0.01	0.23±0.01	0.20±0.02	0.32±0.01	0.34±0.01	0.30±0.01
$max(SW_{DA})$	0.40	0.25	0.27	0.33	0.34	0.31
SW_{+}	0.37±0.01	0.14±0.01	0.22±0.02	0.23±0.01	0.24±0.01	0.22±0.01
$max(SW_{+})$	0.38	0.15	0.27	0.24	0.24	0.23
SW_{*}	0.19±0.01	0.03±0.04	0.13±0.01	0.05±0.01	0.05±0.01	0.05±0.01
$max(SW_{*})$	0.20	0.03	0.16	0.05	0.05	0.05
150 req.						
SW_{DA}	0.35±0.01	0.22±0.01	0.20±0.02	0.27±0.01	0.28±0.01	0.25±0.01
$max(SW_{DA})$	0.36	0.23	0.26	0.27	0.28	0.26
SW_{+}	0.33±0.01	0.13±0.01	0.21±0.02	0.19±0.01	0.20±0.01	0.19±0.01
$max(SW_{+})$	0.33	0.14	0.26	0.20	0.20	0.19
SW_{*}	0.17±0.01	0.02±0.05	0.12±0.01	0.04±0.01	0.04±0.01	0.04±0.01
$max(SW_{*})$	0.17	0.03	0.15	0.04	0.04	0.04

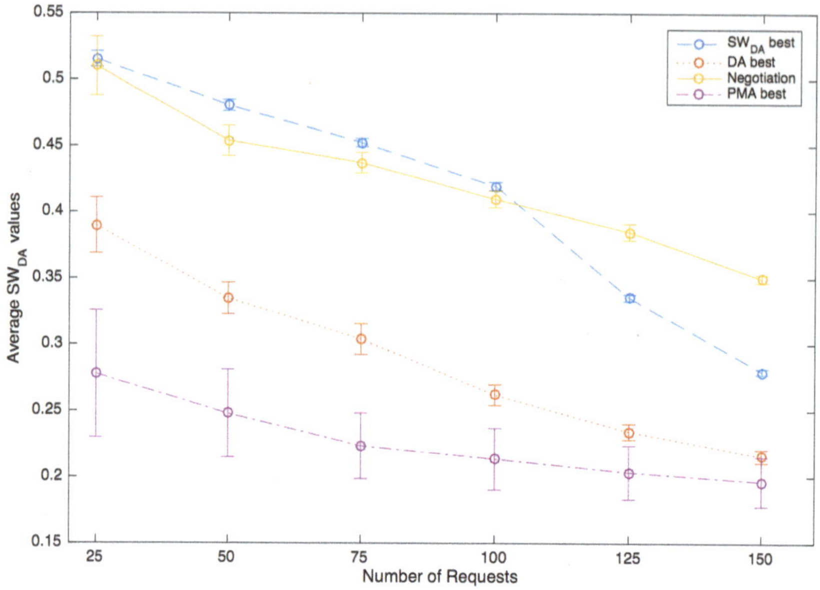

Fig. 2 Average SW_{DA} values for the negotiation and the baseline cases varying the number of requests

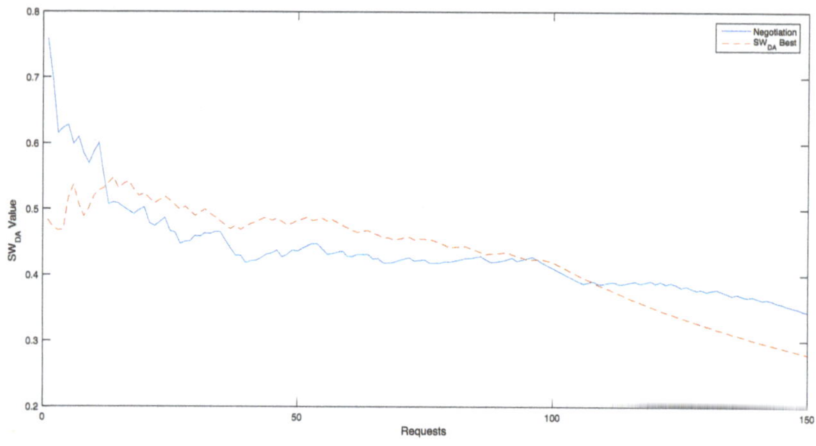

Fig. 3 SW_{DA} values for a single run of 150 queries

considered settings, the percentage of allocated parking spaces is higher for drivers belonging to the Flexible-Business class since they are, on the whole, the ones more available to accept parking spaces with a higher price, while it is very low for drivers belonging to the Strict-Tourist class, according to the weights set for the different parameters of the DA utility function. These results suggest that when requests come from drivers uniformly distributed in different classes, as in the experiments

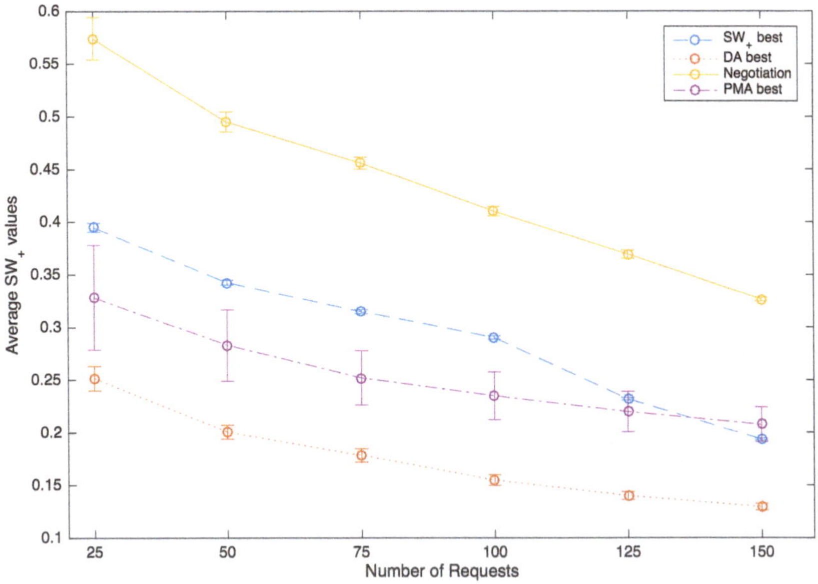

Fig. 4 Average SW_+ values for the negotiation and the baseline cases varying the number of requests

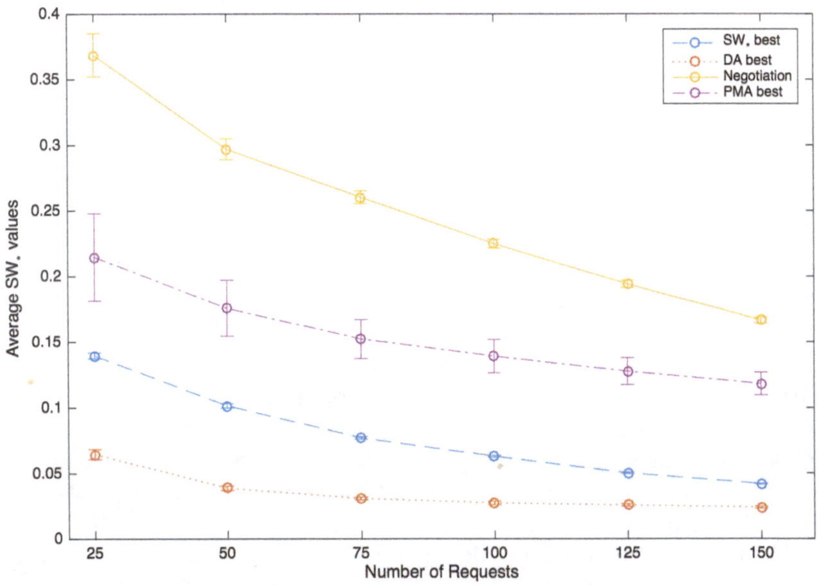

Fig. 5 Average SW_* values for the negotiation and the baseline cases varying the number of requests

Table 4 Negotiation outcomes for each driver class with 100 requests

	%all	U_{DA}	U_{PMA}	SW_{DA}	SW_{+}	SW_{*}
Flexible-business	95±1	0.52±0.01	0.45±0.01	0.52±0.01	0.48±0.01	0.24±0.01
Flexible-tourist	45±1	0.22±0.01	0.27±0.01	0.22±0.01	0.24±0.01	0.13±0.01
Strict-business	53±1	0.33±0.01	0.22±0.01	0.33±0.01	0.28±0.01	0.14±0.01
Strict-tourist	5±1	0.04±0.01	0.04±0.01	0.04±0.01	0.04±0.01	0.03±0.01

reported earlier, the more flexible drivers give room to the less flexible ones for an allocation. That means that by increasing the number of parking requests, the variability of drivers allows to improve the percentage of allocated parking spaces when not flexible drivers are considered.

7 Conclusions

Usually, smart parking applications provide drivers with dynamic information on parking availability within controlled areas, and direct them to vacant parking spaces by taking into account their preferences that, as reported in literature, mainly regard parking cost and location. However, finding a parking space in densely populated urban areas cannot depend only on drivers' needs, but also on needs coming from parking owners, trying to maximizing their profit, and city managers trying to consider the global benefits for the city limiting traffic congestion, or car circulation in specific city areas (e.g., pedestrian areas, or car prohibited areas for special events).

The adoption of software agent automated negotiation allows to model the problem of allocating parking spaces as the process of finding an agreement among different and sometimes conflicting needs. In this way, it is possible to consider not only the usual drivers' needs, in terms of parking cost and location, and car parks owners' needs, in terms of revenues, but also the well-being of the city that is impacted by traffic congestion, overbooking of specific and better located car parks, or unforeseen traffic disruptions. Negotiation occurs among Driver Agents acting on behalf of drivers requesting to reserve a parking space that satisfies their own criteria, and a Parking Manager Agent acting on behalf of a city authority that tries to allocate parking spaces belonging to different car parks by accommodating city needs, so leading to a parking allocation that is beneficial from a city point of view.

In order to provide a measure of the social benefit of the overall allocation problem that takes into account different needs, the negotiation is evaluated in terms of the obtained social welfare of the global outcome of the negotiation for a set of received parking requests and for different sets of users characterized by different profiles. Different types of social welfare were evaluated by taking into account: the distribution of parking spaces with respect to only drivers needs, the same distribution with respect to both drivers and city manager needs, and finally the same distribution with respect to how the drivers and city needs are balanced. The results of the experiments carried out confirm that, also when considering the global parking allocation problem for a set of requests, negotiation leads in average to better allocations and utilities for all the adopted measures when compared to base cases without negotiation. With this approach, we show that a negotiation process is more effective, in terms of social welfare maximization, than a one-sided utility maximization, since it allows a mediation of the conflicting needs of drivers and the city management. In addition, the system does not force an allocation as in typical centralized resource allocations where a single entity decides on the final allocation of resources with respect to all the agents' requests, possibly after having elicited the agents' preferences over alternative allocations [6], or a reallocation of resources locally taken by each agent, as in distributed resource allocations. In the presented approach, no reallocation of resources is considered, but the final global allocation is the outcome of a global negotiation composed of a sequence of negotiations for each parking request.

As future work, we plan to work on the design and implementation of a distributed version of the system allowing for the possibility for DAs to negotiate concurrently with different PMAs. In this way, we will be able to evaluate a more realistic social welfare composed of the different outcomes of the concurrent negotiations.

References

1. Abidi, S., Krichen, S., Alba, E., Molina, J.M.: A new heuristic for solving the parking assignment problem. Proc. Comput. Sci. **60**, 312–321 (2015)
2. Barile, F., Di Napoli, C., Di Nocera, D., Rossi, S.: Evaluating the social benefit of a negotiation-based parking allocation. In: Advances in Practical Applications of Agents, Multi-Agent Systems, and Sustainability: The PAAMS Collection. Lecture Notes in Computer Science, vol. 9086, pp. 15–26. Springer International Publishing, Berlin (2015)
3. Bellifemine, F., Caire, G., Poggi, A., Rimassa, G.: Jade: a software framework for developing multi-agent applications. Lessons learned. Inf. Softw. Technol. **50**(1–2), 10–21 (2008)
4. Bonsall, P., Palmer, I.: Modelling drivers car parking behaviour using data from a travel choice simulator. Transp. Res. C Emerg. Technol. **12**(5), 321–347 (2004)
5. Di Napoli, C., Di Nocera, D., Rossi, S.: Agent negotiation for different needs in smart parking allocation. In: Advances in Practical Applications of Heterogeneous Multi-Agent Systems. Lecture Notes in Computer Science, vol. 8473, pp. 98–109. Springer, Berlin (2014)
6. Endriss, U., Maudet, N., Sadri, F., Toni, F.: Negotiating socially optimal allocations of resources. J. Artif. Intell. Res. **25**, 315–348 (2006)
7. Geng, Y., Cassandras, C.: A new smart parking system based on optimal resource allocation and reservations. In: 2011 14th International IEEE Conference on Intelligent Transportation Systems (ITSC), pp. 979–984 (2011)
8. Geng, Y., Cassandras, C.: New smart parking system based on resource allocation and reservations. IEEE Trans. Intell. Transp. Syst. **14**(3), 1129–1139 (2013)
9. Haklay, M., Weber, P.: Openstreetmap: user-generated street maps. IEEE Pervasive Comput. **7**(4), 12–18 (2008)
10. Jonkers, E., van Noort, M., van der Veen, J.: Parking guidance: modelling, simulation and impact assessment. In: 2011 14th International IEEE Conference on Intelligent Transportation Systems (ITSC), pp. 858–864 (2011)

11. Longfei, W., Hong, C., Yang, L.: Integrating mobile agent with multi-agent system for intelligent parking negotiation and guidance. In: 4th IEEE Conference on Industrial Electronics and Applications, pp. 1704–1707 (2009)
12. Mei, Z., Tian, Y., Li, D.: Analysis of parking reliability guidance of urban parking variable message sign system. Math. Probl. Eng. **2012**, 10 pp. (2012)
13. Meir, R., Chen, Y., Feldman, M.: Efficient parking allocation as online bipartite matching with posted prices. In: Proceedings of the 2013 International Conference on Autonomous Agents and Multi-agent Systems, pp. 303–310 (2013)
14. Mejri, N., Ayari, M., Kamoun, F.: An efficient cooperative parking slot assignment solution. In: The Seventh International Conference on Mobile Ubiquitous Computing, Systems, Services and Technologies, pp. 119–125. IARIA (2013)
15. Pan, B., Crotts, J.C., Muller, B.: Developing web-based tourist information tools using Google map. In: Sigala, M., Mich, L., Murphy, J. (eds.) Information and Communication Technologies in Tourism 2007, pp. 503–512. Springer Vienna, New York (2007)
16. Polycarpou, E., Lambrinos, L., Protopapadakis, E.: Smart parking solutions for urban areas. In: 2013 IEEE 14th International Symposium and Workshops on a World of Wireless, Mobile and Multimedia Networks (WoWMoM), pp. 1–6. IEEE, New York (2013)
17. Ramezani, S., Endriss, U.: Nash social welfare in multiagent resource allocation. In: Agent-Mediated Electronic Commerce. Designing Trading Strategies and Mechanisms for Electronic Markets. Lecture Notes in Business Information Processing, vol. 59, pp. 117–131. Springer, Berlin (2010)
18. Rosenschein, J.S., Zlotkin, G.: Rules of Encounter: Designing Conventions for Automated Negotiation Among Computers. MIT Press, Cambridge, MA (1994)
19. Teodorovic, D., Lucic, P.: Intelligent parking systems. Eur. J. Oper. Res. **175**(3), 1666–1681 (2006)
20. Wang, H., He, W.: A reservation-based smart parking system. In: IEEE Conference on Computer Communications, pp. 690–695 (2011)

Sustainable Farming Behaviours: An Agent Based Modelling and LCA Perspective

Tomás Navarrete Gutiérrez, Sameer Rege, Antonino Marvuglia, and Enrico Benetto

Abstract The paper is focused on the application of ABM (Agent Based Models) to simulate the evolution of the agricultural system of the Grand Duchy of Luxembourg, which aims at the evaluation of the potential environmental impacts arising from policy implementation, following the methodology known as Consequential Life Cycle Assessment (CLCA). The novelty of our approach is on the multi-modeling consideration of the problem of how to evaluate potential environmental impact of farmer's behaviours. We consider the coupling of a computational model (ABM) and a matrix-based LCA model. The paper only presents preliminary results, exploring the influence of farmers' environmental awareness on the environmental impacts linked to farming activities. This is possible thanks to the attribution to the agents' profiles of one specific feature which simulates their "green consciousness level".

1 Introduction

Life cycle assessment (LCA) is a standardised methodology used to quantify the environmental impacts of products across their whole life cycle [12] which is nowadays recognised worldwide, although with some persisting limitations and certain barriers, as witnessed by the survey of Cooper and Fava [6]. It has been used at the fine grain level (at the product level) or at a more global level (policy).

In this latter case (policy evaluation) a more pertinent methodology that has nowadays gained recognition is the so-called Consequential LCA (CLCA). CLCA aims at evaluating the direct and indirect environmental consequences of a strategic decision taken in a moment t_0, which can be in the past, present or future. For example, in the specific case of agro-systems, the land use changes and related consequences following a bio-energy policy. CLCA expands the system boundaries

T. Navarrete Gutiérrez (✉) • S. Rege • A. Marvuglia • E. Benetto
Luxembourg Institute of Science and Technology (LIST),
41 rue du Brill, 4422 Belvaux, Luxembourg
e-mail: tomas.navarrete@list.lu; sameer.rege@list.lu; antonino.marvuglia@list.lu; enrico.benetto@list.lu

A. Alonso-Betanzos et al. (eds.), *Agent-Based Modeling of Sustainable Behaviors*, Understanding Complex Systems, DOI 10.1007/978-3-319-46331-5_9

to account for short and long-term changes in supply and demand capacity on the market, sometimes implying the use of economic equilibrium models [15].

The ultimate aim of the study is then evaluating the difference (delta) in terms of environmental inventory (and related impacts) between the two time marks set for the simulation (t_0 and t_{0+n}). In the specific case of agro-systems, economic modelling approaches, assuming supply-demand equilibrium, are increasingly applied to derive the linkages between agricultural operations and economic market, and their consequences in terms of environmental damage costs and social costs. However, behavioural criteria and adaptive social processes (e.g. imitation) are not taken into account in top down economic models and basing the modelling exercise exclusively on economic criteria is likely to turn out inappropriate to forecast structural changes, simply because economic decisions are never purely rational.

The MUSA (MUlti agent Simulation for consequential LCA of Agro-systems) project http://musa.tudor.lu, aims to simulate the future possible evolution of the Luxembourgish farming system, accounting for more factors than just the economy oriented drivers in farmers' decision making processes. We are interested in the behavioural aspects of the agricultural systems, including the green consciousness of the farmers. The challenges facing the farming system are manifold. Dairy and meat production have been the financial mainstay of the national agricultural landscape with production of cereals and other crops as a support to the husbandry. This sector is fraught with multiple rules and regulations including restrictions on milk production via quotas as dictated by the Common Agricultural Policy of the EU. There is also a complex set of subsidies in place to enable the farmers to be more competitive, while preserving the environment and the landscape. The quotas have disappeared as of March 2015, but new measures concerning greening are now in place. Just as a chain is as strong as its weakest link, any model is as robust as its weakest assumption. Model building is a complex exercise but modelling behaviour is far more complex. Statistical and optimisation models fail to account for vagaries of human behaviour and have limited granularity. Preceding the MUSA project, in [19] we have built a partial equilibrium model for Luxembourg to conduct a CLCA of maize production for energy purposes (dealing with an estimated additional production of 80,000 t of maize) using non-linear programming (NLP) and positive mathematical programming (PMP) approaches. PMP methodology [11] has been the mainstay of modelling methodology for agriculture models relating to cropping patterns based on economic fundamentals. This approach converts a traditional linear programming (LP) into a NLP problem by formulating the objective function as a non-linear cost function to be minimised. The objective function parameters are calibrated to replicate the base case crop outputs. This approach is useful as a macro level such as countries or regions where one observes the entire gamut of crops planted at a regional level. Increasing the granularity to investigate the impacts of policy on size of farms is still possible provided each class of farms based on size exhibits plantation of all crops in the system. When the granularity increases to the farm level, the crop rotation takes priority for the farmer and then one observes only a subset of the entire list of crops in the system. It is at this level that the PMP approach fails as the objective function is calibrated to the crops observed at a specific point in time. This limitation is overcome by the agent based model (ABM) approach. To investigate the possible set of outcomes due to human responses to financial and natural challenges, ABM are a formal mechanism to validate the range of outcomes due to behavioural differences.

LCA can produce as outputs different levels of environmental assessments. The levels are termed mid-point and end-point in the LCA jargon. They correspond respectively to assessments given at a semi-aggregated and aggregated level. An example of midpoint result from an LCA assessment is the popular "Carbon Footprint".

We consider that LCA is a valuable tool to inform farmers on the potential (in the LCA terms) impact of their activities. Starting from this, we focus on the question of modelling the inclusion of LCA results in the behaviour of farmers in ABM model.

In the work we report here, we investigate how to introduce the results of LCA into ABMs to simulate the evolution of an agricultural system. The main question remains how to use the specific results of an initial LCA within the decision process of farmers. For our work, we shall use our own initial agent-based model that will allow agents to perceive the results of an LCA (for the full agricultural system) and use them in the farm planning.

In order to understand the thought process and potential responses to different situations, it was necessary to obtain inputs from the farmers and try to understand their perception of the threats and opportunities of their practice. For this reason a survey was designed in collaboration with: CEPS/INSTEAD,[1] a public Luxembourgish Institute specialised in Socio-economic research; CONVIS, a consortium of farmers based in Luxembourg and the Walloon Agricultural Research Centre (CRA-W).

There are in fact several factors leading a farmer to make a choice related to his/her farm. Some of these factors are sometimes linked to traditions, personal orientations, family motivations, etc. The survey contains questions related, amongst others, to the farm (size, crops planted, type and amount of fertilisers used, rotation schemes adopted, animals reared, milk and meat produced, etc.), the size and nature of the changes introduced in the past in the farmer's activities or planned for the future (new cultures, expansion of cultivated area, etc.); the main factors playing a role in the decision to implement changes (e.g. public subsidies); the farmer's risk proneness or aversion (in terms of economic investments); the attachment to family traditions in the way the farm is managed.

The survey was distributed by CONVIS to 1191 farmers out of the more than 2100 farms existing in Luxembourg, with a response rate of 14 % (168 respondents) and respondents located in 97 different areas. The survey is available for download at the following website: http://musa.tudor.lu/surveyresults. The information collected in the questionnaire will be analysed by the project team and will be used to achieve a better definition of the simulation model that is currently being built.

The survey included 79 questions (some of which were optional) and was divided up in four parts: Part I about the farm; Part II about the farmer's land use choice

[1]Now known as Luxembourg Institute of Socio-economic research http://www.liser.lu.

(with Sect. II.1 about a previous change; II.2 about a planned change; II.3 about a previous attempt to change which failed; II.4 about no change); Part III about farmer's inclination/aversion to risk and Part IV about the composition of farmer's household.

In Sect. 2 we present a brief tour of different agent-based models dealing with agricultural systems as well as LCA research done using agent-based models. Then, in Sect. 3 we describe our model as well as our proposition to link it with LCA. We present the initial results of the simulations in Sect. 5 and present our conclusions in Sect. 6.

2 Literature Survey

Happe et al. [10] use an ABM called AgriPoliS, for simulation of agricultural policies. The focus is on simulating the behaviour of farms by combining traditional optimisation models in farm economics with agent based approach for the region of Hohenlohe in southwest Germany. The paper classifies farms into size classes and assumes all farms in a specific class size to have the same area. The variation in the farms is on account of individual asset differences (both financial and physical). Farms are classified only as grasslands or arable lands with no further classification by crops or rotation schemes. Happe et al. [10] has high explanatory power with regard to farm evolution or competitive market settings, *but* in a rational behaviour context. Agents are often designed following the hypothesis of actions in economically rational way, see for example: [8, 18].

Culture and traditions play also an important role in land use decisions. Humans when they find themselves spurred with overwhelming influences for a single decision, employ strategies in land use that go beyond the simplistic maximisation of profits and opportunity cost and risk minimisation. Berger [1] and Bonabeau [5] are other applications on the lines of Happe et al. [10] with an integration of farm based LP models under a cellular automata framework, applied to Chile. The model explicitly covers the spatial dimension and its links to hydrology with a policy issue of investigating the potential benefits of joining the Mercosur agreement. Le et al. [14] is an application of agent based modelling framework to assess the socio-ecological impacts of land-use policy in the Hong Ha watershed in Vietnam. The model uses in addition to farms as agents, the landscape agents that encapsulate the land characteristics. Additional sub models deal with farmland choice, forest choice, agricultural yield dynamics, forest yield dynamics, agent categoriser that classifies the households into a group and natural transition to enable transformation between vegetation types. Berta et al. [4] with it's Pampas model is a similar study to simulate structural and land use changes in the Argentinian pampas. The model is driven by agent behaviour that aims to match the aspiration level of the farmer to the wealth level. The farmer will change behaviour until the two are close. In addition to [10] equal sized plots in the farm, the farmer has a choice of planting two varieties of wheat, maize and soybean. The model studies the potential penetration of a particular crop and find that soybean is being cultivated on a much larger scale. Bert et al. [2, 3] proposes the LARMA Model, embedded into the Pampas model, with a strong focus on land exchange and includes detailed transactions associated to the land market and includes feedback with natural ecosystems. The different information related to land rental prices is generated endogenously.

The work of Murray-Rust et al. [16] and its modelling framework Aporia, is unique in dealing with simulation of land use decisions and their impacts ecosystem services, because of the explicit inclusion of vegetation models and because it is released as open source software.

In [20], a model based on constrained optimisation is built to simulate farm decision making. The uniqueness of the model relies on the use of a micro economic modelling approach and a choice of alternative biophysical modules that can be directly coded in the software or injected using a third party library.

Despite most of the previously cited related works include a high number of economical aspects in their models, they lack the inclusion of aspects related to green consumption and production. In general terms the side effects on the environment, as considered by life cycle impact assessment are missing from them.

3 Data and Model Structure

In the model, the farmers are represented via entities called "agents" who take decisions based on their individual profiles and on a set of decision (behavioural) rules defined by the researchers on the basis of the observation of real world (e.g. the results of the questionnaire) and on the interactions with other entities. The model currently includes one reactive rule regarding the change of crops planted at the farm.

Luxembourg provides statistics about the economy of the agricultural sector through STATEC [22] and Service d'Economie Rurale [21]. The statistics deal with the area under crops over time, farm sizes and types of farms by output classification, use of fertilisers, number of animals by type (bovines, pigs, poultry, horses) and use (meat, milk). The latest year for which one obtains a consistent data set across all the model variables was 2009.[2] In 2009 there were 2242 farms with an area of 130,762 ha under cultivation that included vineyards and fruit trees and pastures amongst other cereal and leaf crops. For 2009, the national statistics indicate a use of arable land by cultures we classified as either `Cereals`, `Leafs` or `Others`. Table 1 shows the specific crops and the number of hectares cultivated for each farm class as of 2009. The first column of the table identifies the crop, the second indicates the type of crop where `C` means a cereal crop, `L` a leaf crop and `O` other kind of crops. After this first column, we have nine different columns ranged from `A` to `I` that identify nine classes of farms. A class of farm indicates the average farm size.

[2]The details of the data are available in [19].

Table 1 Area (ha) under crop by farm type (A to I)

Crop	Type	A	B	C	D	E	F	G	H	I
wheat_winter	C	0	4	25	81	124	355	684	1590	3712
wheat_summer	C	0	4	27	85	129	370	715	1660	3877
spelt	C	0	0	2	5	8	22	42	97	226
rye_winter	C	0	1	7	31	27	90	141	221	585
barley_winter	C	0	8	39	118	200	573	792	1375	2759
barley_spring	C	0	4	23	70	119	343	474	823	1651
oats	C	0	3	5	36	65	169	253	314	539
mixed_grain_winter	C	0	1	2	3	3	9	15	32	59
mixed_grain_spring	C	0	0	1	2	2	9	14	31	58
grain_maize	L	0	2	5	9	9	31	49	106	198
triticale_winter	C	0	5	29	108	124	450	580	1047	1712
other_forage_crops	L	0	1	7	29	59	388	1196	3148	6515
maize_dry_matter_BG	L	0	1	8	32	66	434	1339	3526	7296
dried_pulses	L	0	1	4	7	6	23	37	79	148
beans	L	0	0	1	2	2	6	9	20	37
potatoes	L	0	3	15	12	5	66	37	138	328
rapeseed	L	0	0	3	11	23	153	473	1246	2577
clover_grass_mix	L	2	9	24	40	39	138	221	481	896
meadows	O	1	38	90	206	196	729	1113	2445	4207
pastures	O	10	249	584	1329	1268	4714	7194	15,804	27,191
vineyards	O	106	203	564	293	47	10	1	16	60
crops_NES	O	0	2	4	7	7	24	38	83	155

The classes are ordered in increasing farm size. Farm class A is for farms with a size under 2 ha, class B for farms with size between 2 and 4.9 ha. The full specification of the classes is provided in the second line of Table 2.

Additionally there is information available by flik,[3] its geometry, crop planted on the flik, the commune or municipality in which the flik is situated. However, for confidentiality reasons, this information is only available in anonymised format that makes it impossible to create a grouping of the fliks by (anonymous) farm manager. The availability of this information would make it straightforward to allocate the initial distribution of the fliks to farms. This forced the model to be initialised with random proportion of crops being planted by each farm, where the random numbers were afterwards scaled to match those of year 2009.

[3]A flik is the smallest geo-referenced land element registered at the cadaster in Luxembourg.

Table 2 Distribution of farms by size (ha) in Luxembourg in 2009

	Farm class	A	B	C	D	E	F	G	H	I
	Total	<2	2–4.9	5–9.9	10–19.9	20–29.9	30–49.9	50–60.9	70–99.9	100+
Number	2242	230	165	217	186	116	246	263	398	421
Area (ha)	130,762	131	598	1533	2667	2890	9956	15,743	33,583	63,661
Average size	58.32	0.57	3.62	7.06	14.34	24.91	40.47	59.86	84.38	151.21

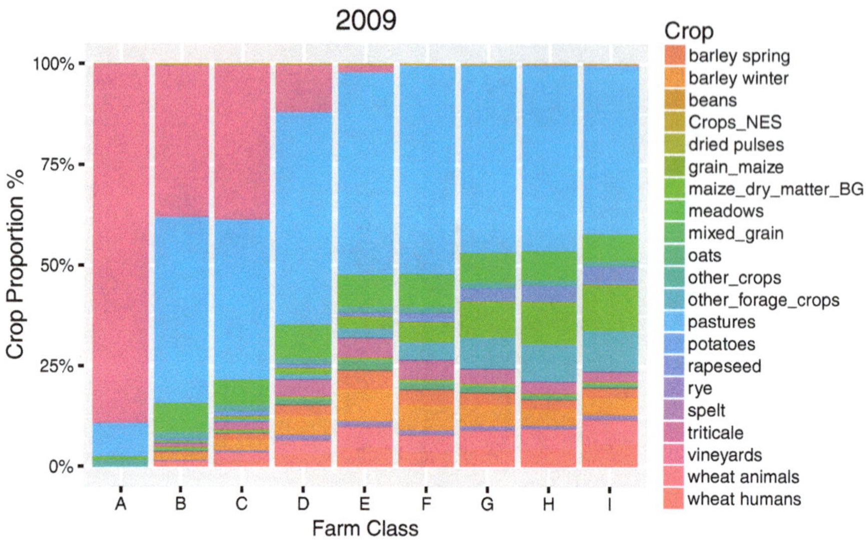

Fig. 1 Initial proportions of crops planted in 2009

Table 2 shows the distribution of farm area by size of farms in Luxembourg for the year 2009. We also know the total area of crops planted under each farm type. Figure 1 shows the initial proportions of each crop present in 2009. This figure shows how farms with small average sizes (`A` to `C` for example) are dominated with the culture of vineyards, in contrast to bigger farms, where cereals are the predominant kind of culture.

From [13] we know the different rotation schemes for crops. Rotation schemes are used by farmers to maintain the health of the soil and rotate cereals and leaves on the same field in a specified manner. We randomly assign a rotation scheme to each farm and then randomly choose a crop from cereals or leaves. As to crops that are neither cereals nor leaves, they are a permanent cultivation such as vineyards, fruits, meadows or pastures. Once the random allocation of crops has been made to the farms, we scale the areas so as to match the total area for the crop under a specific farm type.

3.1 Model Structure

The model is intended to be used as part of a LCA. The LCA measures the environmental consequences (at a global scale and in a life cycle perspective) of different decisions made by farmers in Luxembourg. The ABM model answers to the question: "What are the (number of) products (crops) being produced in Luxembourgish farms, at a given time?". More specifically, we would like to answer this question under the condition that a certain policy-driven scenario is put in place. In order to do so, the model provides as output the changes (in terms of hectares of land planted with each crop) in the land use occurring as a consequence of the exogenously imposed change (for example the decision to reach an additional production of 80,000 t of maize to destine to the production of biogas). In a future enhanced version of the model, it will also be able to provide the changes in terms of cattle heads, milk and meat production. Once these changes are computed, they are used as inputs for the LCA calculations. The answers to this question can then be used to perform an environmental assessment taking into consideration the changes over time of the products of each farm.

The ABM represents the agricultural system of Luxembourg, it is built around the activity of farmers in the country and includes the following entities.

Farmers. The autonomous agents of the system.

Farms. This is a container object used to organise the model. Farms in the model are instantiated using the above mentioned statistics.

The current model of the behaviour of the farmers is to consider that they follow a rational expected maximisation behaviour. This means that a farmer will decide to change the crops to plant, based on the yield of a given crop, the price for the last season of the crop and the costs of the crop. The farmer agent, will decide to change a crop in his farm, that suits the rotation scheme, taking the list of available crops from the environment.

Each farm has an associated rotation scheme. Along with initial allocation of a rotation scheme per farm, we set up an initial allocation of crops planted in a given proportion in the farm.

For clarity purposes, we have decided to make an explicit distinction between a farm and a farmer in our model. Also, although we cannot imply from the information we currently have that multiple farms are managed by the same farmer, this is a possibility that only this separation of entities allows. Another reason for this separation is on the technical side: when programming following the object oriented paradigm (in particular in the java language) classes to instantiate objects must be defined, and for good practice a separation of concerns clearly suggest that Farms and Farmers are two different classes and must be defined independently. Finally we considered all of the entities in the *modelled* system corresponded to an autonomous agent in the system, this is the case for the abstract notion of *farm*.

Product buyers. In the model, a buyer will offer to buy the production for the farmers. There is only one buyer that will offer to buy every product of the whole country. The prices the buyer will use, will be given in the scenario. The specification of the prices for the moment is static: one price for each "year" in the simulation.

Crop. Crops can be either cereals or leafs. They have an associated yield in tonnes per hectare.

In every kind of OTE (technical and economical orientation of the farms as classified in the European context [7]) a farm has a "majority" of it's activity characterised. This means that a farm with "Field Crops"[4] OTE, will not only have planted cereals or leafs, but may also have Grazing Livestock (see [7] for the details).

For the current formulation of the project, we will focus on the area (in ha) planted with each Crop, for each farm. The model accounts the following different scales.

Description levels. The model is given at the individual level, for farms and farming resources. For the prices, we will consider a global representation where all prices are identical for all farmers and are given by the market.

Time. The model will consider a time step being equal to a year. In one time step, the agent sows at the beginning, then harvests, sells the production, and finally decides to change or not a specific crop for this rotation scheme (substitute a **C**ereal or a **L**eaf).

Space. Each farm has a size (in ha) of arable land. No GIS information is included in the current version of the model.

Designing Concepts Agents follow a reactive architecture. They observe the prices in the system and change parameters of their main behaviour accordingly. Since there are no interactions between agents, we consider a sequential application of behaviours.

Implementation In order to keep a maximum flexibility during the development of the model as well as to be able to use the outputs of the model as the inputs for the LCA, we have built our own simulator from scratch using java as the programming language.

3.2 LCA Model

In broad terms, an LCA model is defined by specifying a functional unit to evaluate. This functional unit can be composed of different inputs from what is called the technosphere, and the biosphere. The functional unit can also specify different emissions associated to it. Functional units have already been defined by the LCA community for a wide range of technosphere processes, like "Electricity production", or "use of a computer". All this information is gathered in different databases like for example ecoinvent [25]. It is of common practice, to define a functional unit by specifying a set of inputs from the technosphere, which in turn, were previously defined (in the database). In this way, an inventory is created, in the form of a matrix. A quick shortcut to define an LCA model, is to consider that the technosphere matrix that describe the inputs and outputs of a functional unit is the model. Solving the system described by the inventory matrix for a given demand will provide an inventory flow results. The environmental assessment part takes place afterwards, after selecting a methodology of interest. A Life Cycle Impact

[4]This specific OTE is defined as $2/3$ of activity is field crops.

Assessment (LCIA) methodology translates the results of the inventory (inventory flows) into potential environmental impacts. The last phase is to solve the system of matrices to identify the impact associated to each element of the inventory.

The functional unit considered for the LCA model in our work is the total crop land of Luxembourg. Based on a "cradle-to-gate" approach, the system boundaries include the direct emissions and resources of the cropping system, as well as the background activities of the used transformed products (e.g. impacts related to the production of fertiliser). Life Cycle Inventory (LCI) data are described in [23, 24] and have been implemented in the SimaPro software. Midpoints categories of the commonly used ReCiPe [9] methodology have been considered for the LCIA.

The interested reader may look at [23, 24] for the details of the LCA model created for the environmental assessment of the Luxembourgish agricultural system.

In the LCIA, in order to calculate the potential environmental impacts, the lifecycle amount of each pollutant emitted has to be converted (using conversion factors called "characterisation factors") in an equivalent amount of a reference substance, acting on the same impact category (for example CO_2 and CH_4 act on the category "climate change" and CO_2 is the reference substance). Midpoint characterisation factors are therefore based on equivalency principles, i.e. midpoint characterisation scores are expressed in kg-equivalents (kg_{eq}) of a substance compared to a reference substance. In other words, the kg_{eq} of a reference substance expresses the amount of that reference substance that equals the impact of the considered pollutant within the midpoint category studied.

From the LCA perspective, it has been considered as a better representation of certain crops (those listed in Sect. 3) to split them as multi-output processes. Wheat winter produces 51 % of fodder and the remaining 49 % of the produce is used for bread making. A similar approach is taken for barley, with 6.2 % for brewing and 93.8 % for fodder in the winter variety and 19.2 % for brewing and 8.08 % for fodder for the summer variety. In the case of mixed grain and triticale, both summer and winter are considered as one single element, "mixed grain" and "triticale" respectively.

3.3 *Model Interconnection*

From a coarse point of view the models are connected following a flow from the ABM model to the LCA model. The ABM produces the number of hectares cultivated with each crop, and the LCA models takes them to produce the LCIA.

A block diagram made with SysML of the previous description is given in Fig. 2. The Systems Modelling Language (SysML) is a general-purpose modelling language particularly suited for dynamical systems. It is a standard published and managed by the OMG [17]. We have used SysML because it allows to clearly represent the interactions between two models. The interest of using a graphical language, is that it can render explicit the relationship between the models. Also, we have used decided to use SysML because it can be used further to describe

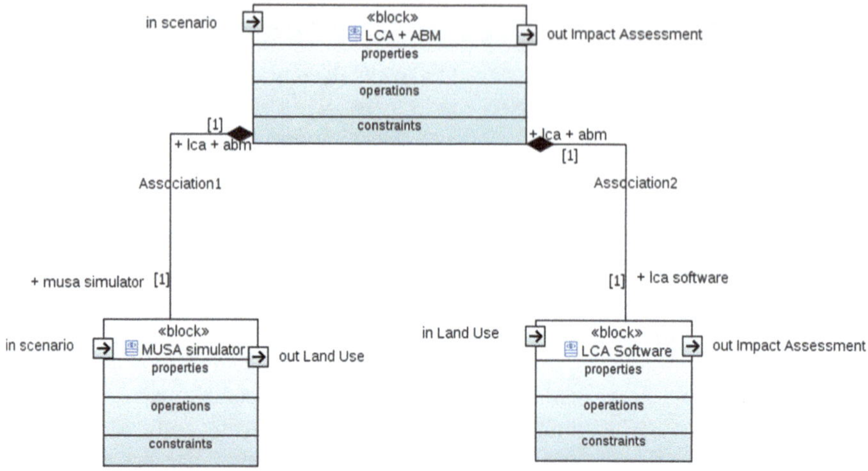

Fig. 2 Block description diagram of the system including the ABM simulator and the LCA software. The block called "ABM+LCA" is in charge of distributing the inputs to the ABM simulator (block "MUSA simulator") and pass the outputs (Land Use) to the LCA software to obtain the final results (Impact Assessment)

the specific software interactions, and since both the ABM and LCA models are in essence computational models, this is necessary in order to engineer a system that makes both models interact. We consider that both models are computational in essence, because to get specific answers that they were built for, they must be executed under the form of a software. Both of them could be solved (that is, executed as to provide answers) by hand, but this is a task that computers are better suited to do.

In LCA, studying a particular functional unit requires some modelling. In principle, any LCA software can be used to perform the environmental LCIA phase, based on the LCA model. However, the authors plan to use the software (Brightway), which being programmed in Python, presents a lot of scripting flexibility and is platform-free (whereas other LCA pieces of software are linked to a specific platform; e.g. the Simapro is linked to Windows OS).

4 Simulation Workflow

4.1 Global Simulation

ABM models are simulation models, and as such they must be simulated before they can actually answer the questions they were built for. The simulator that implements the ABM has the following workflow.

1. Initialise the farm sizes.
2. Assign to each farm a random rotation scheme.
3. Assign to each farm a set of crops being planted (depending on the type of rotation scheme).
4. Scale the sizes of the farms and the fields being planted with each crop to match the statistics of 2009.
5. Apply the behaviours associated to each agent in a "pre-market" phase.
6. Let the agents sell their produce (the result of the yield per hectare per crop) in a market with prices estimated form historical records and future forecasts made using Holt-Winters time series technique.
7. Apply the behaviours associate to each agent in a "post-market" phase.

4.2 Behaviours

We currently have implemented three different behaviours for the agents. Two basic ones and a third based on the first two. The basic ones are: one that represents a use of exclusively economic drivers in the decisions (termed profit maximisation) and another one where the most important criterion for taking decisions is the environmental impact of planting a given crop (in terms of CO_2 emissions).

The initial conception of the model was built around market-driven factors. In the simulations that the model can currently perform all the agents would take the decision to modify their current rotation scheme (change each of the specific cereals or leaf crops they currently grow) based on the prices the agents perceive from the environment. The prices are now set artificially as part of the simulation. That is, we currently specify manually the different values of costs and prices for each crop, based on information publicly available so that the values we assign are close to reality.

The focus on CO_2 is presented here as a mere example, given the fact that the environmental impacts we calculate are the entire set of midpoint impacts obtainable with the LCIA methods commonly used in LCA. In particular, we applied the consensus method ReCIPe [9]. LCIA results of midpoint categories can be represented in two ways: 1. in *absolute* units (kg_{eq}-substance) or 2. in *relative* units (percentage of impact compared to one reference scenario; for more details see [9].

A third behaviour termed "green consciousness" is introduced to add heterogeneity in the behaviours. An agent (farmer) has an internal green consciousness (a value between 0 and 1) that dictates the strength of the environmental considerations for a farmer. If the value is lower than 0.5 the farmer will actually use the same criteria as behaviour "profit maximisation" to change the crops for next year. On the other case (green consciousness higher or equal to 0.5) the criteria to use to select a substitution crop will be the environmental impact of the list of available crops. Algorithms 1 and 2 specify the behaviours.

Algorithm 1: Profit maximisation behaviour implementation

```
Estimate potential economic gain;
for Each crop i in the potential scheme do
    gain[i] ← AreaUnderCrop[i] * Last Selling Price of Crop[i] - Cost of Crop[i];
    total gain += gain[i];
end
```

Algorithm 2: Generic behaviour for substituting crops in the rotation scheme. All crop types, **C**ereal or **L**eave is taken into account. For the green consciousness behaviour, the highest value function uses CO_2 emissions as the criterion (the lower the emissions, the higher the value) and for the profit maximisation, the potential economic gain is the criterion used

```
l ← perceive available Crops in the Environment;
for Each Crop c in l do
    c.profit ← calculate potential value (c);
end
for Each crop i currently planted in the farm do
    s ← select highest value(l,currentplantation, i);
    if ∃s then
        substitute i with s;
    end
end
```

In this version of the model, we have included a post-market phase for future use, as it currently does not include any specific behaviour for the agents.

4.3 *Scheduling*

At each phase of the simulation (as described in Sect. 4.1) the agents are scheduled to be active in a random and synchronous order: one agent at the time is randomly selected and the corresponding behaviours (pre-market, market and post market) are applied to each agent.

5 Experiments

To obtain a wide-ranging perspective of the environmental impacts and damages linked to the described production system, and to check the consistency of the results, several different LCIA methods were applied. The analysis of midpoint impacts, as well as endpoint damages was calculated using the ReCIPe [9] methodology. Just for exemplification purposes, we ran five simulations with the

ABM model and we computed the corresponding variations in the cultivated area of each crop. These variations (deltas) were fed to the LCA software, which allowed the calculation of the corresponding variations in the cultivated area of each crop between 2009 and 2010. A comparison of the results obtained with the average values resulting from the five simulations, and normalised, resulted in a situation where all the different midpoint indicators worsen, except for one.

The simulations have an identical set of initial parameters and behaviours assigned to the agents. The different elements that change for each simulation are the initial sizes of the farms, and crops assigned. For example, in simulation one, the first farm of class A may have an initial size of 1.5 ha, completely allocated to vineyards, while in simulation two, the same farm may be initialised with a size of 1.9 ha, with 80 % vineyards and 20 % other crops. The choice for only five simulations was arbitrary.

We have initial results obtained by simulating the model using the following parameters.

- Each farmer has a random green consciousness that is taken form a uniform distribution between 0 and 1.
- Each simulation is executed only over one time step (each time step representing a year).
- All agents perceive in perfect conditions the available crops in the environment as well as the LCA results.
- Each agent has only one post-market behaviour: green consciousness.

5.1 Results

To ease the initial analysis, we averaged the variations (the deltas) of the outputs. The average is done over the results obtained after five simulations. Figure 3 shows the average cultivated area for each crop, after five simulations. We analysed the percentage changes for each of the different categories obtained from the LCA assessment based on the average changes between the initial proportions of crops planted in the model, and those resulting after applying the behaviours to each agent. For each impact category, we assigned the 100 % value to the simulation for which the impacts are the highest, and the impacts arising from the other simulations were scaled accordingly. The analysis indicated that only the "agricultural land occupation" indicator worsens.

In the simulation results, the only impact category that did not worsen was "agricultural land occupation". This means that for all the other impact assessment indicators we saw a worse condition of the system after 1 year of simulation, except for "agricultural land occupation". To understand this change, we would need to look at the different impacts per crop in Fig. 4, to see that the crops with the highest impact for that category are the ones with the highest average delta.

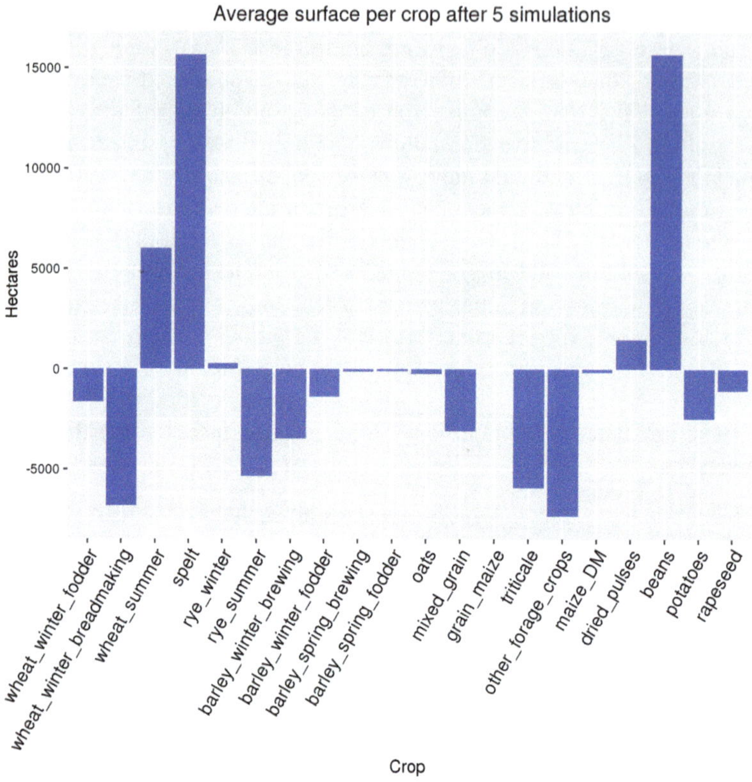

Fig. 3 Average deltas for hectares cultivated with each crop along the five different simulations

6 Conclusions

This paper presents the first version of an ABM which aims at simulating a plausible evolution of the Luxembourgish agriculture and farming system under different conditions (status quo and externally imposed scenarios) at the aim of evaluating the potential environmental consequences of policy driven actions.

We have managed to implement an ABM that lets the agents take decisions based on information they perceive from the environment that is available to all of them. The information was the carbon footprint of different crops cultivated in 2009 in Luxembourg. This is an initial solution to the question of how to include the LCA, results in an ABM model.

Our initial results, concerning the LCA, suggest that if a farmer solely focuses on CO_2 as a criterion and has a "green consciousness" that compels him to make changes in his rotation scheme, he may end up improving his carbon footprint, but may worsen other environmental impacts that could be of higher importance for his specific activities (e.g. those related to soil and water).

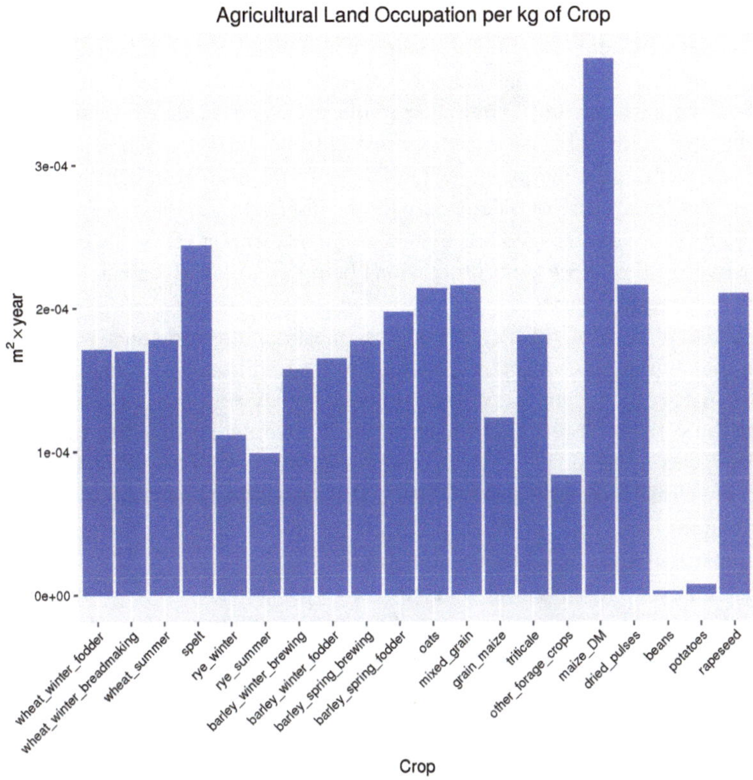

Fig. 4 "Agricultural land occupation" impact per crop

For the moment, we consider only the crop type of resources that a farm may have, but from previous experience [24] and from continuous discussions with different stakeholders of the Luxembourgish agricultural sector, we know that for highly integrated farming systems, animal breeding is also a very important factor in the decisions at farm planning level. Therefore efforts are ongoing to include this element in the model as well.

From the computational side of the ABM simulation special attention was paid to randomly initialise *and then scale* the different characteristics of the farms. Although we only ran five simulations, we can consider that the calibration procedure that we devised, would normally yield very similar initial conditions regarding the farm sizes and crops assigned to each farm. However there is important uncertainty on the type of rotation scheme assigned initially to each farm. Regular uncertainty analysis techniques exist that could allow to better quantify this issue and we shall apply them in further work, provided that we have better access to this kind of information. This would require for example to consider rotation schemes that are not as simple as a following an identical path every 3 or 5 years. However a finer definition of this elements remains out of the scope of this initial model version.

The survey we initially designed was ran in parallel to the development of the work we present here. Our initial model does not take into account the specific elements of information that can be gathered from the results of the survey. However, we have already identified key questions for further refinements of the different behaviours associated to the farmers. Initial analysis of a small sample of the full answers to the survey indicate that for farms with size smaller than 70 ha, most farmers are aged 50 or older, while for the big farms, (sizes greater than 100 ha) the proportion of farmers aged less than 40 is equally important as that of the farmers aged between 40 and 50, both proportions being significantly greater than that of farmers aged more than 50.

A partial sample of the Part III of the questionnaire indicates what share of farmers has willingness to make changes in the traditional way of managing the farm. In particular, when we asked, supposing they had the financial means to do that, how likely (on a scale from 1 to 10) would be that the farmer could consider investing in biogas production in a time horizon of 10 years, the currently available sample indicates about 50 % of the respondents are not at all likely to make changes, with 25 % of the respondents that did not answer, and the remaining equally distributed in the remaining options.

If this tendency in the sample is confirmed with the full results from the survey, the results could be used to characterise the incentives that could be used to influence the farmers. For example, we could consider that without a proper incentive, it would be very difficult getting to the objectives set for 2020 in terms of maize dry matter production for biogas.

6.1 Perspectives

In this first version of the model, we are not taking into consideration the subtleties of the OTE. Including this would allow us to give a more generalised (at least at the European scale) point of view of our results.

Although we implemented a solution that includes a so-called green consciousness, on how to let the agents react using information from LCA results, the question for future work is how to characterise and let evolve this consciousness. For the matter of the agents profile definition, we will use the conducted survey to better characterise the decision making processes in farm planning. As for the evolution, we shall include in further refinements of the model, the use of social networks to whom the farmers would belong and let the interactions with members of the networks influence the evolution of the green consciousness. We can at least think of the professional networks (most farmers are professionally advised by local associations) and the proximity networks (geographical neighbourhoods). So far the model fixes the green consciousness of the farmers at the initialisation phase, but it is to be expected that this consciousness evolves over time. Further refinements could include social contagion and imitation as mechanisms to let green consciousness evolve as agents interact with other agents.

Our inclusion of the LCIA results for the moment is limited to executing simulations over one time step and then calculating the LCA results form the outputs of the simulation. In future work, we envision using the results from the LCIA of each time step, and re-inject them in the environment, so that agents can perceive them and also let the green consciousness and decisions evolve based on the evolution of the LCA results. We can think of this as a feedback loop. However, for this we need to develop LCIA tools that can directly be called from the simulator. Our current tools, like for most LCA practitioners, have limited functionalities that support interaction with other software.

Acknowledgements Luxembourg's National Research Fund (FNR) is acknowledged for the financial support of project MUSA with id: C12/SR/4011535.

References

1. Berger, T.: Agent-based spatial models applied to agriculture: a simulation tool for technology diffusion, resource use changes and policy analysis. Agric. Econ. **25**, 245–260 (2001)
2. Bert, F.E., Rovere, S.L., Macal, C.M., North, M.J., Podestá, G.P.: Lessons from a comprehensive validation of an agent based-model: the experience of the pampas model of argentinean agricultural systems. Ecol. Model. **273**, 284–298 (2014)
3. Bert, F., North, M., Rovere, S., Tatara, E., Macal, C., Podestá, G.: Simulating agricultural land rental markets by combining agent-based models with traditional economics concepts: the case of the argentine pampas. Environ. Model. Softw. **71**, 97–110 (2015). http://dx.doi.org/10.1016/j.envsoft.2015.05.005. http://www.sciencedirect.com/science/article/pii/S1364815215001528
4. Berta, F.E., Podestá, G.P., Rovere, S.L., Menéndez, A.N., North, M., Tatarad, E., Laciana, C.E., Weber, E., Toranzo, F.R.: An agent based model to simulate structural and land use changes in agricultural systems of the argentine pampas. Ecol. Model. **222**, 3486–3499 (2011)
5. Bonabeau, E.: Agent-based modeling: methods and techniques for simulating human systems. Proc. Natl. Acad. Sci. **99**(Suppl. 3), 7280–7287 (2002)
6. Cooper, J.S., Fava, J.A.: Life-cycle assessment practitioner survey: summary of results. J. Ind. Ecol. **10**(4), 12–14 (2006)
7. European Commission: Commision Regulation (EC) No 1242/2008 of 8 December 2008 establishing a Community typology for agricultural holdings. Official Journal of the European Union (2008)
8. Filatova, T., Parker, D., Van der Veen, A.: Agent-based urban land markets: agent's pricing behavior, land prices and urban land use change. J. Artif. Soc. Soc. Simul. **12**(1), 3 (2009)
9. Goedkoop, M., Heijungs, R., Huijbregts, M., De Schryver, A., Struijs, J., van Zelm, R.: Recipe 2008. A life cycle impact assessment method which comprises harmonised category indicators at the midpoint and the endpoint level. Report I: Characterisation, 1 (2009)
10. Happe, K., Kellermann, K., Balmann, A.: Agent-based analysis of agricultural policies: an illustration of the agricultural policy simulator AgriPoliS, its adaptation, and behavior. Ecol. Soc. **11**(1), 329–342 (2006). http://www.ecologyandsociety.org/vol11/iss1/art49/
11. Howitt, R.: Positive mathematical programming. Am. J Agr. Econ. **77**, 329–342 (1995)
12. ISO: Environmental management—Life cycle assessment—Principles and framework. ISO 14040:2006, International Organization for Standardization, Geneva (2010)
13. KTBL: Faustzahlen für die Landwirtschaft (in German). ISO, Kuratorium für Technik und Bauwesen in der Landwirtschaft, Darmstadt (2006)

14. Le, Q.B., Park, S.J., Vlek, P.L., Cremers, A.B.: Land-use dynamic simulator (LUDAS): a multi-agent system model for simulating spatio-temporal dynamics of coupled human–landscape system. I. Structure and theoretical specification. Ecol. Inform. **3**(2), 135–153 (2008)
15. Marvuglia, A., Benetto, E., Rege, S., Jury, C.: Modelling approaches for consequential life-cycle assessment (c-lca) of bioenergy: critical review and proposed framework for biogas production. Renew. Sust. Energ. Rev. **25**, 768–781 (2013)
16. Murray-Rust, D., Robinson, D.T., Guillem, E., Karali, E., Rounsevell, M.: An open framework for agent based modelling of agricultural land use change. Environ. Model. Softw. **61**, 19–38 (2014). http://dx.doi.org/10.1016/j.envsoft.2014.06.027. http://www.sciencedirect.com/science/article/pii/S136481521400200X
17. OMG: OMG system modeling language. http://www.omg.org/spec/SysML/ (2012)
18. Parker, D.C., Hessl, A., Davis, S.C.: Complexity, land-use modeling, and the human dimension: Fundamental challenges for mapping unknown outcome spaces. Geoforum **39**(2), 789–804 (2008)
19. Rege, S., Arenz, M., Marvuglia, A., Vázquez-Rowe, I., Benetto, E., Igos, E., Koster, D.: Quantification of agricultural land use changes in consequential Life Cycle Assessment using mathematical programming models following a partial equilibrium approach. J. Environ. Inform. **26**(2), 12–139 (2015)
20. Schreinemachers, P., Berger, T.: An agent-based simulation model of human-environment interactions in agricultural systems. Environ. Model. Softw. **26**(7), 845–859 (2011). http://dx.doi.org/10.1016/j.envsoft.2011.02.004. http://www.sciencedirect.com/science/article/pii/S1364815211000314
21. SER: http://www.ser.public.lu/ (2016)
22. STATEC: http://www.statistiques.public.lu/en/actors/statec/index.htm (2016)
23. Vázquez-Rowe, I., Rege, S., Marvuglia, A., Thénie, J., Haurie, A., Benetto, E.: Application of three independent consequential LCA approaches to the agricultural sector in luxembourg. Int. J. Life Cycle Assess. **18**(8), 1593–1604 (2013)
24. Vázquez-Rowe, I., Marvuglia, A., Rege, S., Benetto, E.: Applying consequential LCA to support energy policy: land use change effects of bioenergy production. Sci. Total Environ. **472**, 78–89 (2014)
25. Weidema, B., Bauer, C., Hischier, R., Mutel, C., Nemecek, T., Reinhard, J., Vadenbo, C., Wernet, G.: The ecoinvent database: overview and methodology, data quality guideline for the ecoinvent database version 3 (2013). http://www.ecoinvent.org/

Agent-Based Simulation of Electricity Markets: Risk Management and Contracts for Difference

Fernando Lopes, Hugo Algarvio, and João Santana

Abstract Electricity markets (EMs) are a relatively new reality and also an evolving one, since both market rules and market players are constantly changing. Market participants can purchase and sell electrical energy through centralized markets and bilateral contracts. The revenue and cost certainty associated with bilateral contracts presents a number of benefits to producers and consumers, notably price stability and mitigation of market power. An agent-based modeling and simulation approach presents itself as a promising approach to model power markets. Accordingly, this chapter presents several key features of software agents able to negotiate bilateral contracts—both forward contracts and contracts for difference (CFDs). The chapter also presents a case study aiming at analyzing the role of CFDs as a financial tool to hedge against pool price volatility. The results of four different simulations involving CFDs and compensations are compared with a "no-compensation" base case involving a forward contract, allowing us to conclude that CFDs can be a useful financial tool for producers, especially when market prices drop to (or below) their marginal cost of production.

F. Lopes (✉)
LNEG—National Research Institute, Estrada do Paço do Lumiar 22, 1649-038 Lisboa, Portugal
e-mail: fernando.lopes@lneg.pt

H. Algarvio
Instituto Superior Técnico, Universidade de Lisboa
Av. Rovisco Pais 1, 1049-001 Lisboa, Portugal
e-mail: hugo.algarvio@tecnico.ulisboa.pt

J. Santana
Instituto Superior Técnico, Universidade de Lisboa
Av. Rovisco Pais 1, 1049-001 Lisboa, Portugal

INESC-ID, R. Alves Redol 9, 1000-029 Lisboa, Portugal
e-mail: jsantana@ist.utl.pt

A. Alonso-Betanzos et al. (eds.), *Agent-Based Modeling of Sustainable Behaviors*, Understanding Complex Systems, DOI 10.1007/978-3-319-46331-5_10

1 Introduction

Electric power companies operated typically as vertically integrated systems having complete control over production, transmission and distribution of energy. Deregulation unbundled the traditional utilities into separate commercial units that can now operate independently of each other. Today, market forces drive the price of electricity and reduce the net cost through increased competition. Two key objectives of electricity markets (EMs) are ensuring a secure and efficient operation and decreasing the cost of electricity utilization [1].

The economic operation of most power systems is usually managed by a market operator (MO). The security of the system operation is normally assigned to an independent system operator (ISO). The MO and the ISO are the leading entities that play a central role in power markets. Generating companies (GenCos) represent business units that own generators—they may own a single plant (and operate like an independent power producer) or multiple plants (and be part of a larger corporate parent that provides other products in the electricity market). Retailers (RetailCos) buy electricity in wholesale markets and re-sell it to consumers, which include residential, commercial and industrial consumers.

Market participants can purchase and sell electrical energy through centralized markets and bilateral contracts [2]. Centralized markets involve no direct negotiations between the parties—all market participants who wish to either sell or buy electricity on a specific delivery day submit their price and quantity offers and bids. The entire market settles simultaneously, usually following a pricing algorithm that selects the lowest priced resources needed to meet load, whether or not transmission constraints are given. Bilateral contracts are essentially agreements between buyers and sellers to trade electricity at specific terms, including duration, price per hour over the length of the contract, variable megawatt amount over the length of the contract, and range of hours when the contract is to be delivered. Most market participants and analysts agree that such arrangements are crucial to the functioning of electricity markets.

There are various types of contractual arrangements that fall under the broad heading of bilateral contracts—two of the most common are forward contracts and contracts for difference (CFDs). Forward contracts are agreements negotiated directly between two parties to exchange electric power under a set of specified conditions, with the terms of the agreements remaining fixed until the time of delivery [3]. They involve a commitment to sell or buy a specific amount of electricity at a certain future time for a specific price. CFDs are bilateral agreements to provide a specific amount of energy for a fixed price called the strike price—they are typically indexed to a reference price, which is often the centralized market price. If the strike price is higher than the reference price, the buyer pays the seller the difference between these two prices times the amount agreed in the contract. Conversely, the seller pays the buyer the difference between these two prices times the agreed amount [2]. In some cases, CFDs can be one way contracts, when the difference payments are made only by one of the parties.

Electricity markets differ from their more traditional counterparts because energy cannot be efficiently held in stock or stored (as tangible goods can). Consequently, market participants are forced to work with consumption prognoses, which, in turn, create a number of risks, notably:

- *over-generation*: energy prices mainly depend on production costs and, as a result, generating more electricity than is consumed is not economical;
- *hypo-generation*: if suppliers cannot match the demand, the lack of energy can cause power cuts (brownouts) or, if prolonged, blackouts;
- *inconsistent-generation*: non-negligible costs stemming from variations in the electricity production volume that most traditional types of energy generators have to face.

Stated simply, financial risk management is often a high priority due to the substantial price and volume risk that markets can exhibit.

Furthermore, renewable generation or variable generation, such as wind and photovoltaic solar power, has increased significantly in recent years. The European Union (EU) has been one of the major drivers of the development of renewable energies (see, e.g., [4–6]). In Portugal, renewable generation is subject to specific licensing requirements and benefits from a feed-in tariff. In Spain, two different types of retribution may be chosen for renewable energy facilities: the regulated or feed-in tariff and the market price plus premium. In UK, there have been two main policy instruments: the Non-Fossil Fuel Order (NFFO), a centralized bidding system that ran from 1990 to 1998, and the Renewables Obligation (RO), that came into effect in 2002. The Government has given "generous" subsidies, and with the banded RO providing a higher payment for immature technologies, there has been a real development of renewables. However, the UK, which was one of the pioneers of deregulating the electricity market in the nineties, has been forced to reform it to meet the challenges of the electricity sector. Contracts for difference are a key element of the electricity market reform—CFDs provide long-term price stabilisation to low carbon plants, allowing investment to come forward at a lower cost of capital and therefore at a lower cost to consumers. A CFD is a private law contract between a low carbon electricity generator and the Low Carbon Contracts Company, a Government-owned company [7].

Electricity markets are, therefore, not only a relatively new reality but also an evolving one, since both market rules and market players are constantly changing. As EMs continue to evolve, there is a growing need for advanced modeling approaches that simulate the behavior of market participants, particularly how they may react to the economic, financial and regulatory changes that can occur in the environment in which they operate. An agent-based modeling and simulation approach presents itself as a promising approach to model power markets. Conceptually, an agent-based approach is an ideal fit to the naturally distributed domain of a deregulated energy market. Software agents can be designed to act with incomplete and uncertain information, limited resources, and may efficiently manage cooperative and competitive interactions with other agents (see, e.g., [8]).

This chapter presents several key features of software gents able to negotiate bilateral contracts—both forward contracts and contracts for difference. The revenue and cost certainty associated with these types of contracts presents a number of benefits to producers and consumers. These benefits include [9]:

- *Price stability*: load serving entities able to acquire a substantial portion of their electric energy requirements through bilateral contracts at fixed, or predictable prices, will be less exposed to the risk of high spot market electricity prices;
- *Support for renewable resources*: utility-scale renewable resources characterized by non-dispatchable, variable output levels that depend on weather conditions, have high up-front costs and low running costs, making bilateral contracts particularly attractive to finance the associated projects and to provide acceptable rates of return;
- *Mitigation of market power*: the possibility to choose between a range of supply acquisition options, including spot market purchases and bilateral contracts, can help to mitigate market power;
- *Support for development of new generation resources*: developers of major new capital-intensive generation resources can obtain a guaranteed stream of future revenues by pre-selling part or all of the energy (and capacity) of such resources under long-term, bilateral contracts.

Certainly, many contractual arrangements may contain terms and conditions that differ significantly.

The work presented here builds on our previous work in the areas of automated negotiation (see, e.g., [10]) and bilateral contracting in electricity markets (see, e.g., [11, 12]). In particular, it refines and extends our previous work on contracts for difference and risk management [13, 14]. This work is also complementary to our previous work on contracts for difference in that it presents a new case study aiming at analyzing the role of CFDs as a financial tool to hedge, at least in part, against pool price volatility. Its main purpose is to compare the results of simulations involving CFDs and compensations with a "no-compensation" base case involving a forward contract.

The remainder of the paper is structured as follows. Section 2 deals with bilateral contracting in electricity markets, focusing on both forward contracts and contracts for difference. Section 3 builds on our previous work on automated negotiation and bilateral contracting and describes some key features of autonomous negotiating agents. In particular, Sect. 3.1 discusses the risk preferences of autonomous agents, Sect. 3.2 describes a computational framework that handles two-party and multi-issue negotiation, Sect. 3.3 formalizes some key terms of CFDs, and Sect. 3.4 discusses one-to-many negotiation. Section 4 presents a case study on bilateral contracting of electricity involving CFDs. The chapters ends with the presentation of concluding remarks and future avenues of research (see Sect. 5).

2 Markets for Electrical Energy

For the supply of real power, two key market models are as follows: (1) pool trading and (2) bilateral trades or contracts. Pool trading is carried out through a centrally operated entity that determines generation levels and prices based on submitted generation bids and purchase offers. Bilateral trades are defined by privately negotiated bilateral contracts that can be either physical or financial obligations. Section 2.1 is devoted to pool trading and Sect. 2.2 presents some important features of forward contracts and CFDs.

2.1 Pool Trading and Marginal Pricing

Electricity is typically bought and sold through a two-settlement system involving a day-ahead market (DAM) and a balancing/real-time market (RTM). The DAM clears to meet bid-in load demand for an entire day, 1 day in advance. The RTM reflects the actual operation of the agents participating in the market. It sets prices and schedules to match the imbalances caused by the variability and uncertainty present in power systems. Many markets also have intermediate scheduling and pricing procedures on the hour ahead or a few hours ahead to facilitate balancing in advance of real time.

The pricing mechanism of most day-ahead markets is founded on the marginal pricing theory. Generators compete to supply demand by submitting bids in the form of price and quantity pairs, for example. Similarly, retailers and possibly other market participants submit offers to buy certain amounts of energy at specific prices. Schedules and prices are calculated from the market-clearing engine, and price-quantity pairs are settled for all market participants regardless of their actual performance. Generators are instructed to produce the amount of energy corresponding to their accepted bids and buyers are informed of the amount of energy that they are allowed to draw from the system [2].

There are two main variations of marginal pricing [15, 16]: system marginal pricing (SMP) and locational marginal pricing (LMP). In SMP, the generation bids are stacked in the merit order, and the market clearing price or spot-price is defined by the intersection of the associated curve with the cumulative load curve. This price is normally determined on an hourly basis and then applied to all generators uniformly, i.e., regardless of their bids or location. SMP does not explicitly take into account transmission constraints, though several revisions and extensions can be considered to address transmission congestion problems. LMP is a more complex variation of marginal pricing—as in SMP, the system operator collects generation bids and purchase offers, and then determines the optimal generation dispatch. However, the optimization process is now subject to different constraints, such as line load-ability and voltage limits, and can include the supply of losses and other ancillary services necessary to support system operation. Typically, the system operator runs an optimal power flow procedure that defines the energy price at each bus of the network.

2.2 Forward Contracts and Contracts for Difference

A bilateral market is a market in which private parties, sellers and buyers, trade directly at negotiated prices and conditions [17]. A bilateral contract is an agreement between two parties for the exchange of electricity under mutually acceptable terms, including starting date, ending date, price, amount of traded energy, and any other terms which may be deemed applicable. Derivatives are contracts whose values depend on (or derive from) the values of other, more basic, underlying variables, typically the prices of traded assets [18]. The term derivative comes from how the price of a contract is derived from the price of some underlying commodity, security or index or the magnitude of some event. It is used to refer to a set of financial instruments that includes forwards and swaps (swaps are also known as contracts for difference [19]). Forwards and swaps are often considered essential to the functioning of electricity markets, because they allow the parties to have the price stability and certainty necessary to perform long-term planning and to make rational and socially optimal investments.

In more detail, forward bilateral contracts are agreements to sell or buy a specific amount of electricity at a certain future time for a specific price [18]. One of the parties assumes a long position and agrees to buy the energy on a future date for a predetermined price, and the other assumes a short position and agrees to sell the energy on the same date for the same price. The payoff from a long position in a forward contract on one unit of electricity is the difference between the spot price (*SP*) at maturity date and the delivery price (*DP*). Similarly, the payoff from a short position is the difference: $DP - SP$. These payoffs can be positive or negative. If enough sellers and buyers are interested in trading electricity in advance of delivery, a forward market for energy will develop. This market is essentially a decentralized market in which electricity is sold using forward bilateral contracts. The delivery time can range from days to years in the future [17].

Contracts for difference are bilateral contracts in which the purchaser pays the seller the difference between the contract price—the *strike price*—and some market price, usually the spot price [17]. The trading parties agree on a strike price and an energy quantity and then take part in a centralized market. They sell and buy their power through the pool, at the pool marginal price, and then in a separate financial transaction compensate each other for the difference between the strike and actual prices. Specifically, CFDs are settled as follows [2]:

- If the strike price is higher than the market price, the buyer pays the seller the difference between these two prices times the agreed amount of energy.
- Conversely, if the strike price is lower than the market price, the seller pays the buyer the difference between these two prices times the amount agreed.

Thus, CFDs insulate the parties from the price on the centralized market while allowing them to take part in this market.

3 Software Agents and Contract Negotiation

An ongoing study is looking at using software agents to help manage both the complexity of centralized markets and the challenges of bilateral contracting of electricity. The agents are computer systems capable of flexible, autonomous action and able to communicate, when appropriate, with other agents to meet their design objectives. They are able to exhibit goal-directed behavior, can respond to changes that occur in the marketplace in which they operate, and can reach mutually acceptable agreements through negotiation.

3.1 Risk Preferences

Software agents representing market participants are exposed to risks associated with price volatility and uncertainties regarding production and consumption. Based on their attitude towards risk, they fit into one of the following categories:

1. *Risk-averse*: agents prefer settings with fixed and certain profits rather than settings where the outcomes may be better but there are chances of getting lower profits;
2. *Risk-seeking or risk-loving*: agents prefer settings with very high returns (although not guaranteed) rather than settings with lower and certain profits;
3. *Risk-neutral*: generally, agents have no preference over the outcomes and take intermediate stances compared to the two above.

Thus, risk-averse agents usually accept profits somewhat worse than they might be able to get later in exchange for the security of getting fixed profits now. On the other hand, risk-loving agents feel usually free to take significant risks in order to secure larger profits.

The preferences of the agents are represented by a utility function $U(\mathbf{x})$ with the following properties:

(1) $U(x_1, \ldots, x_n) > U(x'_1, \ldots, x'_n)$ if agents prefer $(x_1, \ldots, x_n)$ to $(x'_1, \ldots, x'_n)$;
(2) $U(x_1, \ldots, x_n) = U(x'_1, \ldots, x'_n)$ if agents are indifferent between $(x_1, \ldots, x_n)$ and $(x'_1, \ldots, x'_n)$.

Considering choice under uncertainty, the utility function depends on the specific level x_i of attribute (or issue) X_i and also on the probability π_i that it will actually occur. Under the utility independence assumption, a particularly convenient form that the utility function might take is the following:

$$U(\mathbf{x}) = \pi_1\, v(x_1) + \pi_2\, v(x_2) \cdots + \pi_n\, v(x_n) \tag{1}$$

which is often referred to as an expected utility function or, sometimes, a von Neumann-Morgenstern utility function. This convenient form says that utility can be written as a weighted sum of some function of the outcome level on each issue, $v(x_1), \ldots, v(x_n)$, where the weights are given by the probabilities $\pi_1, \ldots, \pi_n$.

The expected utility function has some very convenient properties for analyzing choice under uncertainty. As a simple and generic example, consider a choice problem involving a uni-dimensional utility function $u(x)$ for wealth [20]. Suppose that an agent a_i has \$10 of wealth and is facing a gamble that gives it a 50 % probability of winning \$5 and a 50 % probability of losing \$5. The expected value of its wealth is \$10 and the expected utility is $1/2\ u(\$15) + 1/2\ u(\$5)$. For a risk-averse agent, the utility of the expected value of wealth, $u(1/2 \times \$15 + 1/2 \times \$5) = u(\$10)$, is greater than the expected utility of wealth, $1/2\ u(\$15) + 1/2\ u(\$5)$. A risk-averse agent prefers to have the expected value of his wealth rather than facing the gamble.

Also, for a risk-seeking agent the expected utility of wealth, $1/2\ u(\$15) + 1/2\ u(\$5)$, is greater than the utility of the expected value of wealth, $u(1/2 \times \$15 + 1/2 \times \$5) = u(\$10)$. Generally speaking, a risk-seeking agent has a convex utility function—its slope gets steeper as wealth is increased. A risk-averse agent has a concave utility function—its slope gets flatter as wealth increases. Therefore, the curvature of the utility function can help measuring a_i's attitude toward risk. For the intermediate case (risk-neutral), the utility function is linear. The expected utility of wealth is the utility of the expected value. The agent does not care about the riskiness of its wealth—only about its expected value [20].

Now, an approach to measure local risk aversion (risk aversion in the small), or local propensity to insure at the point x under the utility function u, is through a local risk aversion function $r(x)$ [21]:

$$r(x) = -\frac{u''(x)}{u'(x)} \tag{2}$$

The absolute magnitude of $u''(x)$ does not in itself have any particular meaning. However, one feature of $u''(x)$ does have a meaning, namely its sign, which equals that of $-r(x)$. A negative (positive) sign at x implies unwillingness (willingness) to accept risks with asset x. Furthermore, a negative (positive) sign for all x implies strict concavity (convexity) and hence unwillingness (willingness) to accept any risks with any assets.

Let $u_1(x)$ and $u_2(x)$ be utility functions with local risk aversion functions $r_1(x)$ and $r_2(x)$, respectively. If, at point x, $r_1(x) > r_2(x)$, then u_1 is locally more risk-averse than u_2 at x. It follows that the corresponding global property also holds—if $r_1(x) > r_2(x)$ for all x, that is, u_1 has greater local risk aversion than u_2 everywhere, then u_1 is also globally more risk-averse in a natural sense [21]. Accordingly, in this work we consider a parameter $\lambda \in [-1, 1]$ correlated with $r(x)$. Given λ, and considering the above discussion, software agents representing market participants are classified according to Table 1.

Table 1 Agent classification according to the attitude towards risk

Level of risk aversion	Value of $r(x)$	Interval for λ
Risk-averse	$r(x) > 0$	$\lambda \in]0,1]$
Risk-neutral	$r(x) = 0$	$\lambda = 0$
Risk-seeking (or risk-loving)	$r(x) < 0$	$\lambda \in [-1,0[$

3.2 Bilateral Negotiation

Let $A = \{a_s, a_b\}$ be the set of autonomous agents and denote the agenda—that is, the set of issues at stake—by I. For each issue X in I, the range of acceptable values is represented by the interval $D = [x_{min}, x_{max}]$. The priority $prt \in \mathbb{N}$ of an issue X included on the agenda is a positive integer that represents the importance of X. The weight $w \in [0,1]$ is a real number that represents the preference of an agent $a_i \in A$ for X. The limit *lim* of an issue X in I is the ultimate fallback position for X, the point beyond which a_i is unwilling to concede on X. The opening x_{opn} is the preferred value for X, the best outcome a_i realistically hopes to achieve.

One of the key aspects of bilateral negotiation is the adoption of a negotiation protocol that settles the rules of trading. We consider an alternating offers protocol [22]. The two agents negotiate by alternately submitting offers (or proposals) at times in $T = \{1, 2, \ldots\}$. This means that only one proposal is submitted in each period, with an agent, say a_i, offering in odd periods $\{1, 3, \ldots\}$, and the other agent $a_j \in A$ offering in even periods $\{2, 4, \ldots\}$. In period $t \in T$, a proposal submitted by a_i to a_j is a vector of issue values:

$$p^t_{i \to j} = (x_1, x_2, \ldots) \tag{3}$$

where x_1 and x_2 are values of issues $X_1 \in I$ and $X_2 \in I$, respectively. The agents have the ability to unilaterally opt out of the negotiation when responding to a proposal made by the opponent.

The negotiation process starts with a_i submitting the proposal $p^1_{i \to j}$ to a_j in period $t = 1$. The agent a_j receives $p^1_{i \to j}$ and can either accept the offer, reject it and opt out of the negotiation, or reject it and continue bargaining. In the first two cases, negotiation comes to an end. Specifically, if $p^1_{i \to j}$ is accepted, negotiation ends successfully and the agreement is implemented. Conversely, if $p^1_{i \to j}$ is rejected and a_j decides to opt out unilaterally, negotiation terminates with no agreement. In the last case, negotiation proceeds to the next time period $t = 2$, in which a_j makes a counter-proposal $p^2_{j \to i}$ (a counter-offer is therefore an offer made in response to a previous offer). This process repeats until one of the outcomes mentioned above occurs.

Negotiation strategies are computationally tractable functions that define the negotiation tactics to be used during the course of negotiation. The agents can pursue several strategies that model typical patterns of concessions. For instance, they can start with ambitious demands, well in excess of limits and aspirations, and concede slowly. High demands and slow concessions, also referred to as "starting high and conceding slowly", are often motivated by concern about position loss and image loss. A formal definition of a negotiation strategy that models this and other existing forms of concession making is presented elsewhere [23].

Also, the agents can have different strengths of preference for the issues at stake—they can place greater emphasis on some key issues and make significant efforts to resolve them favorably. Hence, they may be more willing to make larger concessions on less important issues. The strategy of "low-priority concession making" involves changes of proposals in which larger concessions are made on low-priority than on high-priority issues (see [23] for a formal definition).

Concession tactics are functions that model the specific concessions to be made throughout negotiation—that is, they generate new values for each issue at stake. Let $X \in I$ designate an issue and denote its value at time t by x. Formally, a concession tactic for X is a function with the following general form:

$$Y(x) = x + (-1)^m \, C_f(x - lim) \tag{4}$$

where $C_f \in [0, 1]$ is the concession factor, $m = 0$ if an agent a_i wants to minimize X or $m = 1$ if a_i wants to maximize X, and lim is the limit for X.

The concession factor C_f can be simply a positive constant independent of any objective criteria. Alternatively, C_f can be modelled as a function of a single criterion. Typical criteria include the total concession made on each issue throughout negotiation [24] and also [25]: the time elapsed since the beginning of negotiation, the quantity of resources available, and the previous behavior of the opponent.

For bilateral contracting in EMs, a useful criterion is the amount or quantity of energy for a specific trading period [26]. The associated "energy dependent concession making" strategy involves individual decisions to make concessions based on the (expected) quantity of energy to be consumed in a given block of time. A different criterion was introduced in [13], namely the attitude towards risk. Risk-averse agents are willing to be flexible to secure a deal, typically making substantial concessions to avoid ending negotiation prematurely without agreement. On the other hand, risk-seeking agents tend to be more rigid and firm than risk-averse agents, typically adopting a tougher, more competitive stance, and thus conceding less throughout negotiation. They are more willing to show a weak concern for negotiation success (and tend to adopt small concession factors). The concession factor $C_f(\lambda)$ may be represented by considering either a polynomial or an exponential function. A generic exponential function follows [13]:

$$C_f(\lambda) = C_{fn} \times e^{c\,\lambda} \tag{5}$$

where λ is the risk aversion parameter, C_{fn} is the concession factor for a risk-neutral agent ($\lambda = 0$), and c is a parameter that shapes the function's curvature.

Equation (5) represents a family of tactics, one for each pair of values (C_{fn}, c). Accordingly, several simulations were made to define appropriate values for these parameters (see Fig. 1). After a detailed analysis, we chose the data series 2 in Fig. 1: $C_{fn} = 0.10$ and $c = 0.55$, which leads to the following exponential function: $C_f = 0.1\, e^{0.55\,\lambda}$.

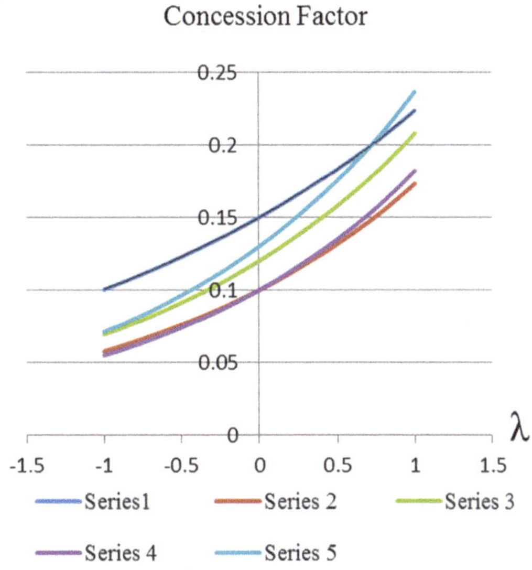

Fig. 1 The concession factor for a given measure of risk aversion (from [13])

3.3 *Contracts for Difference and Bilateral Negotiation*

Contracts for difference are agreements to provide a specific amount of energy for a strike price and, as noted earlier, are typically indexed to a reference price. In this work, we consider that CFDs may also specify the provision of different amounts of energy for different blocks of time, at somewhat different prices. This generalization accounts for time-varying rates (TOU rates) that reflect the value and cost of electricity in different blocks of a day—that is, 2-rate tariffs (peak/off-peak), 3-rate tariffs (peak/medium/off-peak) or even 24-rate tariffs (hour-wise tariffs).

Without loss of generality, consider that the set of issues to negotiate include only the most important issues, i.e., the prices and quantities (or volumes) of energy.[1] In other words, the agenda I includes n strike prices and n energy quantities:

$$I = \{P_1, \ldots, P_n, Q_1, \ldots, Q_n\} \tag{6}$$

where P_k is a strike price and Q_k is an energy quantity (for $k = 1, \ldots, n$ and some $n \in \mathbb{N}$; for the purposes of this work $n \leq 24$).

Let p_k denote the value of P_k for quantity q_k of Q_k. Also, let rp_k be the value of a reference price RP_k associated with a specific block of a day (as noted, CFDs are typically indexed to reference prices, which are often the spot market prices). The financial compensation associated with CFDs—buyers (sellers) pay to sellers (buyers) the absolute value of the difference between the strike and reference prices

[1]This framework may be readily adapted, refined and extended to account for other issues (e.g., duration and time of delivery).

times the agreed amount of energy—can now be formalized. Specifically, when the strike prices are smaller than the reference prices, sellers pay to buyers. The total amount is given by the following expression:

$$C_s = \sum_{k=1}^{n} (rp_k - p_k) \times q_k \tag{7}$$

Conversely, buyers pay a financial compensation to sellers when the strike prices are higher than the reference prices. The total amount is as follows:

$$C_b = \sum_{k=1}^{n} (p_k - rp_k) \times q_k \tag{8}$$

3.4 Multilateral Negotiation

Negotiation may involve two parties (bilateral negotiation) or more than two parties (multilateral negotiation). In bilateral negotiation, a particular market participant (e.g., a generation company or GenCo) can negotiate directly with other market participant (e.g., a retailer or RetailCo) a mutually acceptable agreement—that is, two isolated individuals can negotiate for their own needs and interests. Multilateral negotiation may be classified according to the number of parties involved. One-to-many negotiation occurs when a single party negotiates with a number of other parties. In this type of negotiation, a RetailCo agent may negotiate with several GenCos to obtain a better agreement. A description of the negotiation process follows (for convenience, we consider a RetailCo agent and several GenCos, although one-to-many negotiation may naturally involve other types of market participants).

A RetailCo agent formulates a request for proposal (RFP) for energy on the basis of the anticipated needs of its customers and its risk tolerance for exposure to pool market price volatility. The RFP is often based on projections of its customers' future energy requirements and thus is subject to uncertainty—this uncertainty can stem from such factors as weather forecast errors, inaccurate projections of the type and number of future customers, unpredictable variability in customers' electricity consumption patterns, or anticipated future prices of short-term markets.

The RFP is sent to different GenCos, who analyze it, formulate bid responses, and submit the responses (offers) to the RetailCo agent. The offers may include prices for all or some portion of the requested energy. They are often based on projected market prices—projections are a function of historical information that acts as a collection of past experiences. Each GenCo weighs the costs and benefits (i.e., the net profits) of entering into a bilateral contract versus selling its energy production on the pool. In addition to net profits, each GenCo may also consider other factors into its final pricing response to an RFP, such as the risk of rejection. The RetailCo agent evaluates the offers and either accepts or rejects them. Typically, this agent accepts the offer (if any) that maximizes corporate utility. The decision is based on the offers received and projected market prices.

At this stage, both the RetailCo agent and the GenCos may revise their marketing strategies and initiate a new round of RFPs—dynamic strategy revision may be based on lessons learned from previous rounds and newly forged bilateral agreements among market players.

4 Agent-Based System for Electricity Markets

The major components of the system under development include a graphical user interface, a simulation engine, and a number of domain-specific agents. The graphical user interface allows users to set agent-specific parameters, specify and monitor trading simulations, and perform a variety of administrative tasks such as saving simulations (see Fig. 2, below). The simulation engine does not rely on any domain-specific knowledge and controls all trading simulations.

The agents represent typical market participants, including generating companies (GenCos), retailers (RetailCos), aggregators, large and small consumers, and a market operator. GenCos may own a single generating plant or a portfolio of plants of different technologies. RetailCos buy electricity in a wholesale market and re-sell it to customers in a retail market (typically, end-use customers that are not allowed, or do not want, to participate in the wholesale market). Aggregators are entities that support groups of end-use customers in trading electrical energy. Large consumers can take an active role in the market by buying electrical energy in the pool or by signing bilateral contracts (e.g., with producers). Small consumers, on the other hand, buy energy from retailers and possibly other market participants. The agents are currently being developed using the JAVA programming language and the JADE platform [27].

The system supports pool trading and bilateral contracting of electricity. A day-ahead market sells energy to RetailCos and buys energy from GenCos in advance of time when the energy is produced and consumed. The pricing mechanism is founded on the marginal pricing theory—both system marginal pricing and locational marginal pricing are supported. Bilateral contracts can be either physical or financial obligations. Buyer and seller agents equipped with the negotiation framework presented in the previous section are able to negotiate beneficial agreements.

4.1 Case Study

This subsection presents a case study aiming at: (1) illustrating how software agents negotiate both forward contracts and contracts for difference in a multi-agent energy market, and mainly (2) analyzing the role of contracts for difference as a risk

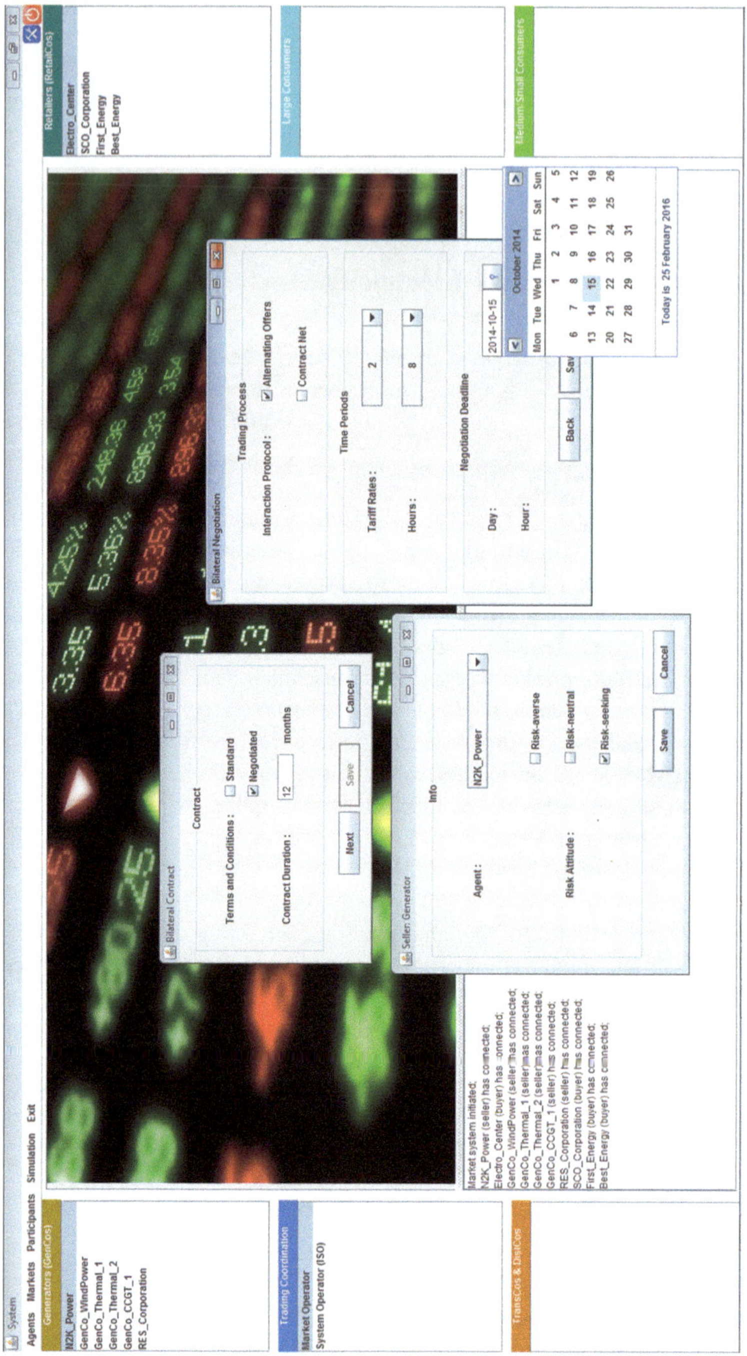

Fig. 2 Snapshot of the agent-based system for simulating electricity markets

Table 2 Details of the five simulations: agents' attitudes toward risk and contract type

	Case 1	Case 2	Case 3	Case 4	Case 5
Seller's attitude	r-averse	r-seeking	r-seeking	r-averse	r-neutral
Buyer's attitude	r-seeking	r-averse	r-seeking	r-averse	r-neutral
Type of contract	CFD	CFD	CFD	CFD	Forward

r-seeking: risk-seeking attitude; r-neutral: risk-neutral attitude; r-averse: risk-averse attitude

management tool. The case study involves five simulations (see Table 2). The first four simulations (case 1–case 4) differ mainly in the attitude towards risk of the two negotiating agents. In these simulations, the agents negotiate a two-way contract for difference. In the base case (case 5), the agents negotiate a forward contract and are assumed to adopt a risk-neutral attitude. The main purpose of the simulations is to compare the results of the cases involving compensations (cases 1 to 4) with the "no-compensation" base case (case 5) to analyze the role of CFDs as a financial tool to hedge against pool price volatility.

The agents are N2K_Power—a seller agent representing a generating company—and Electro_Center—a buyer agent representing a retailer. N2K_Power should decide how to operate its production units and its involvement in bilateral contracting (and the pool) on a daily basis. Likewise, the retailer Electro_Center should consider signing one or more bilateral contracts to procure the electrical energy it needs to supply its clients. Accordingly, we consider that the agents decide to negotiate a 12-month contract for the following period: January 2015 to December 2015. The negotiation process is assumed to start on October 1st, 2014 and end on October 15th, 2014—that is, the negotiation deadline is set to 2 weeks.

The agents decide to consider a rate with different unit prices for usage during different blocks of time, defined for a 24 h day. Specifically, they consider the following two-rate tariff:

1. Peak hours: 8:00 a.m.–12:00 p.m.
2. Off-peak hours: 12:00 p.m.–8:00 a.m.

This rate reflects the average cost of generating and delivering power at different times (i.e., higher prices during the day—day rate—and lower prices at night—night rate).

The generation technologies of N2K_Power are assumed to be renewable (wind) and combined cycle (natural gas), the maximum capacity 1250.00 MW, and the marginal cost 50.00 €/MWh (we consider a near-zero variable production cost for wind generation). We also consider that the retailer Electro_Center wants to sign a contract for the purchase of the following daily volumes: $Q_1 = 830.00$ MWh (peak) and $Q_2 = 500.00$ MWh (off-peak). These two quantities are assumed to be constant during the course of negotiation. Accordingly, the negotiating agenda includes two strike prices, P_1 and P_2. The two reference prices are set as follows: $RP_1 = 53.99$

Table 3 Initial offers, price limits and energy quantities

Agent	Period	Initial offer (€/MWh)	Price limit (€/MWh)	Quantity (MWh)
Seller	Peak	67.00	62.00	–
	Off-peak	52.00	50.00	–
Buyer	Peak	55.00	70.00	830.00
	Off-peak	45.00	55.00	500.00

€/MWh (peak) and $RP_2 = 43.28$ €/MWh (off-peak). These prices were obtained from the Iberian Market, considering the time period of the contract.[2]

The initial offers and price limits for both agents are shown in Table 3. The prices for N2K_Power were obtained by considering the marginal cost of production. The prices for Electro_Center were defined by considering averages prices from the Iberian Market for the period between June 1st, 2014 and September 30th, 2014. Specifically, the buyer agent sets the initial prices 0 % (peak) and 2 % (off-peak) lower than the market prices, and the limits 27 % (peak) and 20 % (off-peak) higher than the market prices.

Since contracts for difference involve only financial settlements, we consider that both agents seek this kind of deal for financial reasons only (and not for satisfying their energy needs). Accordingly, their attitude towards risk in cases 1–4 is non-neutral (i.e., either risk-averse or risk-seeking). In particular, the attitude of N2K_Power changes from risk-averse (cases 1 and 4) to risk-seeking (cases 2 and 3), mainly to analyze the potential consequences of considering different attitudes towards risk. The retailer Electro_Center adopts a risk-seeking attitude in both cases 1 and 3. This is mainly because this agent can "search" the market for a better deal—that is, if it fails to reach an agreement with N2K_Power, there is the possibility to look for another seller to negotiate a fair and reasonable settlement. This agent adopts a risk-averse attitude in cases 2 and 4, again to simulate different attitudes towards risk and analyzing their potential consequences.

Negotiation involves an iterative exchange of offers and counter-offers. Both agents prepare offers according to the negotiation strategy presented in the previous section, referred to as *risk-preference concession making strategy*—it models successive concessions during the course of negotiation based on both the risk aversion parameter λ and the concession factor C_{fn} for a risk-neutral agent (but see Eqs. (4) and (5), above). Thus, we consider that a risk-averse agent shows typically more flexibility to secure a deal, and therefore, tends to concede more to avoid that negotiation ends prematurely without agreement. On the other hand, a risk-seeking agent tends to be more rigid and firm, typically conceding less than the opponent. By engaging in this behavior, negotiation may end without an agreement being in place. Despite this, if negotiation ends successfully with agreement, a risk-seeking agent will probably benefit more than a risk-averse agent in similar situations.

The five simulation processes ended successfully, i.e., the agents reached agreement after exchanging a moderate numbers of proposals. The results are presented

[2]http://www.mercado.ren.pt/EN/Electr/MarketInfo/MarketResults/OMIE/Pages/Prices.aspx (accessed on February 2016).

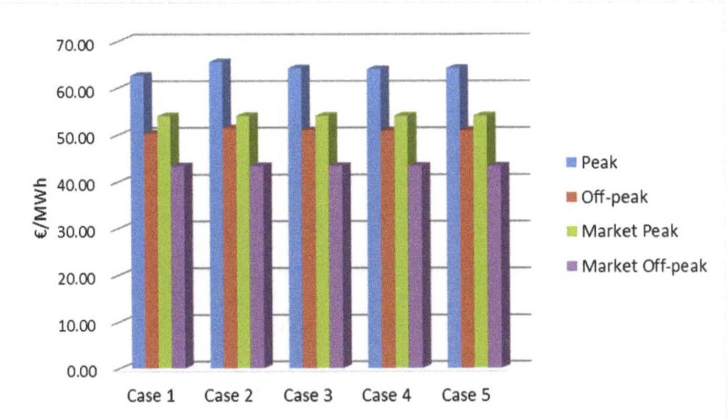

Fig. 3 Simulation results (strike prices) and market reference prices

Table 4 Simulation results: strike prices (€/MWh)

Period	Case 1	Case 2	Case 3	Case 4	Case 5
Peak	62.62	65.50	64.18	63.93	64.15
Off-peak	50.25	51.40	50.87	50.77	50.86

in Fig. 3 and Table 4. Since the prices in 2015 were higher than in 2014, the strike prices negotiated for all cases were higher than the reference prices. Therefore, the retailer Electro_Center needs to pay the difference to the producer N2K_Power (the compensations are shown in Fig. 4).

It is possible to draw some interesting conclusions from the results. Firstly, CFDs seem to be a useful financial tool to hedge against pool price volatility. The producer N2K_Power has (effectively) protected from the drop in prices by receiving payments from the retailer Electro_Center. Secondly, N2K_Power can "maximize" its profit by adopting a risk-seeking attitude during negotiation. This can be shown by comparing the prices in cases 2 and 3 (N2K_Power adopts a risk-seeking attitude) with the prices in cases 1 and 4 (N2K_Power adopts a risk-averse attitude). A third and final conclusion is also worth to mention. CFDs can be a useful financial tool for producers, especially when market prices drop to (or below) their marginal cost of production, since they will receive a compensation from buyers (even without operating their plants).

5 Conclusion

This chapter has presented several key features of software agents able to negotiate both forward contracts and contracts for difference—the agents represent typical participants of electricity markets (e.g., generating companies and retailers). A risk-preference concession making strategy was also introduced. Agents pursuing the

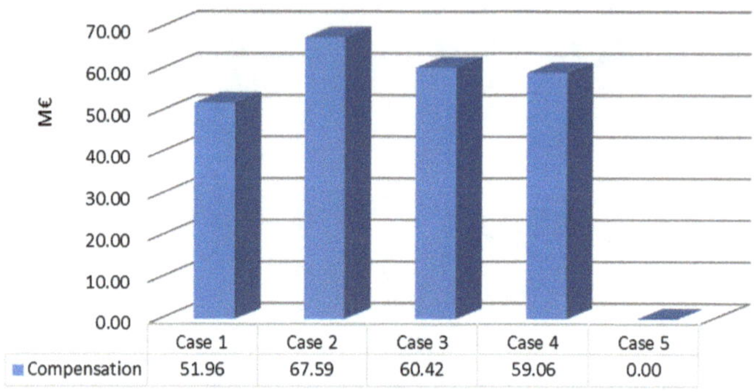

Fig. 4 Compensations (perspective of the producer N2K_Power)

strategy—and their associated negotiation tactics—were able to prepare offers and counter-offers according to their attitude towards risk (either risk-seeking, risk-neutral, or risk-averse), and also to reach mutually beneficial agreements.

A case study was also presented to illustrate how agents negotiate in a multi-agent energy market, and mainly to analyze the role of contracts for difference as a risk management tool. The case study involved five simulations. In the first four simulations, the agents adopt different attitudes towards risk and negotiate a two-way contract for difference. In the base case simulation, the agents negotiate a forward contract and are assumed to adopt a risk-neutral attitude. The results shown that CFDs can be a useful financial tool to hedge against pool price volatility.

Acknowledgements This work was performed under the project MAN-REM (FCOMP-01-0124-FEDER-020397), supported by both FEDER Funds through the program "COMPETE—Programa Operacional Temático Factores de Competividade" and National Funds through "FCT—Fundação para a Ciência e a Tecnologia".

References

1. Shahidehpour, M., Yamin, H., Li, Z.: Market Operations in Electric Power Systems. Wiley, Chichester (2002)
2. Kirschen, D., Strbac, G.: Fundamentals of Power System Economics. Wiley, Chichester (2004)
3. Khatib, S., Galiana, F.: Negotiating bilateral contracts in electricity markets. IEEE Trans. Power Syst. **22**(2), 553–562 (2007)
4. European Union: Energy for the future: renewable sources of energy. White paper for a community strategy and action plan, COM(97) 599 (26 November 1997). http://europa.eu/documents/comm/white_papers/pdf/com97_599_en.pdf. Cited 09 Feb (2016)

5. European Union: Renewable energy road map renewable energies in the 21st century: building a more sustainable future. communication from the commission to the council and the European Parliament, COM(2006) 848 (10 January 2007). http://eur-lex.europa.eu/legal-content/EN/TXT/PDF/?uri=CELEX:52006DC0848&from=EN. Cited 09 Feb (2016)
6. European Union: Directive 2009/28/EC of the European Parliament and of the Council on the promotion of the use of energy from renewable sources and amending and subsequently repealing Directives 2001/77/EC and 2003/30/EC (23 April 2009). http://eur-lex.europa.eu/legal-content/EN/TXT/?uri=CELEX:32009L0028. Cited 09 Feb (2016)
7. Department for Energy and Climate Change: Implementing electricity market reform (June 2014) https://www.gov.uk/government/publications/implementing-electricity-market-reform-emr. Cited 09 Feb (2016)
8. Wooldridge, M.: An Introduction to Multi-agent Systems. Wiley, Chichester (2009)
9. Hausman, E., Hornby R., Smith, A.: Bilateral contracting in deregulated electricity markets. Report to the American Public Power Association, Synapse Energy Economics (April 2008)
10. Lopes, F., Coelho, H. (eds.): Negotiation and Argumentation in Multi-agent Systems. Bentham Science, The Netherlands (2014)
11. Lopes, F., Novais, A.Q., Coelho, H.: Bilateral negotiation in a multi-agent energy market. In: Huang, D.-S., Jo, K.-H., Lee, H.-H., Kang, H.-J., Bevilacqua, V. (eds.) Emerging Intelligent Computing Technology and Applications, pp. 655–664. Springer, Heidelberg (2009)
12. Lopes, F., Rodrigues, T., Sousa, J.: Negotiating bilateral contracts in a multi-agent electricity market: a case study. In: Hameurlain, A., Tjoa, A., Wagner, R. (eds.) 23rd Database and Expert Systems Applications (DEXA 2012), pp. 326–330. IEEE, New York (2012)
13. Sousa, F., Lopes, F., Santana, J.: Contracts for difference and risk management in multi-agent energy markets. In: Demazeau, Y., Decker, K., Pérez, J., De la Prieta, F. (eds.) Advances in Practical Applications of Agents, Multi-Agent Systems, and Sustainability: The PAAMS Collection (PAAMS 2015), pp. 155–164. Springer International Publishing, New York (2015)
14. Sousa, F., Lopes, F., Santana, J.: Multi-agent electricity markets: a case study on contracts for difference. In: Spies, M., Wagner, R., Tjoa, A. (eds.) 26th Database and Expert Systems Applications (DEXA 2015), pp. 86–90. IEEE, New York (2015)
15. Schweppe, F., Caramanis, M., Tabors, R., Bohn, R.: Spot Pricing of Electricity. Kluwer Academic Publishers, Boston (1988)
16. Hogan, W.: Competitive Electricity Market Design: A Wholesale Primer. Harvard University, Cambridge, MA (1998)
17. Stoft, S.: Power System Economis: Designing Markets for Electricity. IEEE Press and Wiley Interscience, Hoboken (2002)
18. Hull, J.: Options, Futures, and Other Derivatives. Pearson Education, New York (2015)
19. Clewlow, L., Strickland, C.: Energy Derivatives: pricing and Risk Management. Lacima Publications, London (1998)
20. Varian, R.: Intermediate Microeconomics: A Modern Approach. W.W. Norton & Company Inc, New York (2010)
21. Pratt, J.: Risk aversion in the small and in the large. Econometrica **32**(1/2), 122–136 (1964)
22. Osborne, M. Rubinstein, A.: Bargaining and Markets. Academic, London (1990)
23. Lopes, F., Coelho, H.: Strategic and tactical behaviour in automated negotiation. Int. J. Artif. Intell. **4**(S10), 35–63 (2010)
24. Lopes, F., Mamede, N., Novais, A.Q., Coelho, H.: Negotiation among autonomous agents: experimental evaluation of integrative strategies. In: Bento, C., Cardoso, A., Dias, G. (eds.) 12th Portuguese Conference on Artificial Intelligence (EPIA 2005), pp. 280–288. IEEE Computer Society Press, New York (2005)
25. Faratin, P., Sierra, C., Jennings, N.: Negotiation decision functions for autonomous agents. J. Robot. Auton. Syst. **24**(3–4), 159–182 (1998)
26. Lopes, F., Algarvio, H. Coelho, H.: Bilateral contracting in multi-agent electricity markets: negotiation strategies and a case study. In: 10th International Conference on the European Energy Market (EEM-13), pp. 1–8. IEEE, New York (2013)
27. Bellifemine, F., Caire, G., Greenwood, D.: Developing Multi-Agent Systems with JADE. Wiley, Chichester (2007)

Energy Management in the Smart Grids via Intelligent Storage Systems

Joelle Klaimi, Rana Rahim-Amoud, and Leila Merghem-Boulahia

Abstract The increased power demand and the renewable energy integration problems have led to the evolution of the traditional electric power grid toward smart grid. In order to permit the interaction among computational and physical elements, the smart grid supports bidirectional information flows between the energy user and the utility grid by integrating Information and Communication Technologies (ICTs). Thus, bidirectional flows between smart grid entities allow energy users not only to consume energy, but also to generate energy and to share it with the utility grid or with other energy consumers. Some researchers have paid attention to the energy management in the smart grids in order to provide an efficient way to maximize the savings of energy bills. However, these researches present some common drawbacks such as: the lack of integration of storage system and the high energy losses. Therefore, this chapter discusses a novel agent-based approach for energy management and control by balancing electric power supply, and minimizing energy bill, while considering residential consumers preferences and comfort level. Simulation results show that our proposal minimizes the energy costs for each energy demand and reduces conventional energy utilization.

J. Klaimi (✉)
EDST, Centre-Azm, Laboratoire des Systèmes électroniques, Télécommunications et Réseaux (LaSTRe), Lebanese University, Beirut, Lebanon

ICD/ERA (UMR CNRS 6281), Troyes University of Technology, Troyes, France
e-mail: joelle.klaimi@gmail.com

R. Rahim-Amoud
EDST, Centre-Azm, Laboratoire des Systèmes électroniques, Télécommunications et Réseaux (LaSTRe), Lebanese University, Beirut, Lebanon

Faculty of Sciences, Lebanese University, Beirut, Lebanon
e-mail: rana.rahim@ul.edu.lb

L. Merghem-Boulahia
ICD/ERA (UMR CNRS 6281), Troyes University of Technology, Troyes, France
e-mail: leila.merghem_boulahia@utt.fr

A. Alonso-Betanzos et al. (eds.), *Agent-Based Modeling of Sustainable Behaviors*, Understanding Complex Systems, DOI 10.1007/978-3-319-46331-5_11

1 Introduction

Nowadays, the dependency level of human on electricity is an increasing phenomenon [1, 2] then the idea of the smart grid has evolved. According to the European Commission Task Force on Smart Grid, the smart grid is defined as "an electricity network that can intelligently integrate the action of all users connected to it generators, consumers and those that do both in order to ensure economically efficient and sustainable power system with low losses, high level of quality, security of supply and safety" [3]. In fact, the smart grid introduces "prosumer" concept. The prosumer is a consumer (residential, commercial, etc.) that is an energy supplier as well as a consumer. Moreover, this grid grants the two-way flow of information and electricity between consumers and electric power companies [4]. In this context, all system components (home, industry, etc.) exchange information and electricity within the smart grid as shown in Fig. 1.

This two-way communication is responsible for transmitting power grid sensing and measuring status, as well as the control messages [4]. So it helps consumers to control their energy usage [5]. To provide near real-time information to utilities and end users, the smart grid introduces several technologies such as ICTs to guarantee reliability, demand management, storage, distribution and transport of electrical energy [3, 6].

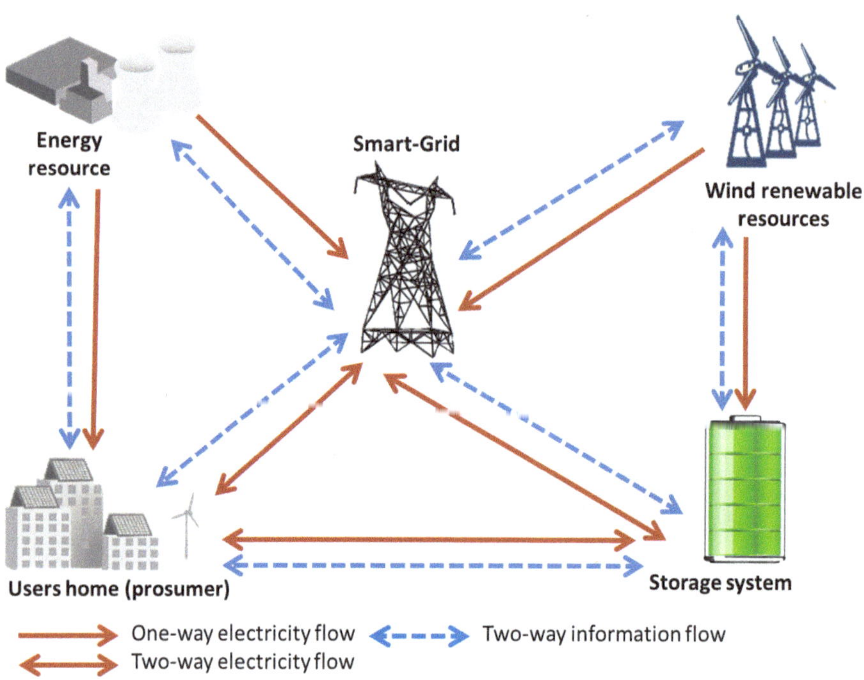

Fig. 1 The two-way flow of information and electricity in the smart grid

Fossil fuels represent the major energy sources in most countries. Absolutely, the fossil fuel resources release toxic gas emissions (e.g. CO_2) that increase the global warming and affect human health. Moreover, these non-renewable resources are exhaustible and no longer available if once used. In this context and to minimize the fossil fuel consequences, smart grid is expected to incorporate distributed renewable resources such as wind or solar energy to offer a green solution compared to traditional energy sources. Although renewable energy is inexhaustible, it is however intermittent and irregular which makes it difficult to balance supply and demand. Thus, energy management algorithms are of great importance to integrate renewable energy use, to meet energy needs, to minimize energy loss and to reduce consumers' energy bill.

The Energy Management (EM) is the set of strategies and functions needed to increase energy efficiency. Furthermore, the EM coordinates several energy sources optimally. It is the process of observing, controlling and conserving electricity usage in a building, a neighborhood, etc. It should provide a solution that optimizes costs and minimizes the risk of production excess loss [7–9].

On the other hand, at certain hours of the day or periods of the year the produced energy can exceed the consumption needs. This energy excess justifies the use of a storage system that allows short-term balance [10].

To be able to make intelligent decisions about energy management, a multi-agent approach has been used by several researchers. Because of its distributed nature, a multi-agent system (MAS) represents an excellent tool for self-control in widely distributed systems whose characteristics are very dynamic [11, 12]. In this paper, we propose a negotiation algorithm to help consumer choose the appropriate producer that provides the energy he needs at the lowest price.

The remainder of this chapter is organized as follows. In Sect. 2, we describe some of the existing approaches that are related to our work. An explanation of the multi-agent concept and storage systems will be given in Sect. 3. We present in Sect. 4 our proposal and we explain our model and its characteristics, the agents' algorithms and the chosen scenarios. We describe in Sect. 5 the simulation results. We conclude our chapter and give our future research directions in Sect. 6.

2 Background and Related Work

Several researchers have paid attention to the use of multi-agent systems and/or storage systems for energy management in smart grids or microgrids. Typically, the microgrid is composed of distributed generators, distributed storage systems and loads [13].

Some approaches have used storage systems in order to help in smart grid energy management [14, 15]. The storage system presented in [14] is composed of many storage agents. Alberola et al. used the predicted data (predicted grid price and predicted users demand) that help to store the needed energy in the storage systems. Moreover, each storage agent chooses the best period to buy energy from the electrical grid. In fact, this period is the one when the forecasted grid price is the lowest during the day. Furthermore, all system users purchase their needs from one or more storage system so they do not request energy from the grid. In this approach, the authors aim to decrease the energy bill. In the other hand, this approach does not take into consideration the use of renewable resources and the case when the consumer can buy the energy immediately from the electrical grid.

Furthermore, an algorithm for energy management in smart grid using a multi-agent system was proposed by Nagata et al. [16]. This algorithm aims to reduce the energy costs. In this algorithm, the grid operator announces two types of parameters: the selling price (SP0) and the buying price (BP0). A centralized entity named the Smart Grid Controller unit (SGC) objective is to negotiate and minimize the buying prices for consumers. Furthermore, the smart grid has several production units and consumption units. These units adapt their set points (SPi and BPi) after negotiation with other units based on the grid price, their operational costs, and the load demand.

In this research work, the system is composed of six types of agents: (1) the Grid Agent (GridAG) that has two principal roles: buying kWh from the smart grid or selling kWh to it, (2) the Generator Agent (GAG) that is responsible for maximizing the total benefit, (3) the Seller Agent (SAG) that is created by the generator agent to trade kWh to the market, (4) the Load Agent (LAG) that aims to create many buyer agents, (5) the Buyer Agent (BAG) that is responsible for purchasing kWh from the grid and finally, (6) the Smart Grid Control Agent (SGC) that aims to optimize its operation by introducing a negotiation algorithm. It is worth noting that, in this research, every SAG and BAG has a limited amount of energy. The centralization and the lack of storage systems are the disadvantages of this research work.

Colson et al. [17] proposed an energy management algorithm for microgrid using a MAS. The authors integrate three basic agent types: the producer agent, the consumer agent and the observer agent. The producer agent aims to monitor power, determine the cost of the power outfitted by the agent and maintain information to other agents. The consumer agent is responsible for monitoring the amount of consumed power, determining the amount of the instantaneous active and reactive power and negotiating the purchase of energy. Finally, the observer agent aims to monitor special parameters within the network and transfer information to other agents according to the state of the node. In this research, consumer agents benefit from negotiation to minimize their bill. The lack of the energy storage system integration is the drawback of this algorithm.

In the research presented in [18], the authors proposed an agent-based algorithm to efficiently use the energy while minimizing cost and taking into consideration the use of renewable energy producers (e.g. solar and wind). In this research, there are three types of agents:

1. Generator Company agents (GC): they are the major sources of electricity and defined by the big generator companies.
2. Prosumer agents: they are small sources of renewable energy distributed near the consumers. They definitely produce and consume energy. The production of the prosumers is negligible compared to the production of the GC.
3. Consumer agents: in this proposal, a consumer agent is a purchaser of energy only.

In this proposal, prosumers prefer to sell the unused quantity of energy. In order to reduce their own bill, the consumer agents negotiate the purchase of energy. The consumers demands are divided into three priorities. The vital loads are satisfied firstly and the consumer negotiates the cost and to buy energy from the source that reduces its own costs. In this proposal, the renewable sources are used to meet the demands firstly. Moreover, when the renewable sources have not a sufficient supply to meet all consumers demands, then the traditional sources are used. In this algorithm, the authors do not use the energy storage systems that would improve outcomes and reduce the consumers bill.

Likewise, the research described in [19] proposed a novel agent-based energy management algorithm for smart grids using a MAS and an intelligent storage system. In this research, the authors argue that the use of storage systems reduces the access to the grid and the consumers' bills. This novel algorithm aims to resolve the problem of generation intermittency and to optimize in real time the consumer bill. Klaimi et al. [19] proposed a multi-agent system composed of four types of agents: (1) the grid agent introduced in the grid, (2) the storage agent introduced in a centralized battery, (3) the prosumer agent introduced in renewable resources and (4) the consumer agent introduced in smart homes. The grid agent role is to satisfy the energy lack and buy the excess of energy produced. The storage agent controls the energy stored. The prosumer agent aims to manage the distribution of its own production. Finally, the consumer agent buys its needs from production units. It is worth noting that this work reduces the energy cost and the grid utilization in the high peak periods where the grid is charged and the energy price is high.

In [20], the authors establish a decentralized energy management algorithm that helps to integrate distributed energy sources in the smart grid context. The main objective of this work is to reduce demands costs. In this proposal, a unit management module (UMM) and a central agent are used. This proposal adopts an iterative approach. At each iteration, the central agent sends information to the UMMs with its chosen strategies. After that, UMMs optimize the proposed strategy and send back their new prices to the central agent in order to maximize their profits. After gathering the bids, the procurement strategy will be modified whenever bids are in line with the cost minimization objective. The mutual interaction ends when no further modification in the procurement strategy is experienced. However, this approach drawback is the centralization.

The research summarized in [21] proposed a model predictive control for energy management in the smart grid that aims to minimize the energy cost function. Based on some parameters (variable cost values, power consumption and generation profiles, etc.), this research plans the battery usage. Unfortunately, this algorithm does not integrate a storage system.

Wang et al. [22] developed a novel algorithm that aims to schedule the energy storage systems according to photovoltaics productions. The researchers believed in predictions to control storage capacity. At each period, this algorithm aims to adjust all its predictions if a perturbation occurs. Based on predicted values, the researchers proposed a storage agent algorithm aiming to plan the future charging and discharging schemes. However, this algorithm does not allow interactions between consumers in order to reduce bill.

Cecati et al. [23] developed an energy management algorithm aiming to minimize the consumers bill and to reduce the renewable resources intermittency effects in the smart grid. This research uses prediction values (generation and load predictions) in order to optimally maintain the network. At the beginning, the system plans the day-ahead predicted data. After that, the real-time optimization process is scheduled. In fact, the predicted values will be modified in real-time according to the process requirements. Furthermore, a neural network learning architecture is used in order to resolve the large amount of data integration into the energy management system. However, this approach does not take into account the intelligent interaction between consumers in order to reduce costs.

An energy management algorithm for the smart grids was proposed by Logenthiran et al. [24]. The main goal of this approach is to develop and simulate a multi-agent system that enables an electricity market for the operation of a microgrid in both islanded and grid-connected modes. In this approach, the microgrid plans to boost its own benefits by minimizing the exchange with the main grid. During the islanded mode, the microgrid role is to satisfy the local energy demand using its local production while reducing its load losses. In the grid-connected mode, the microgrid trades energy to its internal load and exchanges energy with the main grid at market price too. In order to reduce the effects of the insufficiency in microgrid production and high energy prices, the microgrid bought energy from the upstream network and sold to the consumers at the same price. However, this algorithm does not allow energy purchase negotiation to minimize the consumers' bills.

The authors in [25] used the demand side management in order to reduce the peak loads in the smart grid system. The authors of this research proposed a MAS in order to minimize demands. Furthermore, they believed in smoothing peak demands to a time slot where the demands are low. However, this technique may reduce the users' comfort.

In [26], the authors proposed a multi-agent based architecture for optimal energy management in smart homes. In this research, the authors present multiple strategies in order to provide savings in energy and costs. They highlight on four optimization strategies: the comfort optimization, the cost optimization, the green energy optimization and the demand side management optimization. In this context, an information exchange between Control and Monitoring Agents, Information Agents and Application Agents occurs. This information exchange consists on the transfer of predicted data and real-time data in order to plan the optimal strategies. However, this approach does not take into consideration the presence of a storage agent that can help in cost and comfort optimization.

To reduce energy cost, Rose et al. [27] proposed a novel scoring rule mechanism called Sum of Others' plus Max (SOM). In this new scheme, the system is composed of a centralized agent that aggregates all predicted information. The centralized agent aggregates all information received. Based on aggregated data, the centralized agent purchases all needed energy to the system. In order to encourage agents to report their true demands to the centralized agent, this last rewards agents from a budget that is equal to the savings made by using the agents' information.

In this algorithm, the centralized agent distributes all rewards in a budget-balanced way. However, the centralization used in this algorithm may call into question the scalability of the approach and constitute bottlenecks.

In order to minimize the inconveniences reached in the literature, we present, in this chapter a novel energy management proposal for smart grids integrating intelligent storage systems. The next section summarizes all tools used in our proposal.

3 The Energy Management in the Smart Grid Context

To highlight the potential benefits of a MAS and an intelligent storage system in energy management, the basic concepts of agents, multi-agent and storage systems need to be known. Thus, we will review and summarize these paradigms hereafter.

3.1 The Multi-agent System and Its Integration to Develop an Intelligent Distributed System

Based on the definitions given in [28–32], a MAS is defined as a collection of intelligent agents that cooperate and coordinate their operation to achieve a common objective. This system deals with modeling of autonomous decision making entities. In addition, a MAS allows intelligent control of microgrids and smart grids while each component is represented by an intelligent and autonomous agent. Theoretically, the fundamental element of a MAS is an intelligent agent which has three typical characteristics: the reactivity, the pro-activity and the sociability. An agent is reactive so it can react to the environment rapidly. Furthermore, this agent is objective-oriented so it is pro-active. Finally, the agent sociability makes it capable to negotiate and communicate with other agents in the system with their own language. Through negotiation, agents solve some questions [33]. Negotiation is the key issue for the multi-agent system to harmonize and solve the conflicts concerning the MAS, knowledge and resources, and it is an agent mutual mechanism based on the communication language.

There are many tools that help the MAS for better energy management. The storage system is a smart grid component that has a main role in energy management. This entity is defined in the next section.

3.2 The Intelligent Storage System and Its Benefits

Due to the high integration of the renewable resources into the electrical grid within the next decades, the penetration of energy storage devices will become more and more essential. The storage system has a meaningful role in ensuring the stability of the electrical system and in minimizing the renewable energy intermittency effects [34]. In an operational context, the energy storage can be used to store energy excess when supply exceeds demand and then dispatch it when it is needed which improves the energy efficiency. From finance point of view, energy storage is treated as an arbitrage instrument: charging storage when energy spot prices are low and discharging it when energy spot prices are high [10, 24]. In order to increase the battery performance and the battery life, an intelligent battery strategy is required. According to Ibrahim et al. [35], the developed intelligent battery strategies plan to follow the predicted values with the help of the battery and energy purchased (respectively sold) from (to respectively) the electricity market.

In the following is detailed ANEMAS, a proposal providing an intelligent energy management solution.

4 ANEMAS: The Proposed Approach

This section describes our proposal "ANEMAS" (Agent-based eNErgy MAnagement in Smart grids) which aims to resolve the generation intermittency problem and to optimize in real time the consumer bill by integrating a storage system and using multi-agent algorithms (including negotiation).

4.1 Problem Formulation

In this work, we highlight on a real scenario to represent the today's energy scheme that integrates the green suppliers like wind and solar in a market mostly brown. In fact, the renewable resources are mostly influenced by external environment like weather (wind speed or sunshine). In this chapter, we develop a novel agent-based energy management algorithm in order to efficiently integrate renewable energy into the system. The goals of this research are summarized hereafter:

Goal 1: the main goal of this research is to resolve the renewable resources intermittency problem.
Goal 2: reduce all users' bill.
Goal 3: minimize grid use in peak hours.

To release our previous objectives, we use a multi-agent architecture which will be detailed in the following sections. Moreover, we will show all the information traded between agents while providing a generic model composed of wind green energy suppliers, consumers, storage and grid.

4.2 The Proposed Multi-agent Model

The agents are introduced in the different components of the smart grid. They must control generation, load, and storage assets primarily from the standpoint of power flows. In our proposal, we define four agents:

1. Grid agent: satisfies the energy lack and buys the excess of energy produced.
2. Storage agent: controls the energy storage and are introduced in batteries.
3. Prosumer agent: controls the distribution of the energy he produces.
4. Consumption agent: negotiates the energy purchase with other consumers and prosumers.

The energy management system in our approach is divided into two layers: the proactive layer and the reactive layer. The proactive layer is responsible for the prediction of energy production and consumption. We choose the time slot in the proactive layer as $\Delta t = 24\,h$. In this layer, the predicted values of consumption and production are those adopted by EDF (Electicité de France) and defined in [36]. The reactive layer is responsible for planning and negotiating consumption at shorter periods $\Delta t' = 1\,h$ [18] and helps to buy energy with a minimal cost. Figure 2 summarize the management steps of our algorithm.

A priority P will be assigned for each user demand according to the user's preferences and the required quantity of energy. A demand's priority can be High, Medium, or Low, and the energy grid prices, at each period, are fixed by the grid. The agent satisfies the highest priority demands at the beginning, then it will satisfy the medium priority demands and finally the lowest priority ones. The agent can delay the medium priority for a given time, that we call Medium priority Delay (MD), and the low priority for a Low priority Delay (LD), such as $LD > MD$. Moreover, the consumption periods are divided during the day as follows: the Off-Peak period (OP), the Low-Peak period (LP) and the High-Peak period (HP). In the Off-Peak period, the energy use and the cost are minimal while in the Low-Peak period, the energy consumption and the purchase of energy start to increase. Finally, in the High-Peak period, the energy use reaches its maximum and the purchase price of energy becomes very high [7].

Within the constraints listed above, an algorithm is proposed for each agent for the purpose of energy management.

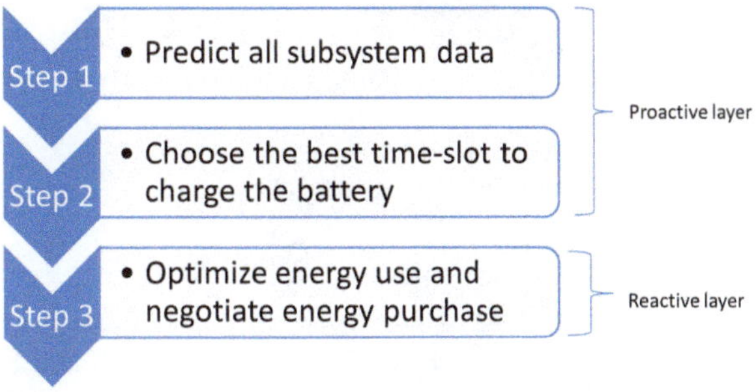

Fig. 2 Management steps

4.3 Agents Algorithms

Each agent has its own algorithm that we will describe in the following.

4.3.1 Grid Agent

The grid agent provides the lack of energy to consumers when the production of renewable energy does not meet their demands. Furthermore, in case of a production excess, this agent can arrange to purchase energy in excess. Hence, the grid agent receives two types of messages: the buying and the selling messages. The buying messages are sent by consumers and storage systems to the grid to buy energy. The selling messages are sent by renewable production units to the grid to sell it their exceeded energy. The grid agent algorithm is described in the Algorithm 1 and its parameters are:

- MaxGridEnergy: maximum capacity of the grid.
- RequestedEnergy: energy requested by a consumer.
- GridEnergy(t): instantaneous grid energy.
- EnergyToSell: energy to sell to the grid by a producer.

4.3.2 Storage Agent

The storage agent is responsible for providing energy to consumers during peak hours or in case of lack of energy. In case of production excess, these agents store energy. This stored energy may be used later in order to reduce energy cost. The storage agent has to respect multiple parameters, presented hereafter, concerning the amount of energy to store in order to meet the users needs:

Algorithm 1: Grid agent algorithm

```
if RequestForEnergy=True and GridEnergy(t) ≥ RequestedEnergy then
    //if there is a lack in the production
    SendEnergyToConsumer();
    GridEnergy(t)− = RequestedEnergy;
end if
if RequestToSellEnergy=True and MaxGridEnergy − GridEnergy(t) ≥ EnergyToSell then
    //if there is an exceeded production and the grid can buy all this exceeded energy from the
    producer
    BuyEnergyFromSource();
    GridEnergy(t)+ = EnergyToSell;
end if
```

- SOC_{min}: minimum state of charge (in kWh).
- SOC_{max}: maximum state of charge (in kWh).
- PR: grid price (in cents/kWh).
- PR_{min}: minimum grid price (in cents/kWh).
- Peak: sum of peak demands predicted for all consumers (in kWh).
- SOC(t): instantaneous state of battery charge (in kWh).
- D_j: requested demand for period j (in kWh).
- RE_j: renewable energy produced on period j (in kWh).
- EN_{st}: energy to be stored (in kWh).
- *tpeak*: time corresponding to the maximum consumption during a day.

The steps of the storage agent algorithm are presented in the Algorithm 2.

In Table 1, we show an example to explain how we calculate EN_{st}. We suppose that $Peak - SOC(tpeak) - RE_{tpeak} = Peak - \sum_{j=0}^{tpeak-1}(RE_j - D_j) - RE_{tpeak}$ and we consider that at period 0 the energy price is minimal so the storage agent will calculate the energy to store. For sake of clarity, we will present only 3 periods. As shown in the example, the High-peak period is the period 1. Normally, the battery should store the lack of energy that can occur, which is 3 in this example but in the peak period the consumers may have a lack of energy that the storage cannot satisfy. In this period (High-peak) the demand may be greater than the total lack, in this example it is 5 which is greater than 3 so the storage system should store 5 kWh (max(5,3)) at period 0.

Algorithm 2: Storage agent algorithm

$EN_{st} \leftarrow max(Peak - SOC(tpeak) - RE_{tpeak}, \sum_{j=i}^{23} D_j - \sum_{j=i}^{23}(RE_j) - SOC(t))$; //Amount of energy to store in order to complement the RE produced and satisfy all the demands (from the current period (i) to period 23)
if $PR = PR_{min}$ **then**
    if $EN_{st} \leq SOC_{max}$ **then**
        RequestCharge(EN_{st}); //The storage agent requests an amount of energy equal to EN_{st} from the grid
    else
        RequestCharge(SOC_{max}); //The storage agent requests an amount of energy equal to SOC_{max} from the grid
    end if
else //PR is not equal to PR_{min}
    if $SOC(t) < EN_{st}$ **and** $PR_i < PR_{i+1}$ **then** //If the storage cannot meet all the demands and the energy price at period i+1 is greater than the energy price at period i
        if $RE_{i+1} < D_{i+1}$ **and** $D_{i+1} - RE_{i+1} \geq SOC_{max} - SOC(t)$ **then**
            RequestCharge($D_{i+1} - RE_{i+1}$); //If there is a space in the storage system ($SOC_{max} - SOC(t)$) to charge the energy needed ($D_{i+1} - RE_{i+1}$), then the storage agent requests an amount of energy equal to $D_{i+1} - RE_{i+1}$ from the grid
        end if
    end if
    GoToNextPeriod();
end if

Table 1 Example of EN_{st}

Period (i)	0	1	2
$D_i(kWh)$	3	12	8
$RE_i(kWh)$	2	8	10
$\sum_{j=0}^{2} D_j - \sum_{j=0}^{2}(RE_j) - SOC(t)$	$(3 + 12 + 8) - (2 + 8 + 10) - 0 = 3$	–	–
$Peak - SOC(tpeak) - RE_{tpeak}$	$12 - (2 - 3) - (8) = 5$	–	–

4.3.3 Consumption Agent

In our proposal, we consider that consumption agents are deployed in smart homes. At each period, the consumption agent divides its demands into three priorities: (1) the priority 1 is the higher priority and its demands should not be delayed or dropped, (2) the priority 2 is the medium priority and its demands can be delayed for a short time, (3) and finally the priority 3 is the lowest one and its needs can be delayed for a time higher than the delay time for the priority 2. In fact, all activities with priority 2 and priority 3 can be delayed respectively with a Medium Delay (MD) and a Low Delay (LD). MD and LD will be chosen in order to maintain consumers' comfort without causing any inconvenience for users. In other words, MD and LD have a maximum value that will be selected so that the activity will be satisfied, this maximum will be set by the system administrator.

Case 1: Decision for highest priority data. In order to maintain users' comfort, data with Priority 1 are satisfied at the beginning.

In fact, each consumer aims to satisfy its demands with the lowest cost. In order to minimize users' costs, we used a negotiation algorithm between consumers. In this context, this negotiation algorithm role is to choose the best combination of consumers and renewable resources that minimize bills. The proposed negotiation strategy is explained below:

1. Consumption agent receives at each period the production and the consumption data from each prosumer and consumption agent.
2. Consumption agent j calculates the utility function f(j) defined in the Algorithm 4.
   Where:
   - f(j): utility function for agent j.
   - NbProsumers: prosumers total number.
   - NbConsumers: consumers total number.
   - P_{ji}: priority that increases if the resource is closer to the consumer, we consider $P_{ii} = 1$ and $P_{ij} = P_{ji}$.
   - E_{ji}: maximum energy that can be bought from prosumer i to consumer j.
   - C_{ji}: renewable energy cost. We consider $C_{ii} = 0$.
3. Consumption agent sends its proposition to all other consumption agents in its network.

Algorithm 3: Consumption agent algorithm

```
if RenewableProduction=true then //If there is a renewable production
  NegociateEnergyPurchase(); //Negotiate energy purchase according to the negotiation
  algorithm
end if
if DemandSatisfied=False then //If there is a demand that cannot be satisfied by renewable
resources
  if PR_Storage ≤ PR_Grid and StorageCanSatisfyDemands=True then //If storage can satisfy
  demands with a price lower than the grid
    StorageSatisfyDemands();
  else //If grid price is lower than storage price or storage capacity cannot meet demands
    if GridCanSatisfyDemands=True then //If grid can satisfy demands
      GridSatisfyDemands();
    else //If the grid and the storage cannot satisfy the demands so the activity will be
    dropped
      if DemandPriority=1 then //If demands priority is 1
        DropDemand();
      else //If consumers demands are with priority 2 or 3
        if MaxDelay=true then //The delay time reaches its maximum
          DropDemand();
        else //If demands can be delayed
          DelayDemand();
        end if
      end if
    end if
  end if
end if
```

4. When consumption agent receives all the proposals, it chooses the one with the highest utility function and purchases energy according to this proposal.

When the negotiation ends, and after buying energy from the renewable resources, the consumption agent follows the steps defined in the following in order to satisfy all its demands with the lowest cost:

- When there is a lack in renewable production so the demands cannot be satisfied only by the renewable sources. In this case, the consumption agent plans to buy the energy with the minimal cost. To achieve this goal, it compares the storage cost and the grid cost in order to choose the minimum between them.
- In this algorithm, if the grid price is lower than the storage price, the consumer agent sends a request to buy energy from the grid.
- When storage costs are lower than grid costs and the available storage energy can meet the consumer demands, then the consumer agent sends a request to the storage system to satisfy its needs.
- When the grid and the storage cannot satisfy the user demands so the activity will be dropped. In fact, this situation is very rare and should be avoided.

Case 2: Decision for medium or low priority data. Data with Priority 2 or 3 are satisfied after the data with priority 1. In this case, if the grid and the

Algorithm 4: Utility function algorithm

```
f(j) = 0; //initialize the utility function
if RE_j < E_j then //The production of prosumer can not meet its demands
   while E_j > 0 do //while demand is not satisfied the consumer or prosumer chooses to
   buy its demands from sources which have the cheapest price
      for all 0 < i ≤ NbProsumers and i ≠ j do
         ChooseMinimal(min(C_ji * P_ji * E_ji))
         f(j)+ = C_ji * P_ji * E_ji;
         E_j = E_j − E_ji
      end for
   end while
end if
for all 0 < i ≤ NbConsumers and i ≠ j do //after satisfying its demands a consumer uses to
meet the demands of other consumers with a minimum price
   if RE_i < E_i then //The production of prosumer can not meet its demands
      while E_i > 0 do
         for all 0 < k ≤ NbProsumers and k ≠ i do
            ChooseMinimal(min(C_ki * P_ki * E_ki))
            f(j)+ = C_ki * P_ki * E_ki;
            E_i = E_i − E_ki
         end for
      end while
   end if
end for
```

storage system cannot satisfy the demands, the activity will be delayed for MD for priority 2 demands and for LD for priority 3 demands. This activity will be dropped if the maximum delay is reached. This maximum is equal to 1 h for priority 2 and to 3 h for priority 3.

4.3.4 Producer Agent

Producer agents are introduced in renewable production units (photovoltaics, wind turbines, etc.). For each period i, the producer agent aims to satisfy consumers' demands. Hence, if the producer is a prosumer agent, it satisfies its own demands at first and then it helps the other consumers. However, if it is not a prosumer, the producer agent will try to satisfy consumers' requests. After that, if the energy it produced exceeds its needs, it will satisfy other consumers demands and sell the energy to the grid or to the storage system. The steps of the producer agent algorithm are presented in the Algorithm 5.

Algorithm 5: Producer agent algorithm

```
if IsProsumer=true then //If it is a prosumer
   MeetSelfDemands(); //Meets its own consumer demands
end if
if EnergyProduced > 0 then //If the produced energy by the production source is higher
than zero
   MeetDemands(); //satisfies requests from consumers
   if EnergyProduced > UsersDemands then //If there is an excess in produced energy
      if StorageNeedsEnergy then //If the storage capacity is not maximal
         ChargeStorage();//sell exceeded production to the storage system
      else
         if GridNeedsEnergy then //If the grid capacity is not maximal
            ChargeGrid();//sell exceeded production to the grid
         else
            GoNextPeriod();
         end if
      end if
   else
      GoNextPeriod();
   end if
end if
```

Table 2 Considered scenarios

Scenario	1	2	3	4	5	6	7	8
Consumers	5	10	5	10	15	20	25	30
Producers	3 prosumers	7 prosumers	[0–10]	[0–10]	15	20	25	30
Storage	1	1	1	1	1	1	1	1
Grid	1	1	1	1	1	1	1	1

4.4 *Description of the Considered Scenarios*

In this proposal, we choose to implement in our system an agent for the grid, an agent for each consumer, an agent for the storage system and an agent for each prosumer or producer. For our simulations, we consider eight scenarios. Table 2 presents the chosen scenarios.

The scenario 1 is simple with 1 grid agent, 2 consumer agents, 3 producer agents and 1 storage agent that can allocate the stored energy to the representing entities. Some complexity is introduced in scenario 2 to test the performance of our proposal when we increase agents number. This scenario has 1 grid agent, 3 consumer agents, 7 prosumer agents and 1 storage agent that has a large charge state to serve all the system. In the scenario 3 and 4, we choose to vary the number of producers from 0 to ten in both cases respectively (five consumers and ten consumers) to evaluate the performance of our approach. Finally, to gauge the efficiency of our approach, we increase the number of consumers from 15 to 30 in scenarios 5, 6 ,7 and 8 and we consider that every consumer is a prosumer; i.e., each consumer has at least one renewable energy source.

Simulation results are presented in the following section.

5 ANEMAS Performance Evaluation

In order to evaluate the performance of our scheme, we simulate the agent behaviors with JADE (Java Agent Development Framework) [37]; "a software framework which is fully implemented in Java language. JADE simplifies the implementation of multi-agent systems through a middle-ware that claims to comply with the FIPA specifications and through a set of tools that supports the debugging and deployment phase" [37].

5.1 Simulation Parameters

The consumption periods during the day are splitted into: the Off-Peak period (OP), the Low-Peak period (LP) and the High-Peak period (HP). Moreover, the day is divided into 24 periods and grid prices vary between 10 and 22 cents/kWh and wind turbine energy (our renewable energy) prices vary between 14 and 7 cents/kWh [36]. The peak hours are represented in Table 3.

5.2 Results

After showing the different scenarios that can take place in a negotiation for energy purchase, the efficiency of ANEMAS will be shown by comparing it to a typical billing scheme. This scheme used in most electric power systems where some or all consumers pay a fixed price per unit of electricity independently of the cost of production at the time of consumption. This scheme is called Conventional-case in the following. In addition, we compare our algorithm to a system that uses the renewable resources but do not use storage systems, the scheme used in this case is called RE-case.

Firstly, we illustrate the simulation results for 1 day of operation. The first simulation test is carried out by using scenario 1 and the results are given in Figs. 3 and 4. Figure 3 shows consumers buying prices with the three defined schemes. During the day, the reduction in consumer bills is caused by the use of renewable energy or by the use of renewable energy and storage systems. During Off-Peak hours the reduction of cost is very high, it is more than 90 % by using ANEMAS and

Table 3 Day peak and prices

Period	OP	LP	HP
Hour	00:00–12:59	13:00–19:59	20:00–23:59
Grid prices (cents/kWh)	10–17	17–21	10–22
Wind power prices (cents/kWh)	7–14	8–10	7–8

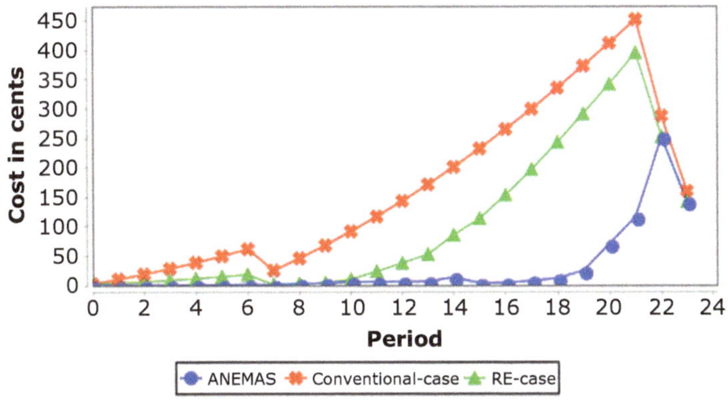

Fig. 3 Consumer bills for scenario 1

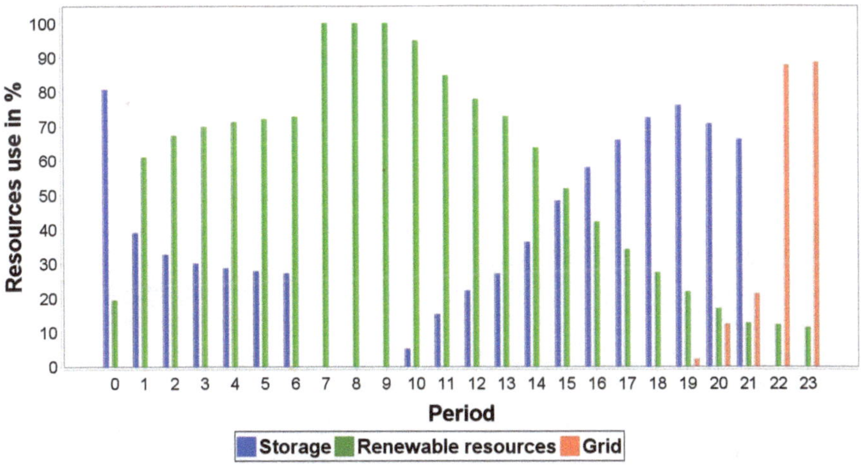

Fig. 4 Resources use for ANEMAS (scenario 1)

more than 60 % by using only renewable resources in comparison to conventional-case. This reduction decreases and reaches 75 % by using ANEMAS in the period of High-Peak (period 21), whereas during this period the decrease in RE-case is 16 %. In both cases (ANEMAS and RE-case), there is no more cost reduction at periods 22 and 23. The decrease in bills is due to the use of renewable resources for RE-case. Increased demand and decreased production cause the decrease in the percentage of price reduction in Low-peak and High-peak hours. Therefore by switching from Off-Peak to the High-Peak, consumer bills increase to reach the grid costs. In addition, we notice that, by using ANEMAS, the costs are less than costs in RE-case. This reduction is due to the use of storage systems with renewable resources, buying stored energy when costs are low, reducing access to the grid and serving the users with a minimal cost even during HP periods (period 21 for example).

Figure 4 shows the use of all resources during the day by ANEMAS. Our algorithm does not use the grid during Off-Peak and Low-Peak periods (grid use is equal to zero). It uses the renewable resources and if these resources do not meet all the demands, it uses the storage. The storage use increases and will be higher than the renewable resources use at period 15. This is caused by the consumption increase and the production decrease at these periods. Furthermore, the use of the grid increases in High-Peak hours and after period 21 the storage use is equal to 0 and the grid use increases. This change is due to the storage charge which reached its minimal value during High-Peak hours.

The results for scenario 2 are shown in Figs. 5 and 6. When we analyze the results shown in these two figures, we can conclude that these results have not much differences from the results of scenario 1. Figure 5 gives the consumers bill for scenario 2, and shows that minimization of costs is the same as that in scenario 1 in both cases (ANEMAS and RE-case). In High-Peak (period 21), the cost minimization reaches 75 % for ANEMAS and 16 % for RE-case.

Furthermore, resources use shown in Fig. 6 demonstrates that the use of the grid increases in high-peak (period 21) when we use our algorithm.

To test the effectiveness and the usefulness of the intelligent storage system proposed in our algorithm, we increase the number of producers from zero to ten in a system consisting of five consumers (10 consumers respectively). Figures 7 and 8 show respectively the energy costs in both studied cases. The cost minimization varies between 0 and 70 % (Fig. 7) and between 0 and 55 % (Fig. 8) in RE-case. Furthermore, by using ANEMAS, the cost minimization varies between 73 and 85 % (Fig. 7) and between 71 and 85 % (Fig. 8). In fact, both figures show that by increasing the number of producers we can minimize the consumers' price.

To test the efficiency of our approach, we increase the number of consumers from five to thirty. We consider that every consumer is a prosumer. Figure 9 shows energy costs in the studied case. The results show that when increasing both of the consumers and producers number together, the energy purchase cost decreases to

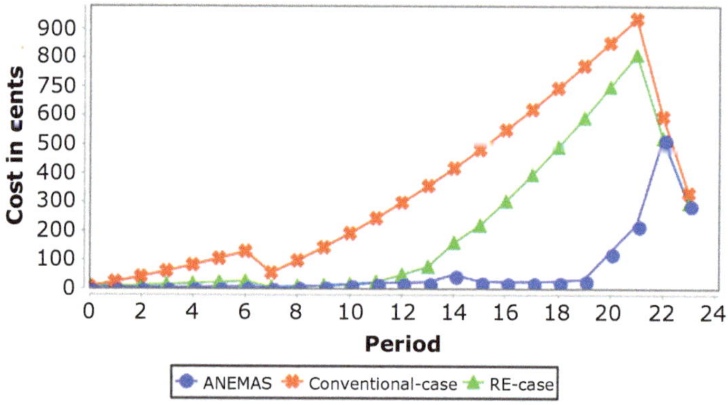

Fig. 5 Consumer bills for scenario 2

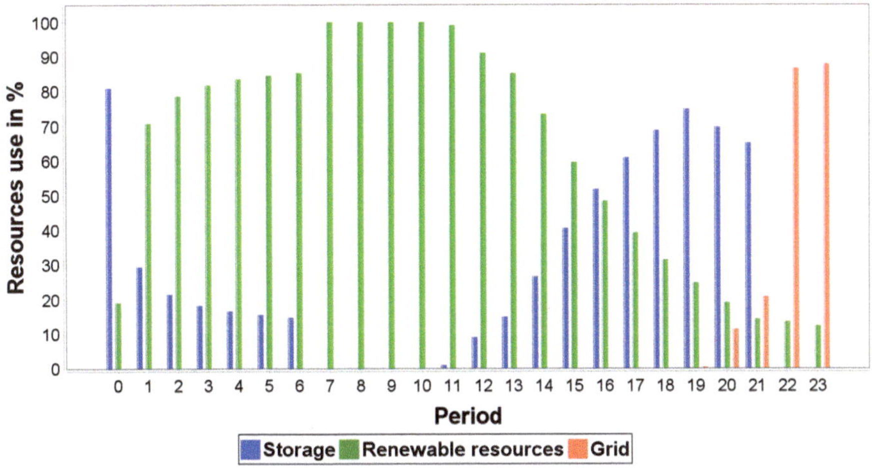

Fig. 6 Resources use for ANEMAS (Scenario 2)

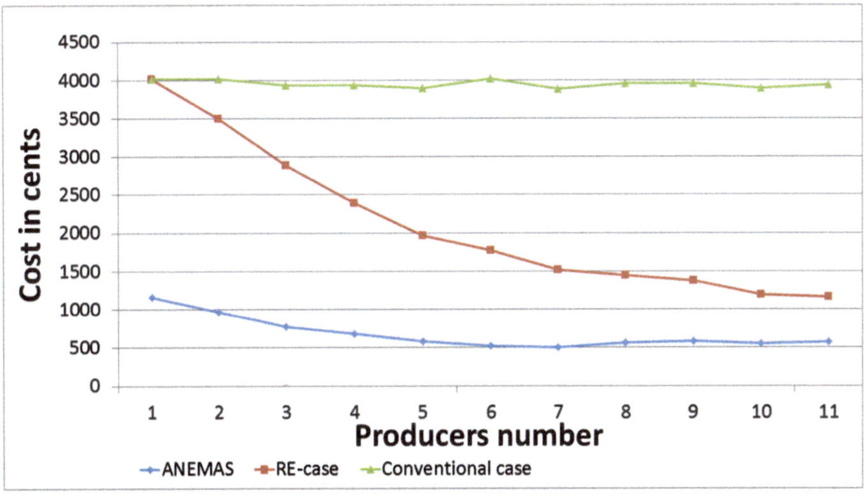

Fig. 7 Day costs when the number of producers increases from 0 to 10 (5 consumers)

about 56 % in RE-case and to 86 % by using ANEMAS. These results show that our approach is an effective approach even by expanding our system.

These results show that when the number of consumers increases, the cost reduction and the grid use remain the same.

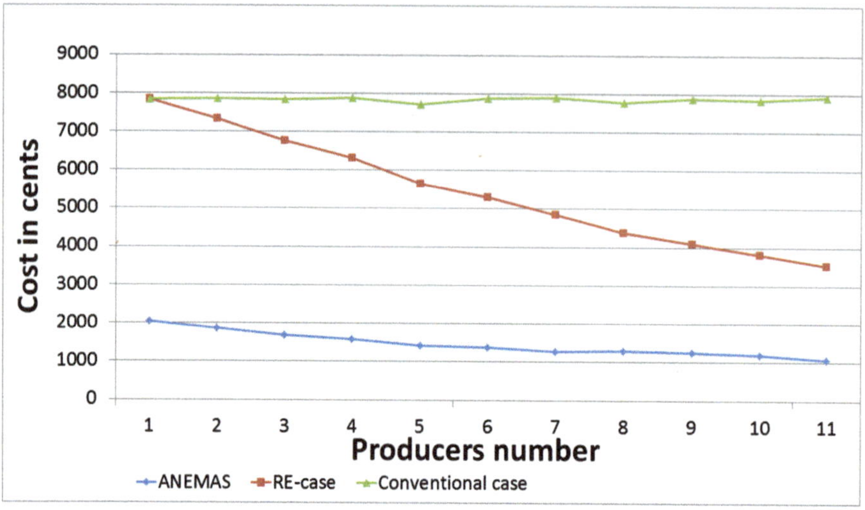

Fig. 8 Day costs when the number of producers increases from 0 to 10 (10 consumers)

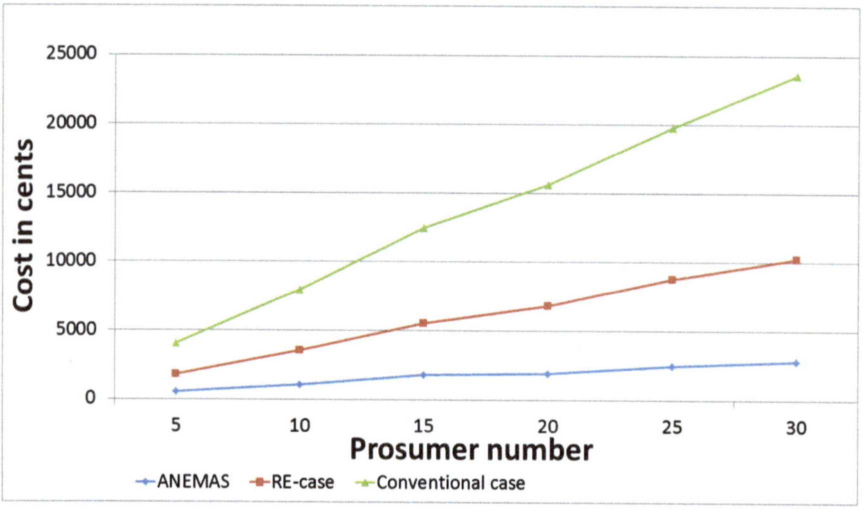

Fig. 9 Day costs when the number of prosumers increases from 5 to 30

6 Conclusion

Smart grid is an important field for the researchers in the near future. The smart grid main purpose is to integrate general and exceptionally flexible strategies for the use of distributed energy resources and/or storage systems. In fact, this grid uses renewable resources in order to minimize toxic gas emissions caused by fossil fuel use. However, introducing renewable energy sources into smart grids have some problems due to the weather and time dependencies. Therefore, there is an evident need to explore the feasibility of using the distributed energy storage systems in the same system with distributed energy resources and their influence on the penetration of renewable energy. Furthermore, it is certainly essential to evolve control strategies for energy management in the smart grid integrating the storage and renewable production side. We considered that the use of energy storage systems and multi-agent approach are good solutions to ensure the energy management, to reduce the consumers' bills and to minimize the access to the grid. In this chapter, we developed a novel energy management proposal using multi-agent systems and intelligent storage systems. Our proposal introduces four agent types into the system: the grid agent, the storage agent, the prosumer agent and the consumption agent. In this research work, we aim to point to more responsible energy consumption while establishing lower contract prices. Furthermore, to evaluate the performance of our scheme, we simulate our proposed approach using JADE simulator. In this context, the simulation results showed that the use of a storage system reduces the energy cost and the grid utilization which will reduce the CO_2 emissions. Finally, we plan in our future works to consider more storage systems which may be in competition and we will study other negotiation algorithms.

Acknowledgements This work is supported by grants from the Troyes University of Technology and the Lebanese University and is partly supported by the Champagne-Ardenne region research project SOLOTEC and the Lebanese-French research project CEDRE Number 32928XK.

References

1. Gao, C., Redfern, M.A.: A review of voltage control in smart grid and smart metering technologies on distribution networks. In: Proceedings of 2011 46th International Universities' Power Engineering Conference (UPEC), pp. 1–5 (2011)
2. Gungor, V.C., Sahin, D., Kocak, T., Ergut, S., Buccella, C., Cecati, C., Hancke, G.P.: Smart grid technologies: communication technologies and standards. IEEE Trans. Ind. Inf. **7**(4), 529–539 (2011)
3. Petinrin, J.O., Shaaban, M.: Smart power grid: technologies and applications. In: 2012 IEEE International Conference on Power and Energy (PECon), pp. 892–897 (2012)
4. Feng, Y., Yi, Q., Hu, R.Q.: Energy efficient self-sustaining wireless neighborhood area network design for smart grid. IEEE Trans. Smart Grid **6**(1), 220–229 (2015)
5. Potter, C.W., Archambault, A., Westrick, K.: Building a smarter smart grid through better renewable energy information. In: IEEE/PES Power Systems Conference and Exposition, 2009 (PSCE 2009), pp. 1–5 (2009)
6. Mets, K., Strobbe, M., Verschueren, T., Roelens, T., De Turck, F., Develder, C.: Distributed multi-agent algorithm for residential energy management in smart grids. In: 2012 IEEE Network Operations and Management Symposium (NOMS), pp. 435–443 (2012)
7. Klaimi, J., Rahim-Amoud, R., Merghem-Boulahia, L., Jrad, A.: An agent-based approach for energy management in smart-grids. In: Highlights of Practical Applications of Agents, Multi-Agent Systems, and Sustainability - The PAAMS Collection, pp. 225–236 (2015)
8. Javaid, N., Khan, I., Ullah, M.N., Mahmood, A., Farooq, M.U.: A survey of home energy management systems in future smart grid communications. In: 2013 Eighth International Conference on Broadband and Wireless Computing, Communication and Applications (BWCCA), pp. 459–464 (2013)

9. Kanchev, H., Lu, D., Colas, F., Lazarov, V., Francois, B.: Energy management and operational planning of a microgrid with a pv-based active generator for smart grid applications. IEEE Trans. Ind. Electron. **58**(10), 4583–4592 (2011)
10. Guo, Y., Pan, M., Fang, Y.: Optimal power management of residential customers in the smart grid. IEEE Trans. Parallel Distrib. Syst. **23**(9), 1593–1606 (2012)
11. Belkacemi, R., Bababola, A.: Experimental implementation of multi-agent system algorithm for distributed restoration of a smart grid system. In: IEEE SOUTHEASTCON 2014, pp. 1–4 (2014)
12. Sardouk, A., Rahim-Amoud, R., Merghem-Boulahia, L., Gaiti, D.: Agent strategy data gathering for long life wsn. Int. J. Comput. Netw. Commun. **2**(5), 71–90 (2010)
13. Kato, T., Takahashi, H., Sasai, K., Kitagata, G., Kim, H.N., Kinoshita, T.: Priority-based hierarchical operational management for multiagent-based microgrids. Energies **7**(4), 2051 (2014)
14. Alberola, J.M., Julián, V., Garcia-Fornes, A.: A self-configurable agent-based system for intelligent storage in smart grid. In: Communications in Computer and Information Science, vol. 365, pp. 240–250 (2013)
15. Vytelingum, P., Voice, T.D., Ramchurn, S.D., Rogers, A., Jennings, N.R.: Theoretical and practical foundations of large-scale agent-based micro-storage in the smart grid. J. Artif. Intell. Res. **42**, 765–813 (2011)
16. Nagata, T., Ueda, Y., Utatani, M.: A multi-agent approach to smart grid energy management. In: IPEC, 2012 Conference on Power Energy, pp. 327–331 (2012)
17. Colson, C.M., Nehrir, M.H.: Algorithms for distributed decision-making for multi-agent microgrid power management. In: 2011 IEEE Power and Energy Society General Meeting, pp. 1–8 (2011)
18. Nabouch, D.E., Matta, N., Rahim-Amoud, R., Merghem-Boulahia, L.: An agent-based approach for efficient energy management in the context of smart houses. In: Highlights on Practical Applications of Agents and Multi-Agent Systems, vol. 365, pp. 375–386. Springer, Berlin/Heidelberg (2013)
19. Klaimi, J., Merghem-Boulahia, L., Rahim-Amoud, R., Jrad, A.: An energy management approach for smart-grids using intelligent storage systems. In: 2015 Fifth International Conference on Digital Information and Communication Technology and Its Applications (DICTAP), pp. 26–31 (2015)
20. Safdarian, A., Fotuhi-Firuzabad, M., Lehtonen, M., Aminifar, F.: Optimal electricity procurement in smart grids with autonomous distributed energy resources. IEEE Trans. Smart Grid **6**(6), 2975–2984 (2015)
21. Ionela, P., Enrico, Z.: A model predictive control framework for reliable microgrid energy management. Int. J. Electr. Power Energy Syst. **61**(0), 399–409 (2014)
22. Yanzhi, W., Xue, L., Pedram, M.: Adaptive control for energy storage systems in households with photovoltaic modules. IEEE Trans. Smart Grid **5**(2), 992–1001 (2014)
23. Cecati, C., Buccella, C., Siano, P., Piccolo, A.: Optimal operation of smart grids with demand side management. In: 2013 IEEE International Conference on Industrial Technology (ICIT), pp. 2010–2015 (2013)
24. Logenthiran, T., Srinivasan, D.: Multi-agent system for market based microgrid operation in smart grid environment. Department of Electrical & Computer Science, National University of Singapore (2011)
25. Logenthiran, T., Srinivasan, D., Khambadkone, A.M., Aung, H.N.: Multiagent system for real-time operation of a microgrid in real-time digital simulator. IEEE Trans. Smart Grid **3**(2), 925–933 (2012)
26. Asare-Bediako, B., Kling, W.L., Ribeiro, P.F.: Multi-agent system architecture for smart home energy management and optimization. In: 2013 4th IEEE/PES Innovative Smart Grid Technologies Europe (ISGT EUROPE), pp. 1–5 (2013)
27. Rose, H., Rogers, A., Gerding, E.H.: A scoring rule-based mechanism for aggregate demand prediction in the smart grid. In: Proceedings of the 11th International Conference on Autonomous Agents and Multiagent Systems (AAMAS 2012), vol. 2, pp. 661–668 (2012)

28. Cintuglu, M.H., Martin, H., Mohammed, O.A.: Real-time implementation of multiagent-based game theory reverse auction model for microgrid market operation. IEEE Trans. Smart Grid **6**(2), 1064–1072 (2015)
29. Logenthiran, T., Srinivasan, D., Khambadkone, A.M.: Multi-agent system for energy resource scheduling of integrated microgrids in a distributed system. Electr. Power Syst. Res. **81**(1), 138–148 (2011)
30. Sharma, A., Srivastava, S.C., Chakrabarti, S.: A multi-agent-based power system hybrid dynamic state estimator. IEEE Intell. Syst. **30**(3), 52–59 (2015)
31. Merabet, G.H., Essaaidi, M., Talei, H., Abid, M.R., Khalil, N., Madkour, M., Benhaddou, D.: Applications of multi-agent systems in smart grids: a survey. In: 2014 International Conference on Multimedia Computing and Systems (ICMCS), pp. 1088–1094 (2014)
32. McArthur, S.D.J., Davidson, E.M., Catterson, V.M., Dimeas, A.L., Hatziargyriou, N.D., Ponci, F., Funabashi, T.: Multi-agent systems for power engineering applications part i: concepts, approaches, and technical challenges. IEEE Trans. Power Syst. **22**(4), 1743–1752 (2007)
33. Dong, X., Chen, J., Hu, Y., Jiang, G., Huang, T.: Multi-attribute negotiation model based on internal factors argumentation. In: 2013 International Conference on Management Science and Engineering (ICMSE), pp. 20–27 (2013)
34. Unger, D., Myrzik, J.M.A.: Agent based coordination of distributed energy storage devices in future distribution grids. In: 2015 IEEE Eindhoven PowerTech, pp. 1–6 (2015)
35. Ibrahim, M., Khairy, A., Hagras, H., Abdel-Rahim, N., Shafei, A.E., Shaltout, A.: Intelligent energy management strategy for decentralized battery storage in grid connected wind energy conversion systems. In: IET Conference on Renewable Power Generation (RPG 2011), pp. 1–6 (2011)
36. EDF: http://france.edf.com/france-45634.html,2013 (2013)
37. JADE: http://jade.tilab.com/,2015 (2015)

Index

A. Alonso-Betanzos et al. (eds.), *Agent-Based Modeling of Sustainable Behaviors*, Understanding Complex Systems, DOI 10.1007/978-3-319-46331-5

FSC
www.fsc.org
MIX
Papier | Fördert gute Waldnutzung
FSC® C083411

Kolibri 360 GmbH
Ferdinand-Jühlke-Straße 7
99095 Erfurt, Deutschland
produktsicherheit@kolibri360.de